D1087743

# PRECISELY

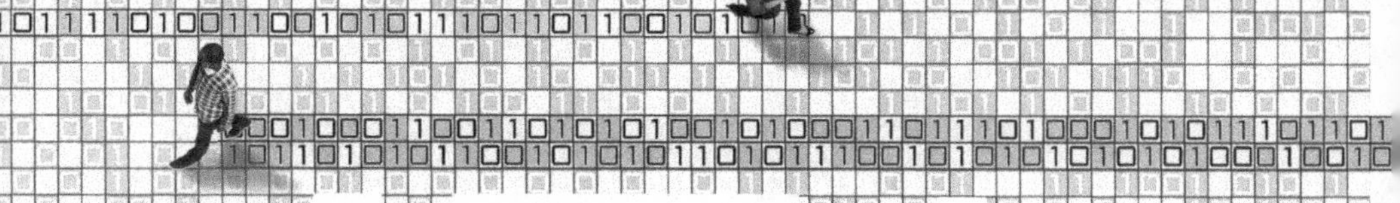

# PRECISELY

## WORKING WITH PRECISION SYSTEMS IN A WORLD OF DATA

ZACHARY TUMIN
and MADELEINE WANT

FOREWORD BY THOMAS H. DAVENPORT

Columbia Business School
Publishing

Columbia University Press
*Publishers Since 1893*
New York Chichester, West Sussex
cup.columbia.edu



Library of Congress Cataloging-in-Publication Data
Names: Tumin, Zachary, author. | Want, Madeleine, author.
Title: Precisely : working with precision systems in a world of data /
Zachary Tumin and Madeleine Want.
Description: New York : Columbia University Press, [2023] |
Includes bibliographical references and index.
Identifiers: LCCN 2022039185 (print) | LCCN 2022039186 (ebook) |
ISBN 9780231200608 (hardback ; alk. paper) | ISBN 9780231553704 (ebook)
Subjects: LCSH: Decision support systems. | Decision making—Data processing. |
Predictive analytics. | Big data—Industrial applications. |
Transformational leadership. | Industrial efficiency.
Classification: LCC HD30.213 .T86 2023 (print) | LCC HD30.213 (ebook) |
DDC 658.4/038011—dc23/eng/20221221
LC record available at https://lccn.loc.gov/2022039185
LC ebook record available at https://lccn.loc.gov/2022039186

Columbia University Press books are printed on permanent and
durable acid-free paper.
Printed in the United States of America

Cover design: Noah Arlow
Cover image: Getty images

*Zach:*
*To my wife Laura,*
*With my love, gratitude, and devotion*

*Maddy:*
*To Nick, who lifted me up and nudged me along countless times.*
*To my parents, who gave me a childhood full of books.*
*To my friends and family who believed I could do it from day one.*

# CONTENTS

# FOREWORD

I am a big fan of precision, so I really like *Precisely*. I confess that I hadn't thought of that term as a way to summarize the desirable data-driven changes taking place in business and society, but I think it's a good one. It connotes accuracy, measurement, efficiency, and attention to detail—all attributes that we want in our decisions and actions.

I like even more the idea of "precision systems," which is really what this book is about. Zachary Tumin and Madeleine Want describe, through many different topics and examples, the problem that many organizations encounter with data, analytics, and artificial intelligence (AI): they don't have the needed "ancillary" systems—processes, attitudes, relationships, and skills—to make effective use of these tools.

The sports analytics examples in the book are great examples of this. Like the authors, I too have marveled at the teams that innovate with data and analysis and make key decisions, not on tradition or intuition but on what plays and players the data and analysis say will be most effective. More and more often, every team—at least in major professional sports—has the same data, the same analytical and AI tools, and even the same types of smart people in analytics or data science roles. But some perform much better than others with similar capabilities. What many lack is close relationships between the data geeks and the coaches, and the sense of trust among decision makers that will ultimately make data-driven decisions, and thus the team, more successful.

Of course, not just sports teams but organizations of all types today are seeking more precision. I primarily work with big companies, almost all of which desire to become more digital, more analytical, more data-driven. But it's not the hardware, software, or data that usually get in the way of progress; it's the wetware. Technology changes rapidly, but most organizations and the people within them change much more slowly. Some companies have achieved great results with digitization, analytics, and AI initiatives, but none of them got there by focusing exclusively on technology.

We can also understand the value of precision systems by observing the absence of them. Perhaps the best example compares South Korea, which the authors profile in the book, to the United States in terms of their COVID-19 response. The United States, of course, had plenty of resources to fight the pandemic, including money, vaccines, computers, health-care workers, and public health experts. However, it lacked a functioning precision system for tracking and responding to the coronavirus. A lack of trust in government and science, politicization of the disease, and a highly fragmented approach to disease response across states and counties contributed to the lack of precision. The outcome has been tragic; over 1 million people have died from COVID in the United States, and it has one of the highest case fatality percentages and death rates per capita in the world.

On the other hand, South Korea, as Tumin and Want describe, had both an early and effective response to COVID-19. The virus was detected quickly, tests were developed rapidly, and an effective tracking and tracing program was introduced. South Korea's fatality rate per capita was less than one sixth of the United States, and its case fatality rate less than one tenth of the United States. Precision systems clearly matter, not just for making more money or winning more games but also for keeping more people alive and well.

At least in the United States, we tend to believe that technology advances are the key to success with precision decisions and actions. This techno-utopianism is deeply embedded in society, and it has nourished a very successful tech industry. But as this book explains, it's only one factor in achieving precision systems, and it's rarely the most difficult to put in place. Creating precision systems is a human discipline, requiring the right beliefs, attitudes, and behaviors from leaders to frontline workers.

*Precisely* doesn't avoid technology by any means, and it's up-to-the-minute in the tools and methods it describes. But it also contains great chapters on leadership, politics, organizational and business design, and other factors that are likely to bring about the "precision enterprise." If you want your organization to be a precision enterprise—and you certainly should—you need to turn or swipe the page and read the rest of this excellent book.

Thomas H. Davenport
Distinguished Professor, Babson College
Visiting Professor, Oxford Said Business School
Fellow, MIT Initiative on the Digital Economy

# PRECISELY

# CHAPTER 1

# INTRODUCTION

From Digitization to Precision

## The Game

In December 2019, fans witnessed two surprising developments in the National Football League (NFL). The unheralded Baltimore Ravens surged to a championship season, while the Dallas Cowboys—"America's Team"—imploded. What made the difference? Both teams were rich with on-field talent. But the Ravens had a secret weapon. In the coaches' box high above the field in M&T Bank Stadium, 25-year old Daniel Stern, a Ravens "football analyst" (holding a degree from Yale in cognitive science) sat elbow to elbow with Baltimore's offensive coordinator, Greg Roman. Play after play, they pumped *Moneyball*-style options down to coach John Harbaugh's headset, each built with artificial intelligence (AI), each a prediction of the game and a prescription for what the Ravens should do next. Harbaugh would pick and choose. Game after game, the Ravens picked up key yardage gains; turned failed, four-and-out possessions into sustained drives; and put points on the board. Games that might have been lost were won.

Dallas coach Jason Garrett had no such advantage, nor did he want it. "Yeah, we don't use those stats within the game," he told reporters. The upshot: precision systems that mix human judgment with analytics can easily outperform gut instinct or data alone. Adapt to the new game in town or perish.

## The Crime Wave

That same month, in the fading light of a winter's day in New York City, Tessa Majors, a nineteen-year-old freshman at Barnard College, was murdered in

Morningside Park, a majestic but decaying jewel. In November, a neighbor had reported twenty-three streetlamps out in the park "for the umpteenth time," a reporter wrote—five in the area at the base of the steps where Tessa Majors was murdered. "Most people know somebody who has gotten robbed there," a victim said. "I was told by police that it was not a very safe place to be." One nearby storeowner had complained to cops at the 26th Precinct station house by phone of harassment by kids. He was tired of it, and he was scared. "A lady told me to tell them to 'move along.'"

If the New York Police Department (NYPD) had no idea this crime was coming, it should have. The knifepoint robbery by three teens was the latest in a neighborhood crime wave. One attacker later admitted that they were in the "habit" of robbing people in the park. Since June, five people reported being robbed near the staircase in Morningside Park where Tessa Majors was killed—including two robberies ten days before. There were twenty-one reported muggings in the park and its perimeter compared to eleven the same period a year prior.

Just in case, for some reason, cops missed these patterns, the NYPD had an advanced AI-enabled precision system called Patternizr for discovering look-alike crimes. A well-managed pattern recognition system—whether AI-driven or manual—should have exposed the attackers' robbery habit sooner, but Patternizr didn't. Cops might have stopped the crime spree sooner, but they didn't. Inevitably, one mugging went wrong. Tessa Majors fought back and lost her life. Lori Pollock, Chief of NYPD Crime Control Strategies at the time, reflected on this: "You could see this coming from a mile away. People weren't paying attention or didn't want to tackle it. Youth crime was becoming an epidemic. No one in the court system or the judicial system seemed to take it seriously. There were no consequences for young people to rob two or three people in an hour—just, roll people. And that's what happened to Tessa Majors." The failure was unusual for New York's Finest. In all things operational and digital, the NYPD holds itself out as the gold standard in crime fighting. In movies, television, and real life, its achievements are legendary, its tools and techniques emulated everywhere.

But when it comes to putting data and AI to work even New York's Finest can stumble. Having a precision system on hand will not flag what matters most if you don't tell it to, nor will it change the world as a precision system should. "Fix broken *process* first," and "Get your *organization* right and ready for precision" are imperatives you will hear throughout this book—before any precision system can work its magic.

In the end, the only pattern matching done here was by Officer Randy Ramos-Luna. On patrol the next day, he recognized a kid on the street by the clothing he was wearing: it matched the Wanted video still he'd seen on his department cellphone. Pursuit, confrontation, and arrest followed. Back at the station house, the kid gave up his two accomplices—just like in the movies.

## Chasing COVID

On December 26, 2019, a Chinese researcher at Weiyuan Gene Technology in Guangzhou was browsing the latest test results from routine disease surveillance. Wuhan had sent over a lung fluid sample taken from a sixty-five-year-old man hospitalized with pneumonia-like symptoms. What the researcher saw alarmed her. The sample, she wrote a colleague, "was brimming with something that looked like SARS." In 2002, severe acute respiratory syndrome (SARS) had killed 774 people worldwide.

"It's no joke," her coworker replied on WeChat. "It's the same level as the plague." Tests the next hour confirmed "bat_SARS_like_coronavirus." "*Worse* than the plague," her coworker replied.

China did not lock down Wuhan for twenty-eight days. The virus, being no respecter of politics, borders, or national pride, seeped out of China by car, train, bus, and plane. It caught governments from Washington to Tokyo unaware. Studies would later show that, had China locked down Wuhan one week earlier, it could have reduced infections across China by 66 percent, three weeks earlier by 95 percent. In the pandemic that followed, COVID-19, as the virus came to be known, went on to kill over 6.5 million people around the world.

Korea was already on the move. In 2015, it had bungled its own Middle East respiratory syndrome (MERS) outbreak—and fixed its playbook to act fast the next time it encountered a pandemic. That time was now: in January, Korea's four confirmed COVID cases weren't exactly a catastrophe—yet it was enough to launch the new protocols. One week in, Korea's Disease Control and Prevention Agency (KDCA) convened twenty test manufacturers and told them to build COVID tests. The government guaranteed their cost reimbursement. KDCA gave them early access to viral samples and standardized the validation process. Korea's Ministry of Food and Drug Safety was ready with authorizations.

When Dr. Jong-Yoon Chun, CEO of the Korean biotech firm Seegene, got word of the pandemic, he responded quickly. Using a supercomputer equipped with AI tools, Seegene's team analyzed the molecular structure of the new virus; worked through billions of possible permutations; and built a COVID-19 test that yielded results within four hours, a test that was standardized so any of Korea's 120 labs could read them

Three weeks in, across Korea 700,000 test kits were ready. Doctors at Incheon Medical Center proposed innovations in testing protocol: drive-through sites would keep drivers in their cars, avoid spreading COVID in hospitals, and triple hourly screening capacity. By February 23, Chilgok Hospital in hardest-hit Daegu had installed South Korea's first drive-through testing center. The invention brought South Korea's testing capacity to 20,000 tests per day. In the opening weeks of the outbreak, 225,000 Koreans were tested. Over that period 8,000 were diagnosed with the disease. Sixty died, but the rapid implementation of a tracing system and social-distancing protocols made possible by the Seegene test and others undoubtedly saved many more lives.

"We are a molecular diagnosis company, specializing in medical testing kits," Dr. Chun told CNN. "We have to prepare in advance. Even if nobody is asking us." The upshot: it takes an ecosystem with all players aligned to make precision happen. With that, precision systems can have global impact, and fast.

## Race on the Water

When the pandemic was brought under more control, and travel again became possible, international sports events resumed. In September 2021, off the coast of Cadiz, Spain, spectators might have caught sight of a flotilla of F50 sailboats racing through the Atlantic at speeds of 100 miles per hour, their ninety-five-foot winglike masts and hydrofoils flying them above the waves. Eight teams of A-list sailors from around the world were making a run for a place in the championship Grand Final of the SailGP campaign in San Francisco the next spring and a prize of $1 million. Every boat was identical. Their crews saw and shared the same data. Each boat was connected to the Oracle cloud, streaming terabytes of data from sensors tracking their location, wind speed, boat speed, wind angle, and a host of other elements.

If each boat was identical and all the data were shared, what would be the key to winning? Talent and experience, certainly. Training, too. During the few months between the Olympics and the America's Cup, these teams had to come together fast with little time to practice. Every minute together mattered. But the difference maker off Cadiz and later San Francisco was the playbook unique to each team. Each choreographed its own moves, starting with a sprint to the video room after practice to see what worked and fix what didn't. That proved decisive on the open waters of San Francisco Bay where the Australia boat defeated all, dodging collisions, capsized competitors, and even a whale wandering onto the racecourse. The upshot: well-managed and built for continuous improvement, world-class precision systems deal with exigency, unleash the power of talent, and take the win.

"You're racing against the best guys in the world," Russell Coutts, SailGP's CEO and five-time America's Cup champion told a reporter. "If you give them more time against you, you're going to get hurt, aren't you?"

Time to the finish wasn't just everything—it was the *only* thing. "Glory," the Australia boat captain said, "lasts forever."

What do these four stories have in common? Massive amounts of data became available, sometimes deluging decision makers. Each was a race of teams against time and nature. Each depended on human-built and -run systems, loaded with AI, but with men and women, not machines, finally making the difference. Some used the data and the tools well, whether in the press of a game or a pandemic and changed the course of history. Others became tangled in the same old work routines and unchanged job designs.

We call this new mix of people, processes, and platforms built to shape the future precisely as managers require *precision systems*. And in today's incredibly fast-paced, competitive, unpredictable world, *precision systems matter*—a lot.

## What This Book Is About

This book provides a new blueprint for today's leader, from a Fortune 10 business executive to a teacher, doctor, or police officer, to convert the enormous potential of today's digital world into powerful change, not just once, but, in the words of Tom Davenport, a global leader in the embrace of analytics, "repeatedly, reliably, and quickly." How? With precision systems—the highly engineered working arrangements of people, processes, and technologies that put big data and AI to work creating exactly the change leaders want, exactly where, when, and how they want it.

From Rosabeth Moss Kanter to John Kotter, change gurus have focused on how to get a person, a team, or an entire organization to start acting differently, one person or group at a time. That's important because a lot of change happens at that level. More recently, Richard H. Thaler and Cass Sunstein's *Nudge* suggests ways to trigger *mass* change. Nudges are blunt-force tools for promoting change. They are passive, imprecise, embedded in a designed "choice architecture," and left to run, always "on." They wait quietly for the next one or a million unsuspecting consumers, users, or readers to stumble across their trip wires and change course. Nudges wait for you at the end-of-aisle grocery store displays, or at the child-height candy shelves at checkout, in yellow warning strips on subway platforms, or in purposefully designed defaults on websites that you either opt out of or fall slave to. Whether for broad public policy goals like climate change, obesity, and addiction, or more modest purposes like painting steps toward a trash bin to encourage its use, the power of a nudge is potent but hit or miss. You have to stumble upon it to get the ball rolling.

Precision systems are different. In their highest form, precision systems don't wait for you to find them—they seek *you out*—one gene at a time or from among millions of people. Political campaigns hunt down crucial single-issue voters and peel them away from parties and party-line voting; railroads scan thousands of engines, boxcar wheels, and track miles searching for a few life-threatening flaws; health authorities desperately seek the handful of truly sick among the millions of well citizens even before the few show symptoms.

If you want to win an election, improve the health of a city or a nation, or sell vastly increased amounts of your company's products, you may have to get millions of people or things, or even just a few, behaving differently, fast and in real time, once or continuously over days, weeks, or months—precisely as you want.

The data is there—*big data*—but it can be a tremendous technical and operational challenge to put that data to work for the change you want, precisely where, when, and how you want it. Giant data sets like GPT-3 for language and ImageNet for images make available billions of examples that make life easier for engineers building everything from chatbots to machines that use images to sort weeds from crops (as we describe in our discussion of John Deere's See and Spray system). Developers don't have to start from scratch.

But these data sets remain unwieldly, costly to use all the same, and flawed—gifted with a "stunning ability to produce human-like text," as one reporter observed of GPT-3, but also "a powerful capacity for biases, bigotry and disinformation."

And there's a catch: the world is rich with events that comprise only a handful of images or words. "I've built AI systems of hundreds of millions of images," AI pioneer Andrew Ng told a reporter. Ng was founder of Google's Google Brain group, chief scientist at Baidu, and now CEO of Landing AI. "Those techniques don't really work when you have only 50 images."

Now new techniques championed by Ng and others can train precision models from *small data*—a handful of images or words. GE Global Researchers recently used only forty-five training sample images to locate kidneys in ultrasound images, for example. These novel techniques open the door to detection of thousands of rarely occurring manufacturing flaws or predictions of low-frequency natural disasters or disease, no matter how scarce the data. In that respect, they expand the reach of precision systems even to those events that are exceedingly rare.

Big data or small, "born digital" tech titans like Facebook, Amazon, Apple, and Netflix make real-time precision look easy—practically automatic. Their AI-driven recommendation systems nudge you to the best next choice, personalized *for you*—or at least for people *like* you—millions of times each day. While by no means perfect, those highly automated precision systems are good enough, and fast enough, to have turned Amazon, Apple, and others into multi-trillion-dollar corporations. It wasn't easy getting there; their battles of years gone by are legendary. But today, at the so-called FAANG firms—Facebook, Amazon, Apple, Netflix, and Google, everyone is a believer; the skill sets, job designs, and organization setups are long proved; and the ability to test, prove, and continuously improve automated decision making built long ago into systems and platforms, operations, and culture. The born-digitals among us still have many issues, but getting to precision isn't one of them.

For those organizations *not* born digital—and most today are not—precision systems require a far different kind of approach by managers. Everywhere, they are putting precision systems to work on a smaller scale, and not just personalized but *particularized* to a specific business problem—to find an elusive pattern, perform a repetitive task, target a payload, or tailor a message. Depending on the nature of the business, the value proposition is entirely different each time: to accelerate mathematics learning among seventh-graders, to prevent workplace injury in mines and manufacturing, or to deepen reader engagement and increase digital subscriptions. Across these diverse missions, the leadership challenge for executive managers looks a lot like the classic work of managers everywhere who bring corporate or organization assets together for a purpose, with results and outcomes that matter. But the nature of precision changes everything, comprising as it does unique technologies, new talents, and prospects for change impossible to achieve otherwise—from the moment managers conceive the precision system, then sell it, make it work, and deliver on its promise.

That's where *Precisely* comes in. This book explores how leaders in every domain are dealing with these challenges and notching wins for their organizations by taking real-time precision systems into the marketplace, the combat space, the political race. We provide insights that will help any leader who is considering the move to precision to choose the system that's right for them, decide which problem to tackle first, sell the importance of precision to stakeholders, power up the people and the technology, accomplish change that delivers precisely what's needed every time—and do it all responsibly.

In this first chapter, we will look at three precision systems, each vastly different but sharing essential traits common to all. The first chapter will give us an early look at some important lessons every manager can learn from.

## Zipline: Delivering Precision, Saving Lives

Since the uncrewed aerial vehicles known as drones were first developed, autonomous delivery has been a dream of innovators for everything from household products to medicines. It turns out to be incredibly difficult, fraught with risk and danger, but when it happens, transforms the world.

Zipline cofounder Keller Rinaudo created the San Francisco–based company with a specific mission in mind: "To provide every human on earth with access to vital medical supplies." One-third of the world has limited access to essential blood products, vaccines, and injectables—none more so, Rinaudo discovered, than in sub-Saharan Africa. In Tanzania, Uganda, Rwanda, Malawi, Ghana, and the Democratic Republic of the Congo, for example, stockouts and expiration rates are high, and cold-chain and short-shelf-life commodities fail against the brutal realities of conflict, heat, and rough access. Rinaudo found the governments of these countries willing to let him experiment with delivery by Zipline. Early networking and hustling found responsive leaders. "We were meeting with the presidents of countries off of a sketch deck," CTO Keenan Wyrobek recalled. "The market 'pull' was there." By 2019, three years after its start, Zipline was delivering 75 percent of Rwanda's blood supply to 2,500 hospitals and health facilities far from Kigali, its principal city, and operating from six distribution centers that covered 100 percent of Rwanda and 50 percent of neighboring Ghana.

Success scaled massively. Through 2022, Zipline operated a network of autonomous drones in Africa, Asia, and North America. Although each instance of success was personal, when massively scaled, it conferred vast social benefit. Thanks to Zipline, for example, an urgent order for blood from a regional hospital for a mother bleeding out during childbirth can be dispatched and arrive in as little as an hour. The drone drops the parachuted package at the destination and flies directly back to home base for reload and the next job. Zipline calls it "last-mile, on-demand emergency medical delivery." We call it a precision system in action.

"I used to see the drones fly and think they must be mad," Alice Mutimutuje, a Rwandan mother said. "Until the same drone saved my life."

If the achievement is astounding, Zipline's end-to-end supply chain is precise to a fare-thee-well. Its two Rwanda distribution centers, for example, each run eight cold-chain storage devices at different temperatures and agitation levels, each fed by three supply chains, among them the national blood bank system. Zipline updates its blood inventory each evening for the next day's deliveries, based on forecasts that Zipline's algorithms built with continuous improvement to 99 percent accuracy.

An optimized pick-and-pack system whittles the time from "order received" to "drone dispatched" down to five minutes on average and to as little as one minute. The drones themselves are remarkable feats of engineering; lightweight and streamlined, their frames maximize efficiency on the ground and in-flight. Every surface is tagged with a QR code for preflight inspections. Technicians point their phones at the QR codes; the plane activates the control surface, technicians image it, and algorithms predict any issues and pass or fail each control surface. Launches are tightly controlled, with each drone launched the same way every time: with an electric catapult that accelerates the plane to 110 kilometers per hour cruising speed in a quarter of a second. "That's why we call it 'zip,'" Wyrobek explains with a grin.

In flight, continuous global positioning system (GPS) signaling and radar allow each drone to "know" its own location in three-dimensional (3D) space as well as the locations of any nearby drones and planes. That's essential for regulators, safety, and scaling into busy airspace. Zipline drones drop their supplies by parachute, taking in terabytes of data in real time, making ballistic calculations and dynamically adjusting for wind speed and direction so that the package drifts with the wind right where the customer wants it—no clinic needs any infrastructure to receive the shipment. Back at home base, a capture system stretches a wire across two goalpost-like towers; as the drone approaches, it communicates with radio receivers, navigates between the posts, extends its tail hook to grab the wire and powers down: mission accomplished. All without a human pilot, all guided by algorithm, the entire voyage mapped to precise three-meter by three-meter coordinates provided by What3Words—the astonishing app that has mapped the globe with 57 trillion such precise coordinates—"the national addressing system of dozens of countries," Wryobek says. "We'd literally just get a GPS coordinate for hospitals and clinics, go visit them, and ask, 'Hey, so where do you want this to drop?' The future of Africa is getting solved by What3Words."

In combination, these elements add up to a precision system dubbed by Wyrobek an "instant logistics infrastructure." Replete with AI and physiotechnical assets managed by doctors, pharmacists, aviators, engineers, and logistics professionals, Zipline delivers exactly what a patient needs, exactly where it is needed, at exactly the right moment—and does it repeatedly, at scale, with utter reliability. It takes a *system* to bring health and safety precisely where and when it is needed most.

Good leadership pivots robust precision systems fast to new possibilities. In 2020, Zipline faced a new test. COVID-19 was spreading in Ghana. In its midst, a presidential election was creating an unprecedented demand for rapid delivery of health-care supplies. "Polls were opening in forty-eight hours," Zipline reported. "Poll workers lacked masks and personal protective equipment. Tens of thousands of face masks needed to be distributed in some thirty-three districts over an area of more than nineteen thousand square kilometers." The Zipline team mobilized, pivoting its operations from health to voting rights. Within twenty-four hours, Zipline flew over 160 drone flights to twenty-nine locations and delivered 18,000 face masks to poll workers. Hands-free drop-off minimized human contact and reduced the risk of transmission of the virus.

Pivots like this emerge unexpectedly as the value of on-demand drone deliveries becomes clearer. Zipline modifies its success metrics accordingly. For example, in its partnership with Novant, a North Carolina–based health and hospital system, Zipline claims an average reduction of $169 for patient costs by telehealth visit, an 18 percent increase in medication adherence with on-demand delivery, and a 28 percent reduction in transportation costs compared to ground transport, all with thirty times fewer carbon dioxide ($CO_2$) emissions per mile than an electric vehicle.

This is the power of precision—not simply an incremental improvement in the way things have usually been done, but a fundamental transformation of a delivery system, changing life on the ground precisely as Renaudo envisioned for millions, one patient, one delivery at a time. No improvement to a regional trucking network could ever produce the exponential increase in efficiency and impact that Zipline achieved. "We set out five years ago with a goal to come as close to teleportation as possible—a goal many people found crazy at the time," he told reporters. Suddenly the way forward "to provide every human on earth with access to vital medical supplies" is within view.

## Zest: Using Precision Systems to Transform the Payday Loan Market

For a different look into the power of precision systems, let's look at another example—a small corner of the financial services industry that has been transformed by a new, precision-enabled business model revolutionary in its day.

"What would you have done if I hadn't answered the phone?" Former Google CIO Douglas Merrill was incredulous. His sister-in-law had just called for help paying for a new tire. Mother of three, with a full-time job, and a full-time student, she was broke. "I would have taken out a payday loan," she said.

At the time, 19 million U.S. households did the same—unbanked, at-risk, they turned to one of America's 25,000 payday lenders to cover needs. Thirty-three percent of Americans couldn't lay their hands on $1,000 when they needed to. They were prepared to pay payday lenders over 400 percent annual percentage rate (APR)—$15 to $30 on a $100 loan. It all added up to about $30 billion in short-term credit each year.

"I thought that wasn't fair," Merrill said. "Isn't there a way to provide loans to these people that's not so expensive?"

By now, millions of gig workers have entered the payday-borrowing market. As more and more people take up new types of work far removed from the traditional, steady, nine-to-five job, many find they need short-term cash to get through periods between gigs or while waiting for the first check to arrive from a new client. This is where access to a reliable source of credit can be a lifesaver.

Until recently, gig workers and their need for credit were sidelined, largely ignored by financial marketplaces that had no way (and not much desire) to understand them, price products and services for them, or connect the buyers with the sellers. But now the markets have discovered these gems and are unleashing their value. How? By listening to millions of new signals that turned these lumps of untouched consumer coal into diamonds.

For decades, the credit scores calculated by the agency originally known as Fair, Isaac and Company and now called FICO have largely shaped the decisions made by loan underwriters. With the invention of standardized credit scores, lenders saw risk better and could increase both credit availability and profit. But FICO scoring was a blunt instrument, freezing out the underbanked and the unbanked—people with scant or no formal financial histories. Their "thin files" gave FICO scorers little good-caliber banking activity to go on. Risk-averse lenders hedged their bets, which meant that a bankruptcy or credit card default could ruin a consumer's record for a lifetime.

Merrill believed that the data needed to calculate fair lending terms for underbanked customers wasn't scarce. "All data is credit data," Merrill said. "We just don't know how to use it yet. If you could get access to thousands of signals, you could correct that."

Merrill created Zest Cash and began scooping up and feeding in millions of signals from online data, building algorithms and models, helping his clients re-predict risk and make loans to the unbanked and underbanked. The more signals, the better the models, the more insight into real risk, and the higher the likelihood of payback from folks who until then were frozen into a group called the "unbankable" and thought of as "uncreditworthy."

Some, true, were awful risks. But they were mixed in with perfectly sound risks. If Merrill could "liquefy" this frozen mass, divide them into segments, and find the gems among them, he could do for them what FICO did for everyone else: price loans according to true risk, making credit more affordable where warranted.

Sifting through all the big data Merrill could get his hands on, Zest added a bevy of small signals to those used by traditional credit scoring: whether an applicant's Social Security number belonged to a dead person, for example; how much time an applicant spent reading the fine print on the Zest website or filling out the application. They also used *contrarian* signals. When customers dialed in to let them know they'd be late on a payment, banks saw risk, but Zest analyzed their subsequent behavior and discovered these folks could in fact be *diligent* and good bets for full repayment. The more signals that Merrill incorporated into his models, the better they predicted borrower risk and the better they could price the loans. "Not 50 percent better," Merrill said, "just a little better. But take a lot of things that are a little better, and you get something that's a lot better."

Zest's machines learned well. Incorporating classic scoring data and then adding 70,000 new variables boosted the predictive accuracy of the algorithms by 40 percent and increased net repayment rates by 90 percent. Add some further analytics by humans and Zest did even better. "The combination of really big data and human artistry is the underlying value here," Merrill said. "We went into this business to save the underbanked billions of dollars," Merrill told an interviewer. "We knew that applying machine learning and big data analysis to underwriting would make a real difference, and it has."

Several key factors helped to explain Zest's success:

- First, a vision, simple in the extreme: "Save the underbanked billions of dollars—and monetize that for profit." That was Merrill's North Star, guiding the enterprise through ups and downs, against a plan sized for action to test and prove his assumptions and discoveries.
- Second, hard resources, including talented people, loaded up with business acumen and technical skills. A platform and software from which engineers and data scientists could extract the data, analyze it, and make it ready for business action.
- Third, soft resources of motivation and leadership for "doing good" but also scoring big paydays. An organization led by a manager who was open to discovery and expected the unexpected, skilled at building models for new products and services that delivered on the promise.

Zest's approach has spawned plenty of competition—and issues. It was good news for the unbanked around the world—*half the world* by some estimates, comprising 800 million people in South Asia alone. A similar approach, for example, has helped Mahindra Finance bring new credit-scoring methods and lower-cost loans to India's nonbanked rural poor. By 2019, reported *Harvard Business Review*, Mahindra Finance had become the largest rural nonbanking financial company in India, serving 50 percent of villages and 6 million customers.

But the application of precision to the financial needs of the unbanked has also served to expand a payday loan industry whose practices and fees are still often labeled as rapacious, exploitative, and exorbitant. *Better* with Zest, but not great. And now more of it. Precision alone doesn't guarantee results that are socially beneficial. Human judgment and ethical values remain as essential as ever.

## The NFL: Precision Systems at Work

In the fall of 2021, ESPN.com's Kevin Seifert wrote this:

> Social media lit up on Sunday night in the moments after the Baltimore Ravens converted a fourth-down run to seal an exhilarating 36–35 victory over the Kansas City Chiefs. Amid the flurry of "WOWs," shock emojis and celebratory GIFs was a tweet from Michael Lopez, the NFL's director of football data and analytics. After watching Ravens coach John Harbaugh make a difficult decision from his team's own 43-yard line, sending quarterback Lamar Jackson up the middle for two yards, Lopez posted the headshot of a man that only his

followers and a handful of other devotees to football analytics would recognize: Daniel Stern.

Lopez needed no words. Stern's smiling child-like visage was well known to the NFL's in-most crowd as one of a handful of twenty-something analysts bringing the rigors of data science and prediction to in-game decisions. Whether for a quick retooling of strategy between halves, or in-the-moment calls to pass or run on first down, to punt or go for it on fourth down, Stern and his cohort were breaking down the historic confines of data from the quiet pastures of recruitment and player development, coming into their own under the bright lights of game time. In the booth, headsets on, game on the line, Stern's mission was plain: lay out win probabilities to Harbaugh for a handful of plays that would have the best chance for a score and a win.

And Harbaugh—how does he gulp down all this data? "I trust my eyes first," he told a reporter. "The information confirms or opens your eyes to something." He will look at it, consider, and make the call.

Come autumn in America, on any given Sunday, fortunes rise and fall for thirty-two NFL teams. It all depends on what twenty-two huge, strong men, eleven on either side of the football, each incredibly fast, all well-coached and -trained, do to get their hands on the ball sitting quietly on a white chalk line, waiting. In a moment, nothing in the world will matter more. They will claw and tackle and pound away for five or six seconds at a time (the average length of a football play), eleven minutes per game (the total minutes of actual contact), for sixty game minutes (on the clock), spread over 3½ hours. Stoked by beers and junk food, millions of fans cheer from the stands in the cold, rain, snow, and heat, or watch at desks, bars, and in man caves around the world. Billions in team revenues, media rights, and fan-wagering dollars are at stake. Competition is fierce, and every move by every player counts. For every coach, on every play, the challenge remains the same: run the data, perfect the models, predict performance, make the call, hope and pray that in the next five seconds the game changes just the way you want it.

For 100 years, NFL coaches have done all this in their heads. Now machines are having their say, loaded up with AI and machine learning, plumbed by analysts (and now offensive assistant) like Daniel Stern for the nugget that will turn "couldn't be" to "maybe" to "probably," all stitched together in precision systems that turn insight to calling plays on the field. The teams' precision systems chow down on data coming from cameras, embedded radio frequency identity (RFID) chips in uniforms and the footballs themselves, and history, all filtered through models that let analysts predict and prescribe and give options to coaches. Built on the promise of gaining the winning edge, they offer deep insight to players on either side of the ball. They do it fast. And the decision "team" of coaches and analysts preps hard.

Harbaugh, his coaches, and Stern start Mondays talking through their approach to the coming game. They settle on principles and rules they'll follow come game time—when Stern will remind Harbaugh of those guardrails in the headset, giving options. On every play, the coach has a few seconds to weigh the numbers and give thumbs up or down to the offensive coordinator's play call.

"A lot of work goes into it," Harbaugh told a reporter. "We always have a plan every week."

Stern learned by watching his bosses communicate with Harbaugh during games—short, terse, to the point. They speak a similar language, making it fast, talking about win probabilities and expected points added (EPAs). The upshot: "Aggressive fourth-down calls have become a weekly occurrence for the Ravens," reported The Athletic's Sheil Kapadia. "They've gotten 10 of 14 fourth-down attempts on the season, averaging 10.5 yards per play. Eight resulted in touchdowns."

Many debate when this modern-day data wrangling all started. Some ascribe it to Ryan Paganetti, of the Philadelphia Eagles, and peg the date not to a win with data, but to a loss *without* it. Paganetti was the Eagles' analyst and game management coach for six seasons, starting in 2015. By 2017 he was sitting up in the booth during games and meeting with Coach Doug Pederson at half times. Like Stern, weeks and weeks of prep, research, and engagement with coaches and players had earned him a seat and a headset. Toward the end of an early season game against the Kansas City Chiefs, Paganetti counselled Pederson to go for a dicey two-point conversion after a touchdown, rather than a sure-bet one-point kick. The coaches game-side and on the headsets freaked out—"borderline, having a confrontation," Paganetti told reporter Liam Fox. Pederson relented, went for the sure bet, and the game was lost.

Reflecting later, Pederson and coaches committed to Paganetti' s analytics. "From that moment on, there was never another time that entire rest of the season where anyone questioned the information I was sharing with him," Paganetti said. It is the stuff of NFL legend that Philadelphia thereafter won nine straight games, and the Super Bowl.

The move to precision systems in the NFL was on. With the right quarterback—a Lamar Jackson, for example, who could run and pass no matter where or when—fourth down became a "go-for-it" proposition. In 2019, the Ravens ran seventeen successful fourth-down tries, a watershed development. Being the NFL, where innovation is on display every Sunday, fast-follower "copycat" NFL teams soon made the shift to a differently-talented playmaker, too—more mobile quarterbacks like Jackson rose to the top of everyone's must-have lists. Through the first two weeks of the 2021 season across the NFL there were 88 "go-for-it" plays on fourth down, more than at any time in league history. "Pass on first down" was gaining new currency, too.

Today, all NFL coaches are competing on precision as never before. But where "old school" competed on who had the best data, all teams now have much the same basic data available, courtesy of the National Football League's data platform and network, dubbed the NFL Player Tracking System and its customer facing product, the NFL's Next Gen Stats. The platform comprises sensors, receivers, and displays throughout every stadium, fueled by nickel-sized RFID chips in every player's pads and the footballs themselves, on the referees, pylons, sticks, and chains, all beaming player and ball data via in-stadium transmitters and antennas ten times per second, precisely locating individuals within inches on the field, instantly calculating their speed, separations, distance travelled, and acceleration. On every play of every game this NFL data "firehose" adds more than two hundred new data points, sent up to the cloud via Amazon Web Services and back down to fans, coaches, and players around the world, everyone putting it to their own best use.

If all teams are equal in data, a new breed of Sunday warrior makes some teams more equal than others. Each team's "data guys" run from kibitzers and Sunday hackers to pros. Top units comprise software engineers, data scientists, data analysts, and product managers. The average ones don't do much more than read out data from commercially available, off-the-shelf algorithms. The best of them pull the data into highly customized models that integrate data for coaches who, like Harbaugh, trust their eyes but keep their minds open. Tested, proved, and fine-tuned to conditions, NFL models are proprietary and well-guarded crown jewels of high performance. But even the best analysts must elbow their way into the booths and onto the headsets and, like any coach, earn their keep with calls that win matchups and games.

It all starts with player drafts and roster construction. Models *describe* the fit of prospects to the coach's strategy, as Lamar Jackson fit John Harbaugh's like a glove. They *predict* outcomes of matchups against opposing teams and players. They *prescribe*, in turn, the general manager's offers of incentive and compensation packages, knowing what models say the team needs from the player to win—numbers of sacks, runs after completions, or interceptions. Those incentives drop down to weight rooms and trainers' tables where players incorporate data for gains in strength, agility, and speed, not just one time in one game but all season long and over a career. "High caliber athletes could always do that [once]," *Sports Business Journal's* Ben Fischer told an audience. "What they're trying to figure out is, 'How do I do that 25 times in the game, and keep that performance going through the entire season?' "

If player longevity is everything, safety is the key. Data drove the invention of vastly improved helmets, for example, precisely configured for the shock of collisions at full tilt, then spurred player adoptions from 40 percent to 99 percent in three years. Data informed NFL rules changes by showing certain plays as the most injurious. It helped the NFL predict impacts on player safety from adding a proposed seventeenth game to the season.

In the stadium, precision systems have transformed the season ticket-holder experience, personalizing it seat by seat, competing successfully with the comforts of the man cave, spurring renewals even for losing teams, whether by showing unique-to-the-stadium replays-on-demand from every angle on massive screens or phones and tablets, or delivering food and drink right to the seats. Precision systems underpin the lucrative Fantasy Football industry and deepen fan engagement with competition unthinkable even five years ago. Betting on NFL games has accordingly soared, with a piece of all the action going to the NFL. In-game systems have opened new vistas for coaches and players, choosing which first- and fourth-down plays to run depending, for example, on the fatigue of opponents in late quarters measured by cameras that time players who trudge back to the sidelines, or sprint.

As new users pull the data forward in astonishing ways, some players, coaches and teams gain further advantage. Every player could, but only some do. It is that unique, distinguishing, and difficult.

Imagine, for example, if all players managed their personal development as Cooper Kupp has managed his. Kupp ran poorly in the pre-draft player showcases and went late in the 2017 draft to the Los Angeles Rams. Riven with curiosity about the science of high performance and driven by personal thirst for success, Kupp transformed his backyard tennis court into a barn and a personal football laboratory. His goal: test and prove a new science of personalized, data-driven football. His gear, laid in for the purpose: stadium-style turf, specialized treadmills, timing gates, towing units, all to measure his speed, agility, and GPS locations as given by imaging and sensors. His team: consultants who helped Kupp winnow his personal data firehose down to "one actual step and football's most critical movement option," *Sports Illustrated*'s Greg Bishop wrote, literally "the first step off the line of scrimmage." Kupp's plan: achieve maximum velocity five yards off the line sooner than any receiver ever had, leaving defenders behind. Train with drills and sprints "not for every day or every session, but for every *rep*."

The upshot? By his fifth season with the Rams Kupp had become a "phenomenal accelerator, aggressive in his bursts," Bishop wrote, "amassing borderline historic numbers." Kupp was the first receiver to have over two thousand receiving yards in a season as both a receiver *and* a running back, on the snap of the ball "shuffling through his repertoire of movement options en route to exactly the right place, at precisely the right time," a gift of the gods to his quarterback and coaches.

As teams and precision systems mature, powerful new consumers like the Rams' Kupp, or the NFL's Health and Safety Committee, or the Ravens' coaches emerge who exert a tidal pull on data, gaining advantage from precision, at least for a time. As novelty wears off and innovation becomes standard issue, new equipment and in-stadium infrastructure anchor the data. New business practices embed it. New models shape it. New outcomes measure and validate it.

If all teams are created (or soon become) equal on data, advantage comes from platforms, processes, and people bringing all these systems together across the enterprise, in time, continuously improving, applying precision solutions exactly where, when and how coaches want the change, achieving highest performance of the parts and the whole, at once and over time. That takes leadership with the mindset and the skills to do it—thinking of precision not as a "nice to have," but as the cornerstone of new strategy. And in that respect, all teams are decidedly *unequal*.

As we have seen, some teams, like the Dallas Cowboys, were slow to embrace the power of data-driven precision. Former head coach Jason Garrett told the press in 2019 that analytics were "not valuable" to the team during the game—an opinion generally received as moronic. Garrett had no Daniel Stern sitting side by side his own coaches up in the booth. That was Garrett's call, and he paid for it, being fired twice from NFL coaching jobs since.

In 2019, after an abysmal 4–12 season, David Gettleman, the New York Giants' general manager finally conceded the game begrudgingly to analytics, at least in the back office. "We have hired four computer folks, software, and we are completely redoing the back end of our college and pro scouting systems." Nothing about *in-game* management, however. In fact, that same year Gettleman hired Jason Garrett as the Giants' new offensive coordinator, then fired him as the Cowboys had, and then lost his own job after the Giants went 4–13 on the season.

Though the times are changing, any coach favoring analytics still must have some steel. Played right, on average, he'll do better following the numbers, but on any given play or game he could lose. Then the "dumb and I'm proud" crowd will descend. When Brandon Staley's LA Chargers "went for it" on five fourth downs but converted just two against the Kansas City Chiefs, and lost in overtime, Howie Long and Terry Bradshaw went after him on Fox. "Neither of us can spell analytics," Long said, "but it took a beating tonight."

By the end of the 2021 season Staley had redeemed himself rising to the top of FootballOutsiders.com's new "Aggressiveness Index," passing the Ravens' Harbaugh on the way, who fell to number sixteen. In his own rearview mirror Staley could see upstarts closing fast, like the Indianapolis Colts' head coach Frank Reich as he stepped over the still-warm football corpse of NFL workhorse coach Bill Belichick. In their game 15 matchup, Reich called for three fourth-down tries, making but two, but ultimately exhausted Belichick's Patriots and their options. Meanwhile, with nine minutes to go in the game Belichick settled for a three-point field goal rather than attempt a fourth-and-goal conversion for six points. Final score: Colts 27, Patriots 17.

"It's crazy how much more the Patriots would win if he made three to four better decisions each game," a league source told Stephen Holder of *The Athletic*. "Despite being perhaps the greatest strategist in the game's history," Holder wrote, "Belichick still has a sizeable blind spot when it comes to analytic decisions."

Being known as a numbers-driven coach can create perverse new *disadvantage*, which other coaches may seize and win on. In a late season matchup between the Pittsburgh Steelers and the Ravens in 2021, Harbaugh and Jackson failed at a game-winning two-point conversion. They went home losers. Steelers head coach Mike Tomlin had studied Harbaugh's game film and was not at all surprised by the Ravens going for the win. "They aggressively play analytics," Tomlin said. "From that standpoint, they're predictable."

As for Tomlin, a Super Bowl Champ, and the only man in history to *not* have a losing season for 15 straight years (and Zach Tumin's favorite NFL coach), he was last on the FootballOutsiders.com "Aggressiveness Index." And still he was finding a way to win.

## Why We Wrote This Book

In this first chapter, we have book-ended an extraordinary decade with accounts like Zest, one of the earliest commercial exploitations of big data primed for market entry; the sophisticated, dexterous use of precision systems for change in Zipline; and the change brought to the venerable NFL. By 2022, as Kara Swisher had predicted, digitization was practically a commodity. As Doug Merrill blazed the trail by *liquefying* (a term coined by venture capitalist Marc Andreessen) credit data, a venturesome few armed with dazzling technologies liquefied whole *markets* and spawned some of the richest and most powerful new firms on the planet—seemingly lifting not a finger in the physical world, providing platforms rich with AI and machine learning, and designed and serviced by brilliant engineers, designers, and product teams. These platforms connect billions of buyers and sellers. "Uber, the world's largest taxi company, owns no vehicles," marketing guru Tom Goodwin wrote. "Facebook, the world's most popular media owner, creates no content. Alibaba, the most valuable retailer, has no inventory. And Airbnb, the world's largest accommodation provider, owns no real estate." For good measure, add Shein—"the biggest fast-fashion retailer in the United States, with no stores," says Benedict Evans.

That's the power and allure of precision systems putting product and service into the hands of billions of new customers for their consideration and choice. But if everyone everywhere has gone digital, what's making the difference between winners and losers in the marketplace, the political race, the battle space? It's the ability to turn all that digital infrastructure into real-world value where people will pay premiums to get the change they want, where, when, and how they want it—an Uber rolling up to a fare on a lonely city street, same-day drone delivery of medicine to a rural hospital, or a vital payday loan in the urban core. Every solution involves a *precision system* built to deliver change exactly—*precisely*—where, when, and how customers want it and managers intend it. If we look, we can we see such precision systems in action *everywhere*:

- *Precision marketing campaigns* uniquely customize offers to millions of end users (voters, shoppers, travelers).
- *Precision pattern search* discovers hidden links across real-world events (crimes, financial irregularities, cyberattacks).
- *Precision customer shaping* suggests the products or services that a target consumer may want to see or buy next (films, music, clothing).
- *Precision sensor systems* monitor the status of machine and human activities via the networked Internet of Things (train and track integrity, oil pipeline safety, health and fitness wearables).
- *Precision product and services* engineers personalized offerings to suit customers' tastes and requirements (athletic gear, automobiles, medical devices).
- *Precision matching* finds best fits to produce specific desired results (matching buyers to sellers, faces to identities, terrorists to locations, sentences to offenders, job candidates to jobs, curricula to students, football plays to game conditions).
- *Precision scheduling* smooths the operations of computer-assisted physical systems (supply chain flows, airline scheduling, and hospital bed utilization).

We've looked at hundreds of such efforts and spoken with hundreds of executives, managers, and teams, from New York–Presbyterian Hospital and BNSF Railways to the *New York Times* and the Biden 2020 campaign. Each used precision systems for their own purposes—whether to boost subscriptions and revenues, win games and elections, speed freight and passengers on their way, or improve health and save lives.

In each instance, we see precision systems comprising a range of techniques and an experimental approach that can be applied in diverse settings. Precision, like math, fits everywhere and can be used for everything. Like math, precision is a collection of techniques, facts, and methods that can be used to answer all sorts of questions—to describe the current world, model and predict the future, prescribe the best step forward.

As this perspective unfolded through our research, we learned a lot about the amazing results that precision systems make possible—the power to create just the change managers want, where, when, and how they want it. That's power *for good*, when used wisely. We've also discovered just how difficult precision systems are to design and implement well, and the risks of doing it badly.

We wrote this book to help managers be successful facing this new onslaught of possibilities and demands. We share what we learned from those in the forefront of precision by studying their tools, techniques, habits, and best moves forward. No matter what kind of organization you may lead or manage, you'll find insights here that can help you see how precision can enable you to do your work better, as well as nuts-and-bolts guidance into the details of making it happen.

## What You Can Expect from This Book

Over the next ten chapters, we'll unpack the technologies, people, and strategies of precision systems to yield lessons that any executive or manager, coach or campaign chief can use right now to get ahead and stay ahead.

Chapter 2, "Six Kinds of Precision Systems," homes in on six precision designs and their essential attributes, technical requirements, and fit to a range of business problems. It explains how the designs work, when to use them, and how to determine what kind of problems your organization currently faces.

Chapter 3, "Design, Test, and Prove: Key Steps in Building a Precision System," brings you inside precision operations. You'll read about and learn from managers as they test whether have found the right audiences for their campaigns, whether their pattern finding works, whether their recommendations are relevant. You'll learn how to determine what counts as a win at the outset—big enough to matter, small enough to work.

Chapter 4, "Getting Started with Data and Platforms," helps you decide which business problems are best suited for precision solutions and how to prioritize opportunities to get the most value possible into the right hands as soon as possible.

Chapter 5, "The Precision Vanguard," explains the roles that organizational leaders play in driving the conversion to precision.

Chapter 6, "The Political Management of Precision," describes the challenges and headwinds facing precision, from status quo bias to the endowment effect. You'll learn how successful leaders have found ways to overcome these barriers.

Chapter 7, "Bootstrapping Precision," examines the kinds of collaboration needed to bring precision to life. You'll learn about the external and internal partnerships that precision design requires, as well as the critical issues of governance for collaboration where traditional lines of authority will be missing.

Chapter 8, "Creating and Managing Teams of Precision," offers insights into the special challenges involved in organizing and leading the kinds of interdisciplinary, cross-functional teams that are needed to design and implement a winning precision system

Chapter 9, "Building the Precision Enterprise," examines the best organizational architectures for precision. We organize the discussion around the critical tasks that precision requires, the formal structures needed to support it, the essential skill sets and work design that make it happen, and the cultural norms and values needed to insert new precision systems into existing business processes.

Chapter 10, "Red Zones of Precision," examines some of the principal technical, ethical, and operating risks that the leaders of any precision initiative must address. It illustrates specific ways of measuring and minimizing these risks.

Chapter 11, "Summing Up: Some Final Lessons for Leaders," summarizes the most important takeaways from the book. You'll learn about the three key leadership roles involved in launching precision systems. The admiral creates a strategy for developing a portfolio of precision projects. The orchestra leader manages a group of self-sufficient professionals who must harmonize their efforts in implementing precision. The pit boss guides precision projects past implementation barriers and pitfalls to reach the end-point delivery of value.

Throughout the book, you'll be introduced to organizations from many industries and economic sectors that have made the big leap to precision systems, overcoming obstacles and challenges along the way. Their stories will help you recognize the enormous opportunities that precision offers your organization—and they'll likely inspire you to begin or accelerate your own journey of exploration into the amazing new world of precision systems.

## CHAPTER 2

# SIX KINDS OF PRECISION SYSTEMS

We use the term "precision system" throughout to refer to the powerfully connected assets of people, processes, platforms, processes and data that managers use to change the behavior of people, places, and things exactly how, when, and where they need to. Whether it's searching out crime patterns, recommending the best next song, or sending the right patient straight to the intensive care unit from the emergency room, precision systems are in use all around us today.

People can be addressed with appropriate messaging (like cohort-specific political campaign emails), presented with thoughtful options (goods and services they might like), or alerted to problems they are concerned about (like an unusually bad night's sleep). These capabilities can be used to influence people to think and act differently—presenting them with a different news article or a different type of clothing in order to influence a vote or a purchase down the line.

Places and objects can be monitored, and the flow of them adapted, like automatically rescheduling hundreds of flights based on predicted bad weather or decommissioning a train whose brakes were detected to have worn thin. Objects are enabled to assess situations and even make decisions autonomously, like recognizing a face in a crowd and alerting law enforcement or identifying a fraudulent transaction and freezing the credit card.

Precision systems are often based on artificial intelligence (AI) and, with machine learning, can take on human-like capabilities, from communicating to planning; reasoning to manipulating objects; and, above all, learning. They are put to tasks ranging from chatbots for lead development to website operation and optimization, to complex imaging of everything from cracks in train wheels to brain slices. While managers can "pop up" precision systems for specific, quite narrow purposes, the great challenge of managers and executives is to stitch systems together for maximum effect across entire "journeys" of people, places, or things, whether the journey is of a patient to good health or a river to clean water or a locomotive to continuous service. "Executives," Tom Davenport wrote in a 2021 *Harvard Business Review* article, "will ultimately see the greatest value by pursuing integrated machine-learning applications" that span the entirety of such journeys, literally from cradle to grave.

And it all starts with technology.

## The Technologies of Precision

One useful way to introduce precision technology is by illustrating how it has transformed preexisting technology types that most people are already familiar with. Consider, for example, the difference between calling a taxi company to request a pickup, as we did only ten years ago, and today's option of opening a smartphone app, requesting a ride, and watching on the screen as the car approaches your location. Ten years ago, the technology used to accomplish this transaction was the telephone. As the customer, your data (your current location) had to be shared with the taxi company verbally. The operator would provide confirmation of dispatch, but after hanging up the phone, you would have no idea where the driver was, how long they would take to arrive, and what the price of the trip would be because the taxi company shared none of this data with to you. This was technology before precision.

Today, the technology used is the smartphone app. It has access to your location (global positioning system [GPS]), your payment method (digital), and your preferences (such as saved favorite locations). You input your destination, and in an instant the GPS locations of all nearby drivers are calculated, they are alerted to a new customer, a price is dynamically calculated, and an estimated time to arrival is offered. This is technology with precision. It involves the purposeful use of data points to offer a customer experience that is simultaneously more customized, more accurate, and more transparent than its predecessors.

Another example of how precision enhances existing technology can be found from the sofa. Not so long ago, the way to find something to watch on TV was to scroll through a list of available channels and, if you were lucky enough to have cable, reading the menu of currently playing and next-up shows. Shows played at set times that were inflexible, although determined customers could record timed shows to watch later. This was a digital experience, but not a precise one.

Today, the browsing experience offered by digital video-streaming platforms is much different. You can pull up an app on your TV, which remembers what you have recently been watching, what you haven't seen yet, and can recommend what you're likely to enjoy watching. These apps collect data about what you click on, the trailers you watch, the shows you binge, and the shows you abort early, and they use all that data to run myriad experiments designed to perfectly hone a profile of what you like and don't like: a quotidian experience powered by unimaginable digital precision.

The term "precision systems" doesn't embrace everything online or everything powered by data. There are plenty of examples of precision operating offline (we'll discuss John Deere's computer vision tractors in chapter 3) and likewise many data-powered systems that aren't precise at all (presenting a viewer with a list of all currently playing shows requires some data).

Precision can be identified by a range of tools, techniques, purposes, and outcomes. Prediction is a recognizable example; something as simple as predicting next month's sales based on last month's sales is an attempt at precision. Another is personalization—launching a product experience unique to each customer. Devices we colloquially refer to as "smart" are often hubs of precision applications: cameras that can recognize the faces of friends and family, speakers that can understand commands and return responses, even refrigerators that can tell our milk is expired.

If a system collects data, and that data is used as an input toward some kind of process, and the output of that process is not predefined by rules, then there's a good chance it's a precision system. Precision systems surround us today.

## Under the Hood of Precision Systems

Under the hood, the actual mathematics powering precision systems can range from sophisticated statistical analyses and advanced machine learning to much simpler, back-of-an-envelope calculations. There's no rule that precision must be AI- or machine learning–enabled. The variety of tools and techniques at the disposal of analysts is vast, and the range of situations they can tackle to solve problems and create value is even more so.

The defining difference between a precision system and any other kind of system, however, is the use of *all* available data by managers to bring about a specific desired change in the status and future of a person, place, or thing—*and no other*. It may be said that *precision systems' entire purpose is to convert the transparency of the world offered by big data into value in the real world, measured by producing benefits with few if any collateral costs.*

Precision systems built on massive new data can confer this advantage because they are more accurate, faster, and more adaptive—more *intelligent*—

than any comparable systems of old. If powered by AI and machine learning, they are *designed* to get smarter with each use—what Jim Collins has called AI's "flywheel effect" and Andrew Ng its "virtuous cycle." The more these systems are used, the smarter they get, the better recommendations they make, and the more they get used—to the point where 85 percent of Netflix customers' choices result from Netflix's recommendations.

It's neither possible nor necessary for every manager to have mastery of all these tools and techniques, and most don't. As long as they grasp the basics under the hood well enough, says Harvard professor Jim Waldo, a Sun Microsystems veteran and recent past chief technology officer at Harvard, they can manage precision systems for success. "The biggest change today is that every manager is now a technology manager," Waldo says. "They either understand that, or they screw up—no matter what kind of management they're doing. Technology is going to be what you're up to and that, in turn, means that whether you succeed as a manager will depend on things that you don't understand." Understanding and mastering the technological problems is only one part of the precision challenge, however, as the following story illustrates.

## Patternizr at the NYPD

In 2018, a December holiday-season robbery wave hit Bronx convenience stores in New York City. It was the Bronx command's turn in the docket of the New York Police Department (NYPD) weekly Compstat, the high-pressure weekly calling-to-account all commands face at some point. *New York Magazine* reporter Chris Smith recalled the excoriation of the Bronx command as Chief of Department Terrance Monahan and Chief of Crime Control Strategies Dermot Shea grilled Bronx detectives.

> "Squad, who is looking at this? This is a commercial gunpoint robbery!" Monahan snaps.
>
> Shea interrupts. "I want to go backwards: We have a robbery and there's a woman in the store. She happens to be on the phone, puts the phone down, opens the door, and the robber comes in?"
>
> "That's correct," says Betania Nazario, a sergeant in the detective squad of the 42nd Precinct, which covers Morrisania and Crotona Park. "Do we believe she's involved?" Shea asks. "Um, we're still working on that," Nazario replies.
>
> "Hold on," Shea says, flipping through sheets of paper supplied by a research assistant. "Seven blocks away, a week earlier, we've got another bodega robbery in the 4-2, four perps, one of them a female. I don't remember that many bodega robberies with female perps. Could she be part of this one too?" "That's

a possibility," Nazario says. Shea's tone turns cutting: "Unless you're telling me it's common for females to be involved in bodega robberies."

"Commercial robs are spiking significantly, all with perps casually walking to a robbery," Shea says. "We've got to backtrack, draw that circle, find what building they went into, and ID them. This is time well spent, this is why we close shootings, because we put the resources in, we backtrack, and we canvass." Then he says it's time for a coffee break, and the entire room exhales in relief.

"I remember sitting in that Compstat," Zach Tumin says. One of the coauthors of this book, Tumin was deputy commissioner at the NYPD on Commissioner Bill Bratton's executive team. Compstat was the weekly meeting of the top 200 NYPD commanders to discuss crime on their beats.

"The Chief of the Bronx was at the podium, being interrogated by Monahan and Shea. Shea was asking the chief how many robberies were *outside* a pattern at that moment. 'About a thousand,' the Bronx chief replied."

"I was astonished," Tumin said. "We know that typically robbers commit more than one robbery. They work out how, and where, and when, and what works. Then they do it again and again. Robberies should be easy to pattern."

Tumin turned to a captain sitting next to him. "Why are there so many robberies *outside* a pattern?" he whispered. "You want patterns to solve crimes more easily, don't you?"

"Yes," came the answer, "but once you 'declare a pattern,' you're on the clock for solving the crime. The commander is on the spot to throw resources at it—resources she may not have. So best to keep it out of a pattern a long as you can."

Tumin realized this might turn out to be more complicated than he thought.

Throughout history, recognizing the similarities among crimes has been the key to solving mysteries and bringing criminals to justice in every police department worldwide. Until very recently, however, this craft has been as much art as science. Television shows and movies showcase the power of genius investigators mulling a seemingly unsolvable crime until—bingo—they use an obscure detail to connect it to an earlier case, the criminal's story quickly unravels, justice is restored, and the city sleeps well that night. It makes for thrilling entertainment; in the real world, intuitive pattern finding is not the most scalable or reliable method of crime solving. Police investigators need either really good memories (which some have, but not all), or detailed systems for record keeping and technical support (often in short supply as well).

Tumin was fascinated by the potential of pattern recognition as a crime-fighting tool. But the exchange at Compstat made it clear that the NYPD's current methods of pattern making had much room for improvement. It was all too iffy and left to chance.

Alex Cholas-Wood, a data scientist on Tumin's staff, also had an interest in patterns. Coming out of the meeting, Tumin chartered Cholas-Wood to launch a project to see whether and how computing could pattern crimes faster. Together with Evan Levine, another data scientist at NYPD, Cholas-Wood set out to bring the power of precision to the age-old problem of recognizing when one recent crime is similar to another and likely perpetrated by the same criminal.

Sitting around his worktable with Cholas-Wood and Levine, Tumin probed three metrics that mattered. First, what proportion of robberies are *not* patterned at any given time? "In a perfect world," Tumin suggested, "that number should be pretty close to zero." The underlying guess here? Most if not all robberies are committed by an individual or group who has done it before. But is that true? The first step would be to verify that guess by drawing a sample—it could turn out that 95 percent of all robberies were repeats. If 95 percent of robberies were committed by one or more or the same individuals, and you had 1,000 robberies not patterned, you had a lot of work to do to make more matches. After all, the data predicts that 950 of those 1,000 are related but not yet matched.

Second, how many crimes are committed before a pattern is declared? "In a perfect world, again," Tumin said, "that number would be two." In this respect, crime patternmaking might be a lot like the TV game show *Name That Tune*—the host reads a clue and two contestants bid the fewest notes they need to identify the song.

Third, how much time elapses from first crime of a pattern to the last, when the perpetrator is arrested? "How long did we let this guy walk around being dangerous? In a perfect world," Tumin suggested, "top performance is exactly the time elapsed between the first and second crime. If we find the pattern in two events, then he's on the prowl only for the time between those two. But if it takes us ten robberies to see the pattern and make an arrest, he could be on the prowl for months while will we figured out his pattern. That's a lot of dangerous time on the loose."

"If we can create a pattern-making system with these three metrics as our guide," Tumin said, "we should never again hear a chief stand up at Compstat and announce he has a thousand un-patterned robberies. That would be an embarrassment."

Cholas-Wood and Levine recognized that this problem had all the characteristics of a good candidate for precision systems treatment. A precision system could turbocharge criminal pattern detection by making those "genius" connections great detectives make between case A and case B much faster and across a much broader range of crimes than any investigator could reasonably cover alone. The automated system would scale well across hundreds of detectives, adding consistency and improving effectiveness. The data required to train the models were available and high quality. Machine learning with new cases would constantly improve the data's accuracy. With thousands of robberies, burglaries, and assaults sitting outside patterns across the city, the need was urgent. No one could dispute the value of stopping crime short: it would save lives.

Although the NYPD was hardly born digital, it did have a long history of innovating crime-fighting tools and techniques. In the 1990s, Bratton's legendary aide Jack Maple had pressed precinct commanders to push pins into precinct maps wherever a crime occurred, updating them regularly. As crimes clustered visually on the map, commanders could respond by putting "cops on the dots," as Bratton famously boasted.

What Maple soon realized was that the locations of the pins on the map were never updated. It was precision in name only. Frustrated, Maple strung wires from the upstairs servers at One Police Plaza headquarters, through his window into a RadioShack computer and downloaded the data himself. NYPD crime fighting had gone digital, presaging digital Compstat—and leading to a reduction in violent crime of 10 percent in two years. A 10 percent year-over-year gain was huge. It suggested either that the cops were much worse than anyone thought at the core business of fighting crime, or perhaps they were great and Compstat just made them that much better.

Bratton came down in the middle, as was his style. He served the NYPD twice as its commissioner, first from 1994 to 1995, when he developed Compstat, and again from 2014 to 2107. Tumin left the NYPD shortly after Bratton did in 2018. Cholas-Wood and Levine were transferred to the command of the new Chief of Crime Control Strategies Lori Pollock. While Chief Pollock was open to change, she recalls being skeptical of the precision project that Cholas-Wood and Levine were bringing to her, which had now been given the name of Patternizr.

"It all depends on how unique the crime is," she explained. She often lectured Tumin on the realities of policing:

> You could have a guy using a brick to smash a window in a candy store between the hours of midnight and 5 a.m. But that's not that unique of an MO—"brick-through-store-window" happens a lot. Can you make a pattern out of that? Yes. you can, but only by using other details—like getting DNA off one brick and seeing if it matched the DNA off another brick. Of course, just because you have DNA, obviously it doesn't identify somebody. It just tells you that it's the same person.

Undaunted, Levine and Cholas-Wood took stock of what they had: the skills to build the model; access to the analysts and detectives who could use the tool; and sets of data covering a decade's worth of grand larcenies, robberies, and burglaries. Their hardware was modest—a standalone Mac computer, a gift from the head of information technology (IT) who thought they would use it to produce the department newsletter.

They got to work using a common machine-learning model framework called random forests. The random forest framework classifies items from a given data set into useful categories. In this case, it examined individual crimes of interest—so-called seed crimes—and asked, "Does this crime fit a pattern with another crime?"

The result was a collection of decision trees used to parse the input data set and separate the crimes methodically according to their various features. Some of these features applied to all the crimes—for example, the date of the crime and its location. Others applied only to specific types of crimes: Was a robbery armed or not? If it was, what type of weapon was used? If a grand larceny was involved, what was the item stolen and what was its dollar value? Each of the many decision trees that make up a random forest sorts through such features and calculates the likelihood that one item in the database is related to another.

"What Patternizr does," Chief Pollock explains, "is it sifts through these large data sets of time and space after you put a seed complaint in, to generate things that closely resemble your seed complaint. Usually, we looked at the top ten because they give you the highest ranked confidence levels. From there, you can decide whether they're part of your pattern or not."

Historical data about established crime patterns were used to train and test the model to be as accurate as possible. They called this rebuilding a pattern—could Patternizr accurately identify the same crimes in a pattern that detectives already had? For example, if a historical crime pattern had five crimes, and Patternizr was given one of those five, could it suggest the other four as related? In a test of 150 patterns, Patternizr correctly rebuilt 37 percent of burglary patterns, 26 percent of robbery patterns, and 27 percent of grand larceny patterns. For those three crime types, it was able to find patterns with a match in the top ten crimes of 82 percent, 78 percent, and 76 percent, respectively. This was a good starting point for a brand-new precision system, but there was still much work to do.

Cholas-Wood and Levine were keenly aware of the need to make Patternizr trustworthy and trusted. They removed from the data sets any feature describing sensitive personal information such as the race and gender of victims and perpetrators, as well as precise details of crime locations. These steps helped to reduce the number of false positives based unfairly on race or gender, and they avoided exposing the precise locations of innocent people. The team also established a multilevel structure of human review so that any detected pattern would be confirmed by police leadership before the arrest of a suspect.

Cholas-Wood and Levine developed the user interface for ease of use and ability to integrate with other NYPD systems. Patternizr proved useful in its early applications. But design elements and work routines affected its use in operations. These are worth enumerating because they impaired its overall usefulness and value.

## The Challenge of Patternizr

Even with technical safeguards, real-world design elements and work routines affected Patternizr's use in operations. First, Patternizr did not run in the background, constantly examining new cases and setting off alerts when a new case might bear similarities to existing unsolved crimes. Rather, it had to be turned on—operated on demand by civilian analysts and detectives for robberies, burglaries, and grand larcenies

Second, the need to turn on Patternizr meant its use had to be prioritized, which could easily move it to the back burner. For example, if the NYPD brass placed a high priority on suppressing a rash of gang shootings in some tit-for-tat turf battle in the Bronx, that could divert detectives' and analysts' attention from other violence, such as stranger-on-stranger muggings.

Third, Patternizr was initially built for only three crimes: burglary, robbery, and grand larceny. This left holes in its usability, making it something of a mysterious "black box"—and potentially unreliable for investigators.

Fourth, machine learning works worst on its first day and better every day it is in use. But that takes engineers, data scientists, and end users looking at results and fine-tuning models. In an organization not set up to sustain AI systems, that requirement puts AI systems like Patternizr at risk of becoming ineffective and thus lowers confidence of its users.

Fifth, metrics did not change—no one was held to account for patternmaking when using Patternizr any more than they were in the good old days prior. That risked letting crimes go unpatterned that should have set off alarm bells instead.

Sixth, perverse incentives could delay "declaring a pattern" deep into a crime spree. Chief Pollock explains:

> If I'm a precinct detective and I'm working a robbery, say, a guy robbed somebody with a gun. I do all this work—I get video, I talk to witnesses, I have all my DNA, I get ballistics. I do all this work, and now I put that crime into Patternizr. And I find out that the same guy did it two boroughs over—that's going to go to a different unit. So all my work is going to get taken out of my queue. I'm still going to be part of it, but all my work is now going to go to Manhattan Robbery, let's say, or Central Robbery, guys who work patterns citywide.
>
> So as you get closer to the investigator, he may or may not want to use Patternizr. He may want to get rid of that case right away. He's like, "Let's put this in Patternizr. Let me see if there's a pattern somewhere in the city that I can go, 'Here you go.'" You know? It's politics, right? It's not that the crime is not going to get prosecuted, hopefully. It's just, there's different reasons why people outside of the crime analysts might or might not use Patternizr.

All of this was sand in the machine, grinding against Patternizr's usefulness in practice compared to its potential. None of the points of weakness listed above constitute fundamental failures. The availability and quality of the data, the performance of the original model, the skills of the team who built it, and the good intentions of the organization all remain unchanged. The NYPD developers carefully built three high-performance models; they rightly excluded sensitive data like gender and race and even kept the data stored on local computers rather than using cloud storage and risking a data breach. In all these ways, the NYPD teams did a fine job in crafting Patternizr. The NYPD's failings here were less obvious, more complicated, last-mile failures that prevented—and likely still prevent—a good product from having the impact that it should have.

The upshot: when leaders invest resources in a risky but potentially powerful precision system but then layer it on top of unchanged everyday workflows, neglect to remetric performance accordingly, and leave it imperfect in development and optional in use, they may unwittingly pave the way for failure. Precision systems require careful preparation and constant care and feeding. As if painting a room or a house, preparation of the underlying surface is everything; regular maintenance is essential to improved performance. Failing either risks a loss of value in the investment and declining performance over time.

## Systems for Finding Patterns

As the Patternizr story shows, implementing a precision system and achieving the results desired is far from simple. Solving the technology challenges is only part of the equation. Equally important are the managerial and social challenges that make most leadership tasks difficult, including those involved in the world of precision.

Fortunately, myriad sources of information can help leaders dive into this world. In the next part of this chapter, we describe common problem types to which precision systems can be applied rather than focus on specific technical tools. By being problem- instead of solution-centered and focusing on the why instead of the how, we can achieve two goals. First, we can help you grasp the value of precision without asking you to master a profusion of technical information. Second, we can provide a range of examples across varied industries and sectors that will encourage you to recognize possible applications to your own challenges.

The challenges that Patternizr was designed to address vividly illustrate one of the most common problems faced by leaders today—the challenge of *finding patterns*. In a world that's awash in a vast, constantly flowing stream of information, it's extremely difficult for a human brain to extract the most accurate, useful signal from the noise. The scale and speed of the data are simply too much for one person or even a team of people to manage. They may know or suspect that within this information lie patterns hidden to the naked eye—correlations between one variable and another, clusters of similar objects, or a reliable prediction that A often leads to B. But most people are unable to put their fingers on the patterns without extra help.

Imagine a baseball manager who wants to know how many base hits his everyday left fielder is likely to have in a crucial upcoming series against the team's toughest rival. He is wondering whether it's time to bench the player, who has been struggling at the plate, in favor of a different outfielder who may be ready to get hot. To predict this, the baseball manager could attempt to watch recordings of every game his team and the opposing team have ever played at this stadium, scrutinizing player profiles, specific pitcher-batter matchups, even weather conditions. Ingesting this volume of data would be difficult or impossible, let alone absorbing it well enough to identify subtle correlations within the games. A precision system designed to watch hundreds of hours of games could do so without trouble, remembering and analyzing every millisecond and independently estimating that, given the variables of the upcoming games, the star player is likely to continue his struggles with the bat. The baseball manager can take that insight and restrategize as he sees fit.

Now consider a marketing manager who wants to know the following: of all the people who arrive at my website, which ones are most likely to end up making a purchase on the website—the people who arrive from a Google search, those who arrive from an email referral, or those who click a link on Facebook? She could trawl through millions of rows of raw click data in a database, looking for visual clues indicating the relative frequency with which a purchase click happened from a user who came to the website via one source or another. She could also use a system that would crunch those totals for her in a moment while also offering her different variables, such as the country of the user and the dollar amount of the purchase, and streaming in new click data to update the result every few minutes. She could then prioritize her marketing efforts accordingly.

These scenarios rely on tools that find patterns in information, an activity sometimes known as data mining. These tools can be instructed to look for a specific pattern or to sift through the data without looking for anything specific, surfacing discovered patterns of any kind.

Data scientists looking for patterns may also use tools trained in skills like *anomaly detection* to catch notable deviations from the norm, *linear and logistic regression* to find associations, or *cluster analysis* to find similar groups. After the insights have been mined from the data, any number of things can be done with them. Regardless of the specific quantitative technique used, what matters is the overall goal or purpose of the application: to sift out the signal from the noise and learn about patterns and insights hidden within the data that would otherwise be invisible to the human eye.

## Systems for Personalization

The concept of *personalization* is a mainstay of ecommerce companies and digital organizations in general. Personalization is the effort to take what is known about an individual customer, prospect, or other target and present to them a version of the digital experience that reflects what they are interested in and likely to engage with. This can result in literally millions of slightly different experiences being presented to millions of people worldwide, by the same company, in the same instant.

Flagship examples of personalization include Google, Amazon, and Netflix—tech giants who make your search results different from the results for the next person, even when you both enter the same search term. When you log into your Netflix home screen, it's not just the "recommended for you" section that reflects what you prefer—it's everything. The order of the sections on the page, the arrangement of program titles within each section, the titles that appear first when you type the first letter of a search term, even the short ten-second clips that play automatically when you hover over a title are all selected just for you, in an instant.

There's a reason for personalization. No longer just "nice to have," it is for many businesses practically existential—"the rod of iron that runs through the foundation of every successful organization," Jenna Delport wrote in a blog post, "without which there is little left to differentiate itself."

"Personalization adds a layer of stickiness to the business," Patrick Reeves at marketing firm Axiz told Delport. "It binds the customer and keeps them in an orbit you can use to further engage and build loyalty."

Many news publishers and aggregators employ personalization techniques that are similar to Google, Amazon, and Netflix. They consider the types of stories you have clicked on before balanced with the top headlines of the day, which results in a package of articles personalized for you. Especially within the context of news media, this kind of personalization can have far-reaching impacts on public perception of truth and facts.

One of the ways these systems achieve personalization at such massive scale continuously is through the concept of *audiences*. Precision systems can bunch together groups of people and objects based on characteristics it knows that they share. For example, millions of people who have clicked on products, movies, or news articles about self-driving vehicles might be grouped together under an audience label like "interest group: machine learning & autonomous vehicles."

The technique often powering these audiences is called collaborative filtering, and it relies on metadata about the content in order to recommend similar content to the people who interacted with it. This means that the next time an article about self-driving cars is published, the recommendation engine will show it to that audience with a higher priority than it shows it to other groups, on the assumption that they'll be more interested.

Sometimes collaborative filtering can't be used, for example, when no one has interacted with new content yet. Without interactions, collaborative filtering can't find similar users. This is known as the cold start problem. To get around it, a technique called the genomic approach is emerging; it aims to understand the actual nature and attributes of the content itself, using content descriptors, and begin recommending that content to people who have liked content with similar attributes.

Precision content management goes beyond just recommendations. The articles, posts, and videos we see are *generated* more and more often by intelligent systems, too. If you were reporting on the Olympics, would you need to write individual and different articles to summarize the results of hundreds of different matches in different sports? The *Washington Post* thinks not. It built a bot to scrape the results of each match and use a *natural language processing (NLP)* algorithm to produce succinct event summaries. They called the system Heliograf, and today it has been expanded to high school sports events and even election coverage, freeing up journalists' time to write more complex pieces involving investigative and editorial expertise.

Other examples of precision systems generating, augmenting, personalizing, and distributing content abound. Their use has exploded along with the internet, social media, and consumer rates of digital consumption. Once the content has been produced, it can then go through the rigors of personalized recommendation systems so that what you see on your screen may been shown only to you and a few "people like you" without having been touched by a human. Advances in speech-to-text and text-to-speech technology allowed Amazon's audiobook arm Audible to sync up the audio reading of a book with its text sibling word-for-word, allowing readers to switch seamlessly between reading on device and audio narration—another form of personalization made possible by precision systems.

## Systems for Predicting Values

Sometimes, all you need is a number. *Predicting values* is one of the most common applications of precision. The challenges involved may run the gamut, from simple, even obvious, to enormously complex.

A business that is forecasting sales numbers for next quarter needs to consider changes to internal factors like the number of salespeople and sales support technology staff, along with external factors like economic strength and average consumer disposable income. It then uses proven predictive techniques to forecast future sales values. They may need to repeat this exercise every quarter.

Another example might be a customer care center manager who needs to predict the number of staff she'll need to schedule for Cyber Monday, the annual shopping day when online holiday sales hit their peak. She could use a probability distribution to blend data from prior Cyber Monday customer call volumes with her own intuition from years of managing the center. The function would spit out a range of predictions with various likelihoods, and she'd be able to create a plan A and a backup plan B for staffing.

For a use case on a massive scale, consider a company running a programmatic advertising campaign for a particular brand that needs to calculate the exact dollar amount it should bid on an *impression opportunity*—that is, a chance to appear on a website visitor's page—in order to win the auction and have its ad appear, while not bidding too much higher than competing brands and thereby wasting money. For large programmatic advertisers, this number needs to be calculated thousands of times *every second*. Settling on the answer depends in part on tracking and measuring the real-world impact of programmatic "spend." Sources of sales data for such purposes vary from Nielsen home-scanning panels, where shoppers record their purchases, to Facebook data that show whether ad exposure brought people to car dealerships, to footfall analysis that links individual store visits to specific programmatic campaigns via in-store WiFi, to classic click-through rates.

But until Instacart came along, it was all ultimately rough recollection and guesswork. marketing's version of "Just So Stories," as one critic lamented, "made up examples of superficially plausible targeting," but lacking the rigor of multiple tests and controlled exposures. Instacart, by contrast, "was the first 'performance marketing' consumer package good makers had ever seen," Ryan Caldbeck of CircleUp told a podcaster. "It gave them direct awareness of every item a consumer ever bought. If they know I love Heineken beer, and if the product manager of Stella wants to try to convert me, they can give me a Stella coupon and then see if I change my behavior over time. Getting that data is the holy grail."

Once settled, precision systems designed to evaluate a multitude of changing input factors and create the optimal output value—be it a sales forecast, a staff number, a bid, or anything else—can range from a handmade one-time analysis on the back of an envelope to an industrial-scale, real-time prediction engine. As with data science in general, the scale of the system needed for a use case depends on the speed, frequency, and volume of decisions the system needs to make. There is no need for a machine-learning algorithm to answer a simple question once a year.

When a complex question must be answered, an array of statistical tools may be called upon. For example, there's *uni- or multivariate linear regression* that predicts future values based on past inputs, economic concepts like *supply and demand modeling*, and advanced data science techniques like *random forest algorithms* that can predict the changing likelihood of an event occurring. All of these can be deployed in predicting important values.

Ben Singleton, a data science leader formerly of JetBlue Airways, led a team that used data science tools called XGBoost and LightGBM to predict the specific dollar amount that customers would be willing to pay for an extra-leg-space seat upgrade for a flight. These tools, which Singleton describes as "statistics on steroids," consider the length of the flight, the time of day (red-eye flights are more painful), and the average ticket cost, among a number of other data points, and outputs the recommended price for each flight. To eke out a decent profit margin from this offering in the brutally competitive airline industry, an airline may need to calculate this for every single flight, every single day—lest it let the price sink too low and lose profit or let it rise too high and lose customers.

This practice is known as *dynamic pricing*, and it's something most airlines have invested heavily in by now. Earlier techniques for price calculation were rule-based, meaning that the airline used "if X then Y" logic to define the price for a ticket or service; for example, "if this customer is booking more than thirty days prior to flight, the price offered should be $300". This was not precise, by our definition, because the prices were neither computed for individuals nor computed on the spot.

In comparison, newer algorithmic pricing techniques take in a wider range of factors: the customer's loyalty status, competitor airline prices in real time, and even the weather, to produce a specific price for each customer at the time of search. Rather than prescribing the price according to rules, the correct price is derived from a continuous process of offering individual prices and adapting them according to consumers' decisions to purchase or not. Altexsoft describes these two approaches as "static pricing" and "dynamic pricing," respectively.

Predicting can be a matter of life or death, too. In early 2020, Peter Chang, physician and director of the Center for Artificial Intelligence in Diagnostic Medicine (CAIDM) at the University of California, Irvine (UCI), saw the rapidly rising numbers of COVID-19 cases and knew immediately that hospital intensive care units could easily be overwhelmed, which in turn could lead to unnecessary mortalities from the disease. Using statistical techniques, he built an algorithm that used a patient's demographic data, physical condition, and lab results to predict the likelihood that the patient will require intubation in intensive care. He called it the Vulnerability Scoring System and was able to roll it out for use in UCI Health facilities within four weeks.

## Systems for Entity Recognition

If you've ever complied with an internet browser's infuriating demand to identify all the trees, cars, or bridges in a photo, you have participated in training an *entity recognition algorithm*. Yes, these so-called captchas are used to weed out humans from bots, but the process also effectively outsources data labeling. Thanks to you and millions like you, the underlying model got slightly smarter at recognizing trees, cars, or bridges in other photos.

Entity recognition is something the human brain is naturally very good at. You don't need to think very hard to recognize your friend's face approaching you on the street or tell the difference between a photo of a dog and a photo of a muffin. Now AI systems (box 2.1) are learning to master entity recognition on a much larger scale than a human can.

Systems for entity recognition are in wide use today. Sky News used entity resolution to recognize the faces of celebrities present at a royal wedding for fashion and style coverage of the event. BNSF Railway uses entity recognition to find microscopic cracks in wheels as trains roll by at seventy miles per hour (more on this case in chapter 9).

University hospital AI departments are now developing clinical support tools for radiologists that can sift through X-ray, magnetic resonance imaging (MRI), and computed tomography (CT) scans, sorting them into normal (healthy) and abnormal (unhealthy), and highlight sections of the scan to

### *Box 2.1 AI*

AI is often a principal technology component of precision systems. Using AI, humans train computers to make humanlike decisions. Making calculations, understanding written text, and performing a trained set of actions are all things that an AI system can do. But before it can learn, humans need to teach it.

Machine learning is a subset of AI in which computers are enabled to learn things without humans explicitly teaching them. Given a huge amount of data and a starting algorithm, the computer is set loose to discover things in the data, learn patterns, and make predictions.

prompt a radiologist's focused attention. Cardiologist and author of *Deep Medicine* Eric Topol told a conference audience that his favorite algorithm outputs to review are heart scans: "I've been doing it for a few decades. I'm afraid to admit, and I would never know if the patient is a man or a woman, but of course you can train algorithms to know that pretty darn well—not to mention their heart function, whether they have anemia and the extent of it—and then to make difficult diagnoses like amyloidosis or pulmonary hypertension, very accurately."

Competitive sports leagues use entity recognition technology to analyze in-game videos, measuring everything from the distance between an offensive player trying to score a goal and a defender trying to block the shot to the angles at which players in a cluster move toward and away from each other in a contact scenario. These advanced, commercialized precision systems create exhilarating slow-motion replays that fans love—and players too—watching it all on stadium jumbotrons. Using these same precision systems, coaches can zero in on specific player weaknesses or decide whether to dispute a referee's call in real time.

Retail stores, casinos, and other workplaces have long been under the watch of video cameras scanning aisles, card tables, and cash registers for known shoplifters, card counters, or sticky-fingered store clerks. Airline check-in and building security scans now match images to authorized passengers and employees. Counterterrorism teams use entity recognition technology for signs that passengers who may be seated separately in airport waiting areas actually know each other and are working in unison—or walking oddly, indicating perhaps a suicide vest or heavy armaments underneath street clothes. Counterespionage teams track foreign agents on their streets and watch for meetings with confidential informants, revealing tradecraft and networks never before known or seen. Facial recognition technologies help governments around the world spy on their own citizens at work, at play, and even in their own homes.

Concerns about privacy and security aren't the only difficult aspects of entity recognition systems. Unlike other precision system types discussed in this chapter, entity recognition has fairly high technical barriers to entry. Rudimentary systems can be programmed to execute specific rules—for example, if color equals red, then fruit equals apple, and if color equals orange, then fruit equals orange. But such systems are quickly overwhelmed by the complexity of reality in which fruit bowl images may contain strawberries, apricots, and other confounding fruits. However, off-the-shelf software systems are gradually learning how to meet these technical challenges. Entity recognition algorithms are now being widely popularized as software-as-a-service offerings with simple interfaces and programmatic access options capable of processing millions of images for a few thousand dollars. Thus, the barriers to entry for people exploring the use of precision systems for entity recognition needs are lower than ever.

## Systems for Geotracking

We know that the tiny supercomputers we keep in our pockets all day—our mobile phones—capture our locations and periodically relay them to satellites, servers, and nearby devices. Precision systems for *geotracking* ingest information like GPS coordinates, maps, Bluetooth beacons, satellite terrain imagery, internet protocol (IP) addresses, and other spatial data. With geotracking, precision systems track and trace, detect and spot, predict and plan. They can re-create past movements and locations of people, cars, and planes, but they can also predict futures: collisions, on-time arrivals, or fuel shortages, and prescribe course changes as needed.

ESRI, a provider of geographic information software, describes two main types of geospatial data: discrete *feature data* such as exact locations, often used to describe the built environment with coordinates and administrative attributes like census information, and *raster data*, which is used to segment chunks of the natural world and apply descriptive characteristics like temperature and precipitation. Using the right tools, analysts of geographical data can perform the same kinds of analyses they would on other data types: hunting for associations, trends, anomalies, clusters, and characteristics. For those willing to dip their toes into the pool of geotracking analysis, the rewards can be powerful.

Geotracking systems are the foundation for many of the personal utilities we've come to take for granted, including the GPS maps we use on highways, finding routes through urban canyons, photo sharing among friends via Bluetooth, detecting changes in a time zone, and the magical appearance of the right boarding pass onto your mobile phone screen as you approach the airport.

The same systems are also used for a growing array of less familiar applications. Satellite imagery is essential for governments and nonprofits in their battles with big-game poachers in wildlife reserves, in tracking the melting rate of the polar ice caps keeping our planet habitably cool, and in monitoring large swaths of agricultural land for quality degradation and crop management. Privately owned Google Earth aims to capture a complete picture of the entire world regularly, indexing it and sharing it widely at no cost.

Geotracking in precision systems helps businesses transform customer experience. The New England Patriots football team pioneered the use of high-bandwidth, in-stadium Wi-Fi and digital "fences" for video replays on thousands of fans' smartphones simultaneously, with in-seat food and beverage service while they watched, all to make the in-stadium experience superior to watching at home, even in the New England winters, and worthy of a season ticket purchase. On a more modest scale, the gym chain Equinox surfaces customers' swipe-in passes on their phone once they are within a few feet of the entry, saving them from digging around for it at check-in.

Entrepreneurs use geotracking for the public good in many ways. As we saw in chapter 1, the Rwandan start-up Zipline has built a fleet of self-navigating medicine drones, which, in the words of the company, "deliver critical and life-saving products precisely where and when they are needed, safely and reliably, every day, across multiple countries." Governments use geotracking for everything from forensics and crime scene investigations to regulatory reforms, to pandemic control. Investigators of the January 6 U.S. Capitol insurrection identified many attackers from logs of their cellphones—all pinging from inside the Capitol building during the hours of attack. Businesses operating within the European Union use location information to know whether a customer falls within the jurisdiction of European privacy law and offer them consumer controls accordingly.

In cases like these, the precision technology is often the easy part—the human challenges of collaboration, adoption, privacy protection, and citizen engagement prove much more complex.

## Systems for Modeling Scenarios

Predicting the future is notoriously difficult but also vitally important—hence the widespread fascination with efforts to sketch potential futures and form a plan for responding to each of them. Such *scenario modeling* involves examining different combinations of factors both inside and outside our control and using mathematical methods to identify which future patterns are most likely to occur. Working with a volume and variety of data never before manageable at scale and with speed, precision systems are now making such exercises more accurate and useful. Budgeting is one of the most recognizable ways we do this. Organizational leaders constantly ask: "Given the monetary constraints we have and the financial commitments we've made, what would the future look like twelve months from now if we spent conservatively, moderately, or liberally? And which of those three scenarios is the most likely?"

A biopharmaceutical start-up concerned with investing its limited resources in the manner most likely to yield a viable product might model the scenario outcomes of various research and development (R&D) strategies. Considering constraints such as dollars, team members, and opportunity costs, how many variants of a potential drug is the company likely to cycle through before it hits something with promise, based on how long it's previously taken in similar situations?

Competitive scenario modeling increases the complexity by including the likely behavior of an adversary or a challenger. Until recently, systems designed to model variants of competitive scenarios had to be run, adjusted, and run again under human supervision. The earliest algorithms to compete with human chess players, for example, required a human chess player to compete against and were developed by data scientists who tweaked and refined them to be better every time. Today, the same processes no longer need to be set up and run by humans. Google AI's fully autonomous, self-teaching chess algorithm AlphaZero learned how to win a chess game after just four hours of training and is now available as a product that runs innumerable game variations against competitors both real and simulated.

Modern applications for competitive scenario modeling span many industries in which a strategy is required and a variety of actions is possible, each with its own ramifications. Merger and acquisition firms use game theory to strategize the most effective way to buy or sell a target firm, hypothesizing what the party on the other side of the table might do every step of the way.

Combat simulation is another application of scenario modeling. The U.S. Navy developed some of the most sophisticated scenario models to coordinate the positioning, targeting, and cyberdefense of aircraft carrier strike groups during attack. The Netherlands Aerospace Center offers a suite of systems for purchase that use machine learning to generate computer generated forces, or simulated combat opponents. The user of the system can simulate various interactions and battle strategies, playing them to their likely conclusion, without ever having to step onto a real battlefield or fire a real weapon. This is positioned as an intelligence imperative because real-world opponents are likely to have invested in such scenario modeling themselves. Such military modeling systems are closely related to entity recognition systems because each of the opponents and objects within the simulated scenario are generated through similar techniques.

## Shared Principles Across Precision Systems

This framework of six types of precision systems is our preferred way to think about high-level categories of precision, but it is not the only way. Other leaders have proposed different lenses to view the same problem. For example, Steven Miller at Singapore Management University proposed only three buckets: cognitive process automation for automating workflows and decisions, cognitive insights for finding information within data, and cognitive engagement for interactive tools like voice assistants and chatbots.

We find the slightly more granular breakdown into six buckets to be more helpful. In the chapters that follow, we will reference each of these six types of precision systems and study examples of their application. They are used again and again to go further than even the cleverest human brains can go in the public, private, and nonprofit sectors; in a range of industries; and by leaders from executive managers of large corporations to individual contributors.

As you can imagine, there are important differences among the many kinds of precision systems being developed and deployed by this array of organizations. But there are also several underlying similarities, all of which offer useful lessons for those interested in exploring the creation of such a system. Next we discuss four of the most important.

## Precision Systems Exist Within a Human and Organizational Context

It's tempting to think of a precision system as a sum of the platforms, code, and data that power it. But a precision system does not exist in a vacuum. It is a living combination of technology, people, processes, and goals—all operating within an organizational culture and a broad social context. People are needed to understand the problems and opportunities characterizing the status quo and to envision systems with the ability to improve it. Processes serve to link the technology to real life and adapt to changes over time. Goals are needed to state the purpose of the system and measure progress toward them, connecting the present to the desired future. And the organizational culture and social context help to define how employees, customers, users, and other stakeholders will respond to the precision system and use it effectively—or not.

## Designing a Precision System Is a Fundamentally Nonlinear Process

Creating a precision system is a form of systems engineering, with all the complexities that work generally involves. But data science—the multidisciplinary field that underlies the developing of precision systems—involves unique difficulties, particularly in software engineering.

In traditional software engineering, a target state of desired features and functionality can be articulated before work begins, and engineers can estimate with relative accuracy the time and complexity required to move from the current state to that target state. For example, a product designer can sketch a new screen for an app, specifying what each button should do when clicked, along with the expected behavior for a variety of edge cases. Engineers can then work collaboratively to assign that vision a complexity versus effort estimation. This process enables teams to plan while retaining the flexibility they need to adapt to changing priorities.

This is not the case with data science development. Much as in traditional software engineering, target states can usually be described; for example, the team working on a precision system for the emergency room in a hospital might decide "the target state is that we can predict with 80 percent accuracy which patients arriving at an emergency department will require intensive care."

However, the problems to be solved by engineers and data scientists to arrive at that target state are much more difficult to describe in advance. Data science development is fundamentally nonlinear—it requires iterations of research and development, trial and error, model optimization, training and retraining. It's simply not possible to know whether or when you'll reach the 80 percent accuracy goal—or whether and when circumstances will change and drive that 80 percent down to 70 and 60 until models are corrected and high performance resumes. For this reason, when a precision system is being built, timelines and outcomes are both uncertain. The sooner leaders understand that, the better.

## When Developing a Precision System, Defining What's "Good Enough" Is a Core Challenge

Because of the inherent ambiguity involved in creating a precision system, improvements can be pursued indefinitely. An accuracy rate of 80 percent may be solid enough to solve most of the problem for most users of the system most of the time. But the team working on the system must consider questions like, "Would 60 percent or 70 percent be just as acceptable as 80 percent? Would the effort to improve to 90 or 95 percent be worthwhile?"

Answers to these questions are rarely obvious. In the pursuit of performance, much can be gained and much can be lost. When to call your precision project "done" may not be clear, and leaders will ultimately need to draw a line to avoid infinite tweaking, seeking certainty that data science inherently resists. The choice between building a simple, good-enough system and an ultra-high-performance system—a skateboard or a Ferrari—will involve a number of factors, including the resources available (such as data, time, and talent) and the cost of errors (human, commercial, strategic, and otherwise).

A simple Excel spreadsheet that took a few hours to design (the skateboard) may be the right answer for some problems. In other cases, a highly complex and massively scaled system (the Ferrari) may be necessary. Both can get you from A to B, but with different costs and benefits. As a leader, knowing whether you need a skateboard or a Ferrari is a make-or-break insight for your precision endeavors as cost of improvement increases against relative value of improvement.

## Precision Systems Are Subject to Several Forms of Bias

The problem of bias built into decision-making algorithms isn't new; it existed long before precision systems did. In 1988, the *British Medical Journal* published a scathing account of a London medical school that had been disproportionately rejecting female applicants and nonwhite applicants because the algorithm that sifted through applications initially had been trained to pick and prioritize applications that most closely resembled previously approved ones. In other words, because most of the student body was white and male, that's what the algorithm was looking for.

Today, we still see the same kind of bias perpetuation, but now the data and the computers are bigger and more powerful than ever. Data science can have the unfortunate effect of cementing bias—which means that those who are developing precision systems need to be on guard against bias at every stage in the process. Here's a simple example of how bias can be perpetuated in the design of a precision system. In 2018, researchers Timnit Gebru from Microsoft and Joy Buolamwini from the Massachusetts Institute of Technology (MIT) proved that some facial recognition products made inaccurate deductions about women and nonwhite men much more often than they did when analyzing the faces of white men. Depending on the purposes for these facial recognition technologies—for example, the identification of criminal suspects—the results could produce significant harm to members of particular demographic groups.

Gebru and Buolamwini found that the cause of this bias was the underlying data set on which the machine-learning algorithm had been trained. The data set contained approximately 80 percent images of white people and only about 20 percent nonwhite people. Because the algorithm was less familiar with nonwhite faces, it was more likely to classify them incorrectly. When the researchers retrained the same algorithm using a different data set that had been deliberately balanced across gender and race, it performed equally well for people of all backgrounds.

As this story illustrates, guardrails need to be put in place at every step of the process of creating a precision system to ensure that bias is avoided. We'll have more to say about this challenge in later chapters, particularly in chapter 10, where advice on avoiding some of the most dangerous pitfalls of precision systems is detailed.

In the chapters that follow, we'll see teams in many different situations using whatever resources, skills, and time they have to envision and realize their precision strategy. Good precision leaders learn how to use the variety of tools in their toolbox and are ready to adjust, retool, and even abandon their agreed-upon strategy when changing circumstances or new discoveries make it necessary. They're prepared to experiment with all the variables—technological, human, process, and beyond—at their disposal, even down to the last mile of the journey. Above all, they know that, unless all the pieces come together at the right time, in the right way, to serve the most urgent human needs their organizations seek to meet, all their effort may be in vain.

## CHAPTER 3

# DESIGN, TEST, AND PROVE

## Key Steps in Building a Precision System

Precision is, first and foremost, a way of doing business that uses established scientific methods to achieve finely tuned results at scale. The technology and the vast data that may be employed are also hallmarks of precision, but precision is nothing without an underlying methodology of *design, test, and prove.* Understanding how organizations achieve transformation through precision begins with understanding the basis of this methodology, which this chapter sets out to introduce and elaborate.

The difference between a product or service that operates simply at scale versus one that operates *with precision* at scale is visible throughout our everyday lives. Facebook serves billions of users, but each user's news feed is personalized based on their networks and interests, and it is designed to become more accurately personalized with each new click. Contrast this to a traditional news publisher, which prints one home page full of articles and one only—no matter who reads it. Or contrast waiting on the street to hail a taxi and hoping one happens to be nearby with requesting a ride via an app like Uber and watching it, tracked by a global positioning system (GPS), approach you instantly. Or contrast a teacher who delivers one consistent lesson to a classroom of thirty diverse students with an instruction platform that enables self-paced learning and unique remediation pathways for individual students based on their quiz results and other performance indicators.

The precision we see and touch increasingly in our daily lives is only the tip of the iceberg. Services we don't notice or have never heard of also employ analogous methods. On the farm, computer vision systems trained by Blue

River and affixed to tractors built by John Deere poison individual weeds; contrast this with traditional methods, in which weed killer is distributed across a whole field.

Just *how* did companies like Blue River, Uber, and Facebook achieve these transformations? How did they begin? How did they proceed? And what ways of thinking allowed the creation and success of these revolutionary products? This chapter will answer these questions.

The methodology of design, test, and prove is a process of creation and improvement that begins, first, with framing a problem or opportunity and a definition of success. Second, it is followed by an initial attempt at a solution, rapid feedback, then iteration with increasing complexity and scale. Above all, it is guided by a mindset of curiosity and discovery, and continuous measurement and improvement. Across many different industries and products, the precision approach of design, test, and prove is consistent. It existed long before digital data did, but it was supercharged by its arrival.

## Blue River, John Deere, and Precision Agriculture

The Blue River Technologies story captures the essence of the design, test, and prove methodology, and it is a good way to begin our discussion. It starts with a vast variety of data in open farm field environments being gathered and then transformed by the most sophisticated computer applications, fitted to infrastructure common to almost every farm around the world: the tractor.

Look at a John Deere tractor and you'll see the familiar, shiny yellow and green machine with six-foot tires—a favorite tool that has helped farmers perform a range of basic agricultural tasks for generations. But when Jorge Heraud and Lee Redden, two young entrepreneurs out of Stanford Business School, first saw a John Deere tractor, they saw a beautiful platform waiting to deliver, with precision and perfection, one of the most extraordinary innovations in American farming in 200 years.

Heraud and Redden are founders and CEO and CTO, respectively, of Blue River Technologies and the inventors of John Deere's See and Spray precision agriculture system. The Deere tractor tows an array of imaging and spray devices behind it mounted on a bar on wheels that is 120 feet side-to-side, covers thirty-six rows of crops at a time, and moves over them at twelve miles per hour. As it rolls, cameras snap images of every plant beneath it twenty times per second and map each plant to its unique, centimeter-square coordinate. Within a second or two the image will be compared with a vast database of similar plants to determine whether the plant is a weed or a crop. If it's a weed, 200 inkjet-like squirters dubbed by Redden "The Jury" will, in the next second, and as the rig keeps rolling, deliver a precise, lethal microdose of herbicide (or overdose of fertilizer) to that centimeter-square location, killing the weed. (This is an example of an entity recognition precision system as described in chapter 2.)

See and Spray automates a task that farmers have performed by hand for millennia. Lee Redden's grandmother was one of those farmers in Nebraska. "She'd walk her 32 acres during harvest," Redden says, "check each plant, harvest the right ones, go through and kill individual weeds." Now the same job gets done faster, easier, and better thanks to supercomputing modules that are mounted right on the tractor itself.

"That's a ton of compute," Redden says, "the highest compute capacity of any moving machine in the world, on par with IBM's supercomputer project, Blue Gene. Ruggedized, in a form small enough to embed on a machine."

It's all about what Chris Padwick, Blue River's director of computer vision and machine learning, calls *visual inferencing*. Each See and Spray towed array is essentially a field robot that must figure out whether it sees a plant or a weed, decide whether to spray it, then dose it or pass.

"As the machine drives through the field," Padwick explains, "high-resolution cameras collect imagery at a high frame rate. We developed a convolutional neural network using PyTorch to analyze each frame and produce a pixel-accurate map of where the crops and weeds are. Once the plants are all identified, each weed and crop is mapped to field locations, and the robot sprays only the weeds—and all within milliseconds so the tractor can keep moving."

The supercomputer manages the whole system, from imaging to weeding, with a core decision engine of purpose-built software making the weed/no weed call for each plant in milliseconds. It embeds automated testing and deep learning so that See and Spray continuously improves, perfecting the accuracy of the weed/no weed decision across more and more crop types, season after season.

The benefits of See and Spray capture the value of precision well. When See and Spray is used, "It sprays weeds, not acres," Deere boasts. This is no blunt-force, one-size-fits-all spray operation of yesteryear (think of the crop duster chasing Cary Grant in Hitchcock's *North by Northwest*), poisoning everything in its mist. See and Spray applies a microdose of weed killer on every plant it calls a weed, and on no other. Some 911 million acres of farmland are broadcast-sprayed with chemicals each year in the United States. When used, See and Spray reduces the total amount of herbicide in use by 80 percent to 90 percent, saving farmers millions of dollars on chemical purchases alone while also reducing soil toxicity and increasing crop yields by up to 100 percent. It generates better yield, less chemical runoff, healthier water tables and rivers, and money saved by farmers, all impact areas of classic broadcast spray operations.

"Taking care of each individual plant unlocks a lot of value," a Blue River executive told a reporter. "Each individual decision is pretty cheap—the costs to take an image and predict whether a plant is weed or crop is low, as is the value you generate. But do that a million times an hour with high density actions across a whole field, coupled with the ability to take actions right there on the spot, and you shift the paradigm and generate much higher value."

In 2017, John Deere purchased Blue River for $310 million. Being backward compatible—taking advantage of existing on-the-farm infrastructure—helped ease this technology innovation onto the farm. To use it, Deere owners simply hook the entire rig onto an existing tractor.

"Traditionally, the agricultural community does not like to be 'disrupted,'" Daniel Koppel, an agricultural technology ("agtech") entrepreneur, told a reporter. "So the fastest way to prove benefits of our technology was to partner with a brand that growers already trust and work with daily. Technology can be smart and innovative. But it's no good unless it's actually used by its intended customer."

Getting to this point wasn't easy. At first, Blue River's computer vision and John Deere's spray machines didn't gel perfectly together. The massive booms of the sprayers wobbled while tractors were moving, preventing Blue River's cameras from taking clear, usable photos. The sprayers' protective covers cast shadows, reducing the robots' accuracy at detecting weeds.

The team had to work directly with the camera manufacturer to build in several enhancements, including much more powerful capabilities for eliminating visual noise, so that cameras could stabilize their own images better. John Deere also had to invent a new version of the boom that wobbled less.

To make sure the machine-learning models distinguish weeds from crops as accurately as possible, they need to be constantly trained with the best possible data. So the combined team has over thirty full-time staff working solely on the collection, preparation, and management of training data. Padwick explains: "We have millions of images collected in thousands of different fields on different conditions. We have images of small cotton and larger cotton, cotton during the dusk with shadow, cotton that has a bit of a disease, cotton that has hail damage and cotton that has just been rained upon and cotton that has lost leaf because it has a fungal infection." The team continuously tests the performance of its image recognition models, more so than any other component of the system. After all, that's the core of See and Spray's value—the "seeing." Heraud explains how the team does it: "We go to a field, collect images, run them through the system as if the system was there and see whether it makes the correct decision or not. We evaluate how many we got wrong, and how many we got right. On top of that we go to customer fields, and we use the system like a farmer would. We have testing that our agronomists do for all sorts of edge conditions. We have all these levels of testing. Testing is a significant portion of the effort on developing this system."

Through years of trial and error and untold hours of work, the team's efforts have created a single, generalized model for each crop. The cotton model, for example, has been trained well enough on so many different variables, that it can now reliably handle real-life complexity, in what Heraud calls the N + 1 scenario: every new field—the "1" in the equation—represents a novel challenge which, when met, becomes part of the "N" generalizable model.

See and Spray demonstrates what precision systems can do at their most powerful, namely, achieve precisely the change intended, no more, no less: every weed destroyed, and no crop. This comprises the true *technical* measure of See and Spray's success. By applying a design, test, and prove framework that is widely applicable across domains, the team eventually built a world-changing product and achieved organizational success. By 2020, there was a 10 percent chance that the lettuce on your fork had survived the roll beneath a John Deere tractor towing See and Spray.

The results are a team success too: they inspire and engage current and future employees. Heraud reflects: "I would say that somewhere between 70 and 90 percent of our employees have joined because of our mission. We're doing something really good. We're making better food: the food that you are going to eat will have less chemicals because of this. My son, my daughter, my parents are going to eat higher quality food. This is farming the way it should be. And food is important to humans, it's life and death. We need to figure out how to do this right." See and Spray is helping to make the age-old, all-important art of growing food into a science that is increasingly safe, heathy, and precise.

## Design, Test, and Prove: The Core Steps

Here are the basic steps involved in the design, test, and prove methodology—steps you should plan to follow whenever you launch a precision project:

- Define the problem and the hypothesis.
- Define success and performance indicators.
- Collect data.
- Build a proof of concept and test it.
- Scale and optimize.
- Launch, maintain, transfer ownership, and reflect.

We'll discuss each step in more detail next.

### Define the Problem and the Hypothesis

Articulating the problem with the current state of things—whatever it is you're concerned with improving—is the beginning of the process. It might be a widely accepted problem or one that only you see; either way, capturing it in detail will be the basis of effective decision making and communications through the entire journey.

Taking Blue River as an example, the problem statement might have been: "Traditionally, farmers distribute weed killer across all crops uniformly even though it is needed only for weeds. This results in massive waste of treatments, environmental harm, and unnecessary cost." This sets out the See and Spray twin precision goals of *producing more goodness*—food—over the same crop land, while at the same time *reducing the harms* of blunt force misapplication of pesticides by traditional means. These are measures of efficiency, effectiveness, and equity, all of which See and Spray seeks to maximize with its precision system and thus to improve overall performance without trading off one for the other, as traditional pesticides applications do.

The beginning of any new project or development is the point in time when ambiguity is highest and concrete knowledge is lowest. Over time, those two positions reverse, and certainty rises. For this reason, articulating the problem itself is a crucial start to any project, especially those involving trial and error, and multiple people and skill sets—as almost all precision projects do. A well-articulated problem statement aligns the team: even when a specific future vision state can't be defined, a clear current problem state usually can be. It ensures that team members are focused on the same challenge and prevents the project from fracturing into as many unique personal visions as there are members of the team.

The cousin of a good problem statement is a well-formed hypothesis. The problem statement is simply a *description* of the current state, while the hypothesis is a *prediction*—it proposes a relationship that can be tested and either proved or disproved. It often begins with "if" and describes a better future state where the problem is solved. Continuing the Blue River example, the hypothesis might be: "If we can selectively apply weed killer only to weeds and spare the crops, we can achieve significant cost savings on weed treatment, improved crop yields for each acre, as well as the ecological benefits from reduced treatment application of weed killer."

A problem statement and hypothesis allow a team to begin working on a solution—a *prescription* of what to do next to change to the desired state away from the predicted one. Team members make sure everybody starts on the same page about what's wrong or could be improved and set an initial direction for research and development to begin pursuing the desired future state. Description, prediction, and prescription: the foundations of precision system change.

## Define Success and Performance Indicators

The next step is to define the desired future state in concrete terms. This could take the form of qualitative statements that must be true in order for the goal to be considered attained, or it could take the form of quantitative goals or key performance indicators. The point of defining success is to clarify at the outset of the project, before any solutions are in development, what it will take to consider the project done. Defining success at the beginning helps you avoid the tendency to move the goalposts while the ball is already in the air and, with it, the tendency to bias the definition of success toward results that have already been met.

Quantitative goals generally provide more value alone than qualitative ones, although a mix of both can often be successful. John Doerr's classic book *Measure What Matters* introduced the concept of objectives and key results (OKRs) and heavily emphasized the value of numerical goals. They leave no room for ambiguity or subjectivity, and they frame the goal in a way that allows it to be either achieved or not achieved. This is important to reduce subjective debate and to calibrate expectations.

Following Doerr's advice in the case of Blue River, precision leaders might have said: "A successful outcome will be to reduce the average farm's cost and distribution of weed killer by 75 percent while maintaining less than 2 percent weed coverage." This proposes that success entails meeting two criteria: reducing weed treatment usage without increasing the presence of weeds in the crops. By specifying a particular amount (75 percent) of intended reduction, the team has a clear impact goal to aim for. Any higher and they succeed; any lower and they fail. But this introduces the risk of selling precision short. The gold standard of *technical* success for See and Spray is *every weed, and no crop, destroyed*. Anything short of that means there is some *systemic* failure in See and Spray's recognition technology or in its application technology rather than a shortfall of effort or a marginal cost calculation that further harm reduction is not worth it. For precision systems, the *total addressable universe* of plants in a field is possible and sets the upper limit of performance at the limits of nature and technology. Alcoa's legendary CEO Paul O'Neill famously quipped about his staff's proposed safety measures (paraphrasing), "Top quartile performance is not good enough. Anyone can be as good as Dupont. We seek, rather, performance at the limits of what God and science permit." See and Spray founders would likely agree.

"I'm not suggesting that all of John Deere needs to follow," Heraud told us. "But on projects that we tackle as the Blue River subsidiary, we don't go for singles. We swing for the fences, for the home run. It has happened several times that we've had to choose—do we want the low hanging fruit or the world-saving problem? We go for the world's biggest problems, which is pretty cool."

Sometimes it is difficult or impossible to measure the actual thing that matters most over and above mere technical precision, and that is the big-picture *transformational* impact. This can happen when the impact seems too far beyond the team's control. Continuing the Blue River example, Blue River has set for itself the goal of reducing water table pollution. "That's why we're called *blue river*," Heraud said. "Right from the start, we decided that one of the biggest impacts of this type of technology was going to be on rivers. And we were we were looking for a name for that company that represented what we were wanting to do. A blue river *is* a clean river."

But myriad factors influence river pollution, near and far; while the promise of See and Spray is precision that *should* deliver clean rivers, no one farm or farmer might control much of that. If upstream farmers continued to pollute, the Blue River team would probably not be able to meet its goal, even if they developed a highly efficient and effective "see and spray" system that achieved 100 percent of its technical goals (every addressable crop saved, every weed destroyed). By their own "blue river" measure, See and Spray would fail in its promised goals of transformation.

This conundrum explains why alternative measures of success known as *proxy metrics* can be useful. Proxy metrics focus closely on the performance of the precision system itself and exclude exogenous factors that are further beyond control. For example, Blue River could measure the system's potential impact on river health by sampling soil for tell-tale pollutants near and far from its farms on the assumption that this would be a good proxy metric for river pollution. If the overall measure of success is whether the system actually reduces river pollution, however, it behooves See and Spray to search not just for proxy measures of river pollution but also for direct measures and to inquire further why some rivers seem to be improved and others not—perhaps to fix See and Spray's methods further.

Proxy metrics measure what you can control, and impact metrics measure whether or not the project had the intended impact overall. Thus, a thoughtful combination of proxy metrics and impact metrics can set a system up well for measures of both technical and transformational success. (We pursue further dimension of precision system impacts in chapter 10 of this book.)

## Collect Data

Modern digital precision systems depart from earlier attempts at precision management in the use of data, often big data, to create analyses and solutions. This makes data collection an essential part of the precision process. Data collection can be a surprisingly arduous and lengthy task; it shouldn't be underestimated when budgeting time. Chapter 4 will explain in more detail how to assess the quality, availability, and accessibility of the data you need. But first, it's important to simply understand which data you need in order to do what you're trying to do.

The Blue River team was developing an entity recognition system using advanced computer vision to distinguish weeds from crops in real time. It should come as no surprise that the team required a huge number of images of both weeds and nonweed plants in order to train their algorithms. By equipping tractors with cameras, they were able to collect these images and form a huge, central database that would power the system's future decisions.

In machine learning, there is a concept of separating data into two different sets: a *training set* and a *testing set*. The training set is used to teach the algorithm what to do. For example, the Blue River training set might have had 50 percent images of weeds and 50 percent images of crops, all of which were labeled accurately. The testing set is then used to evaluate the performance of the algorithm: The Blue River testing set would have included a mixture of crops and weeds, none of the labeled—the system would have to label them itself. Data scientists are accustomed to handling data sets very carefully and making sure that no training data and test data ever mix, which would spoil the validity of the tests.

Similarly, digital marketers are fluent in *A/B testing*, a scientific process of optimization that brings the core tenets of experimental design online. For example, if you were running a website and wanted to test the hypothesis that "Showing website visitors a 15 percent off purchase coupon will cause them to make more purchases," you could test that by showing the coupon to half the visitors and not showing it to the other half. Comparing the two groups' purchase rates would tell you whether or not the hypothesis was true (more on this methodology shortly).

Those who aren't data scientists, digital marketers, or other precision natives can still benefit from understanding the core approach to data. Data is the fuel that powers accurate and precise systems, and the better the quality and handling of the input data, the better the results.

## Build a Proof of Concept and Test It

One of the most important ideas in the domain of product management is the value of learning early, iterating rapidly, and failing fast when developing a new product. Jim Brickman of Y Combinator, one of the most successful seed-stage investors and start-up incubators in the world, explains the reason this way: "When you build a product, you make many assumptions. You assume you know what users are looking for, how the design should work, what marketing strategy to use, what architecture will work most efficiently, which monetization strategy will make it sustainable, and which laws and regulations you must comply with. No matter how good you are, some of your assumptions will be wrong. The problem is, you don't know which ones." Using real-world testing is the best way to identify and correct those flawed assumptions. This is why building and testing a proof of concept is the next step in precision development. Just like any new product, a precision system benefits from early and frequent feedback and testing.

Defining the scope of a proof of concept requires careful thinking about what is needed now in order to capture useful feedback and what can be delayed until later. There is no hard and fast rule that simplifies this decision across all projects. It can be helpful, however, to assume that the vast majority of features and most of the functionality and sophistication you envision for the ideal precision system are *not* necessary for the proof of concept. All that matters is the most rudimentary skeleton that allows you to test the hypothesis.

Another way of making the same point is to recall the concept of the Ferrari versus the skateboard described in chapter 2. The proof of concept for a Ferrari is not a miniature, functional car—it is a skateboard. After all, the core purpose of a Ferrari is to get you from point A to point B. Testing that ability is the purpose of the proof of concept—and for that, a skateboard is perfectly adequate.

In the case of Blue River, the team behind See and Spray could not wait until they had built a fully functioning computer-vision tractor with variable treatment-dispensing capabilities to find out whether their hypothesis was correct or not. They had to start much smaller and first prove whether or not they could correctly distinguish weeds from crops. Their proof-of-concept tests proved that this was possible before they invested any further. Testing proof of concept early in the project has another benefit: it establishes a testing system that you'll be able to use, perhaps with slight variations, throughout the rest of the development process. Continuous testing will be one of the tools that will help keep the entire project on track.

It merits emphasis that not all testing for a precision product should be quantitative, or model performance, testing. Precision products are used by people, and their success or failure depends just as much on whether people like and trust and would use the product as it does on how accurately the product functions. For example, testing the interactions between human and machine on a voice-assistant product will yield important insight into whether users are satisfied with and likely to adopt the assistant into their day-to-day lives, or whether it's too clunky, patchy or weak to get traction with users. This information comes alongside model performance insights like error rates, accurate response rates, and false positives and negatives. Quality assurance engineer Debjani Goswami explores the voice-assistant testing example further and proposes three considerations for usability testing: defining learning objectives for the assistant, compatibility between user's needs and the assistant's solutions, and continuing learning after real-world deployment.

Usability testing for precision projects can help to prevent a risk scenario described later in chapter 10: the last mile problem, where a performant system is launched that still somehow fails to get traction in the real world.

## Scale and Optimize

Once your proof of concept has passed its initial tests, the next step is to begin adding incremental functionality and complexity to your precision system piece by piece, developing the product toward the desired end state and testing, retesting, and testing again all the way through. In some cases, this process will involve making the system available more widely—to more users, or (in the case of Blue River) across more acres. In other cases, it means introducing new features that will be needed eventually—like the weed treatment dispensing system that leverages the weed/no weed decision and turns it into action.

This step can often be the longest one in a precision project because it's where the rubber meets the road in a new way. A precision system can't simply work; it needs to work well enough to be fully launched and meet the impact goals set at the beginning of the entire project. Thus, as the team continues to improve the system by optimizing its performance and readying it for significantly larger scale of operation, the question becomes, "When will it be ready?"

By specifying success metrics in advance, you have already provided yourself with a way to answer that question. Your precision system will be ready to launch whenever it meets the bar you originally set. For the team behind See and Spray, the system was ready for launch and sale once it reduced cost and usage of weed treatment by 75 percent while maintaining a consistent level of weeds among the crops.

The scale-and-optimize stage is also noteworthy for requiring the broadest array of multidisciplinary people and skills. While earlier steps in the precision system process may rely almost entirely on the work of data scientists, analysts, and researchers, the process of scaling and optimizing requires many other inputs. Systems engineers rebuild the product with prime-time scale in mind, turning it from a frail experiment into a robust, world-ready system. Product managers facilitate user feedback testing, which helps them prioritize future improvements that will be needed before they ultimately approve the launch. Marketers and leaders ready the sales and awareness engine, preparing to get this product out to whoever needs it—in the Blue River case, the intended buyers of an intelligent tractor.

## Launch, Maintain, Transfer Ownership, and Reflect

When the scale-and-optimize process has been completed, your precision system is ready to launch. Once it is live and in the hands of real users, the problem has, for all intents and purposes, been solved. Of course, if unexpected problems emerge after launch, you'll need to take a step back. Another round of scale-and-optimize work will probably be needed to achieve the level of performance you mistakenly thought you'd already reached. Having to withdraw your system from use, or tolerate less-than-adequate performance while you scramble to improve it, can be embarrassing and perhaps politically damaging. But it's usually not the end of the world. Once the problems are fixed, the imperfect initial launch is apt to be forgotten quickly.

When your precision system is up and running as intended, it's time to shift into maintenance mode. This means finding a long-term organizational home for the system itself, assigning roles and responsibilities for its future maintenance and improvement, and winding down the project without loose ends. Most teams approach this stage haphazardly, often under the ill-advised direction of leaders who would like to hurry on to the next problem as quickly as possible. This approach robs the team of the opportunity to reflect and distill learnings that would help them to do better next time. What working assumptions did the team make that proved incorrect? Did they have the right talent, data, and tools at their disposal to succeed? What mistakes could have been avoided? How could the work have been completed more quickly, more easily, and more accurately?

Dedicating time to winding down and reflecting on your success and failure is a standard practice of the best teams who produce the most successful products. Many frameworks are available that teams can use to facilitate structured reflection and yield insights (such as the agile software development methodology's concept of a *retrospective session*), but the particular framework chosen matters less than that a framework is chosen at all and that sufficient time is given to work through it. One such framework from the Lean Startup methodology (figure 3.1) serves as a good example.

***

The design, test, and prove methodology underpins the approach of precision teams around the world. No matter what industry they're working in or what product they're working on, this straightforward approach can be seen again and again in successful precision investments.

Douglas Merrill used this methodology when he revolutionized the payday loan market by creating Zest Cash. He saw a problem in which people who were underbanked couldn't access fairly priced loans and resorted to payday loan sharks, who charged predatory interest rates. He hypothesized that by collecting a raft of other signals beyond the traditional credit score, he could reasonably predict whether underbanked customers were at low or high risk of defaulting on a loan, and offer them more fairly priced short-term loans. He collected the data—regular or sporadic phone calls to customer service, presence or absence of a Social Security number, thorough or hurried reading of the Zest terms of service—and used them to build a new model for risk prediction.

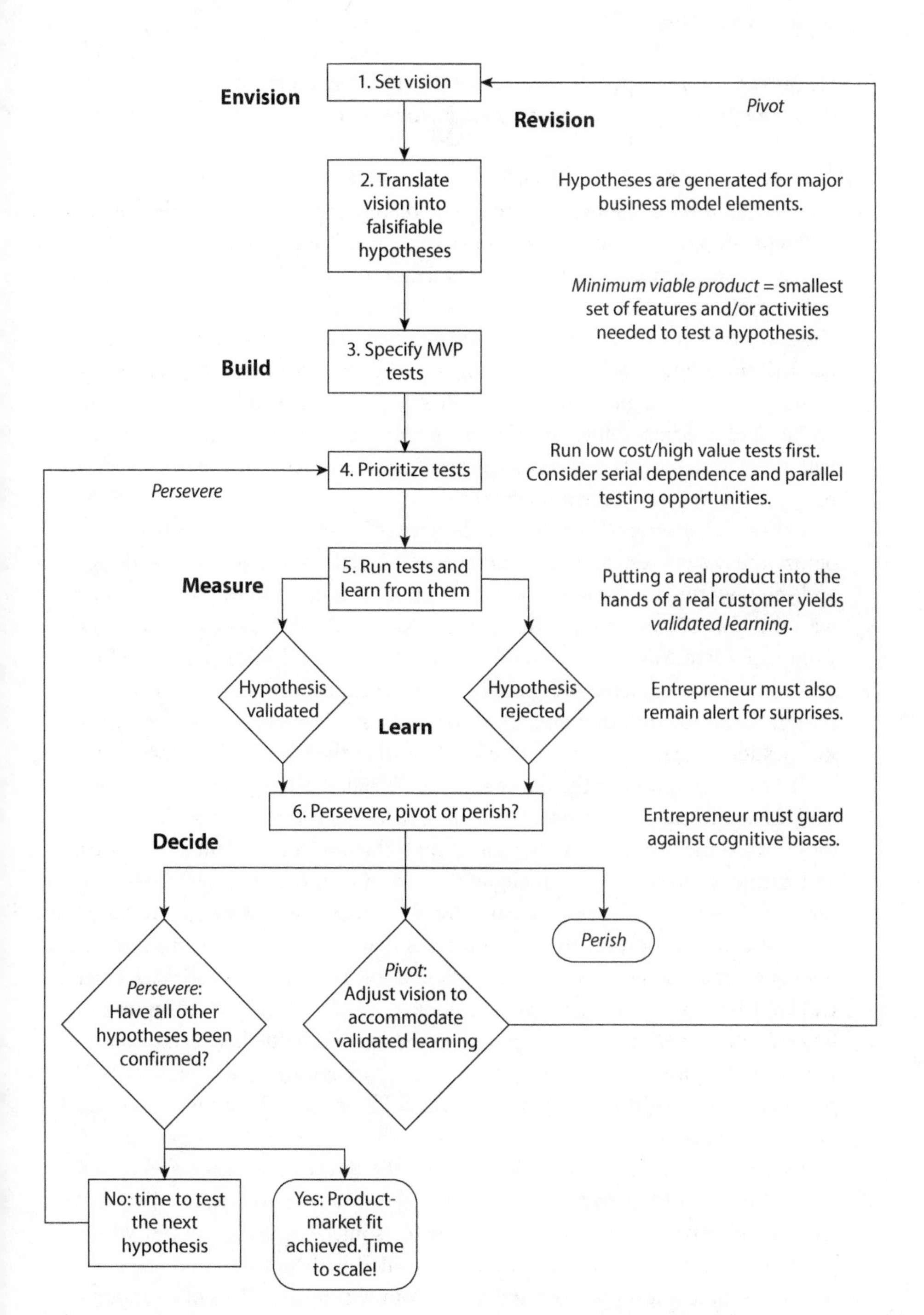


**3.1** A framework for the steps of a hypothesis-driven entrepreneurship process. From Thomas Eisenmann, Eric Ries, and Sarah Dillard, "Hypothesis-Driven Entrepreneurship: The Lean Startup," HBS No. 812-095 (Boston: Harvard Business School Publishing, 2013).

He proved the concept and then scaled it, launching a new service targeted to that demographic and offering a more effective alternative to traditional payday loan sharks.

Alex Cholas-Wood and Evan Levine at the New York Police Department (NYPD) used the same framework when creating Patternizr. They identified a problem—too many crimes going unprevented and unsolved due to similarity patterns unrecognized. They hypothesized that they could group related crimes together algorithmically using historical data about known crime patterns to build a new prediction system for future crimes. They collected the data they needed, including thousands of historical crimes both in patterns and stand alone. They proved that they could effectively detect relationships between crimes and use that ability to find new crime patterns. They scaled that into Patternizr, which managers could (in theory) use to decide how best to deploy police resources and prevent further crime.

If you've ever seriously studied math, science, or statistics, you probably recognize the design, test, and prove process as a variation of experimental design and the scientific method. Those who work at tech companies are so familiar with this way of working that it's second nature and barely needs explanation. Companies like Amazon, Facebook, Google, LinkedIn, Netflix, Airbnb, Uber, and others have mastered the art of embedding experimental design so deeply into the corporate mindset and ethos that even staff members who don't work in quantitative teams know to test and prove what they do, however they can.

For those who do run the numbers, a book called *Trustworthy Online Controlled Experiments: A Practical Guide to A/B Testing* was written by three leaders from the world of tech and big data: Ron Kohavi (vice president at Microsoft and Airbnb), Diane Tang (Google fellow) and Ya Xu (head of data science at LinkedIn). It offers a comprehensive dive into digital experimentation based on their combined experience at companies that each run over 20,000 experiments a year. The book opens with a quotation from Jim Barksdale, former CEO of Netscape: "If we have data, let's look at data. If all we have are opinions, let's go with mine." It's a wry reference to the age-old battle between observation and intuition—and the way that, in any organization, the highest paid person's opinion (HIPPO) tends to dominate decision making unless actively disproved by data.

As for A/B testing, it existed long before the internet, big data, and digital marketing ever did. Amy Gallo at *Harvard Business Review* explains that "A/B testing is a way to compare two versions of something to figure out which performs better. While it's most often associated with websites and apps, the method is almost 100 years old and it's one of the simplest forms of a randomized controlled experiment." Biologist Ronald Fisher honed the specifics of A/B testing on farmland in the 1920s, when he tested and documented the effects of applying more or less fertilizer to certain areas of land and comparing the results. When A/B testing is applied to product development, it supports the design, test, and prove methodology that yields precision systems that perform well, are wanted by users, and have massive impact.

Specifically in the world of models and algorithms, A/B testing or experimental design is also used to compare models or versions of models to each other. For example, two versions of a recommendation algorithm for a music-streaming service may be different: they might have been trained independently on different training data sets, or they might be leveraging entirely different types of algorithms (think collaborative filtering versus the genomic process suggested in chapter 2). They can both be run simultaneously so that 50 percent of music-streaming users hear recommendations coming from the first model, and the other 50 percent hear recommendations coming from the second. Measuring the two against each other to see which produces the happiest users and highest engagement is effectively executing an A/B test and leveraging the experimental design framework within a longer initiative to design, test, and prove the best possible music recommendation service.

Amazon Web Services suggests a protocol for comparing versions of models in this way, which includes gradually rolling out new versions, monitoring performance metrics, and keeping track of versions for control. This introduces the idea of a "champion" model (the original version that most users are experiencing) and a "challenger" model (whose task is to outperform the original and win the main share of users).

Historically, quantitative and scientific expertise resided in pockets within big organizations, the domain of analysts in ivory towers who were occasionally called on by executives for their input. Not anymore. In today's most successful precision organizations, the mindset of design, test, and prove has permeated so deeply that it has become an invisible core value. Everyone, from the intern to the CEO, needs a level of understanding of experimental design because it's the way decisions big and small are made. Just like John Harbaugh in the NFL, those who understand how to use data for decision making at the highest levels pull ahead, and those executives who wave it off miss out.

Whoever you are and whatever you do, if the game you're playing is precision, then you need to grasp design, test, and prove, and begin to operate by it. All the sophisticated technology and rich data in the world are worthless if you aren't equipped to work with them intelligently. The methodology of design, test, and prove is an essential tool for the task.

# CHAPTER 4

# GETTING STARTED WITH DATA AND PLATFORMS

The methodology of design, test, and prove underpins precision. But there are two other key components to precision: the data and the platforms. Data are the fuel that gets shoveled into the fire of experimentation. Platforms shape and empower it all. In combination, the methodology, the data, and the platforms form the three-legged stool of technical precision. (In later chapters, we draw the important distinction between *technical* and *transformative* precision and describe what you must do to achieve the latter.)

## The First Step in Getting Started: Let the Right Data Flow

Just as a chemist must have the right chemicals, beakers, and lab glasses before beginning an experiment, so too must anybody embarking on a data-driven precision project have the right data and tools for analysis. Where are the data coming from? Are they continuous or batched or manually entered and error-prone? Do they go stale and need refreshing? The stumble-and-fall of data projects frequently points to the same culprit. "Data-intensive projects have a single point of failure," GSK computational statistician George Krasadakis asserts. "Data quality."

Whether in the chemist's lab or the analyst's data flows and stores, hygiene is everything. "The obvious challenge," writes Krasadakis, "is to query heterogeneous data sources, then extract and transform data towards one or more data models. The *non-obvious* challenge is the early identification of data issues, which—in most cases—were unknown to the data owners as well."

This is the part of the journey where the generalist manager from departments like strategy, marketing, and product development needs to get into the weeds. Most incumbent companies that are considering a shift to precision have access to large amounts of data. But that doesn't necessarily mean the data exist in a form that makes them easily available and of high quality.

As a rule, the further away your work is from the mechanics of technology systems and the data they produce, the hazier becomes your knowledge of the true state of the data, and the likelier you are to think that your organization has access to data that it does not. As a result, it's easy for business managers to fall prey to "optimism bias" regarding the quality of the data on which their own decisions are based. A great example of this optimism bias comes from Google's "prediction markets," built by Bo Cowgill and colleagues to study information flows among Google office workers. Google employees could "bet" on the outcomes of everything from serious stuff, like Google projects or new products, to goofy stuff like which film would win the Oscar for Best Picture. Cowgill's team collected data on 80,000 of these so-called bets. Among their many fascinating discoveries: executives high up in the organization consistently underperformed on bets made on the serious stuff—questions like, "Will Google open its Russian office on schedule?" to "Will Gmail users use search more?" Cowgill's conclusion: centrally positioned executives were biased toward optimism for Google's new products and services by comparison with managers "at the edge," who had a more realistic view of the company's prospects.

Data are subject to the same bias. From the executive who assumes that the customer personas developed by her product marketing team are founded on rigorous analysis to the middle manager who hasn't touched a line of code since he was promoted but assumes that everything still works just as it did back then, distance from the details correlates reliably with unfounded optimism about the feasibility of a precision project. No matter where you sit in the organization, you will need to bridge the chasm between you and your data. "Know your data" is a key imperative for any manager and should come early in any project in the form of a *data quality assessment.*

## Jorge Bernal Mendoza at Pemex

For Jorge Bernal Mendoza (a composite portrait of senior information executives at Pemex, Mexico's energy utility) data issues were rife, and all pointed to one of the biggest problems encountered by Pemex CEO Jose Antonio Gonzàlez Anaya and his executive team: massive nationwide fuel theft. It was 2016. Organized crime was stealing its oil, costing Pemex thousands of barrels a day that were siphoned from pipelines around the nation and sold on the black market.

The criminal ecosystem was set in motion by gasoline prices inflated by the government to provide a steady and vital stream of tax revenue. Oil-rich Mexico was gouging its own citizens, who paid 25 percent more for a gallon of gas than Americans. Economic inequality and a Robin Hood culture supportive of "redistribution" opened the spigots on illegal pipeline taps everywhere—doubling from almost 7,000 taps to 14,910 in two years as organized crime discovered the wealth within the pipelines. With a distribution system that was informal but organized and effective, looters could make up to $50,000 per month. On Highway 150D, which originates at Pemex facilities in Veracruz and runs through the state of Puebla and on to Mexico City, you might count *hundreds* of people standing on the roadside selling gasoline for half what the Pemex gas stations charged. If you went into the towns of Puebla, chances were good that in twenty-five of them you would find groups standing around a pipe someone had slashed open, splashing around in oil puddles filling up containers. *Hauchicol*, as the contraband fuel was known, was everywhere.

The situation was extremely dangerous. That was *gasoline* they were sloshing around in, after all, which from time to time enveloped a crowd in massive fireballs. Days before Christmas 2010, in San Martin Texmelucan, a city about the size of New Haven, Connecticut, a breached pipeline wreaked havoc. "Several streets were flooded with fuel. With a spark, there was a river of fire," the interior minister reported—killing twenty-eight people, and destroying thirty-two houses.

And through the systematic hijacking and diversion of fuel from refineries and pipelines, the organized criminal enterprises known as *huachicoleros* were costing Pemex upwards of $2 billion each year. But that was just a guess. With the pipeline stretching like a spider web over Mexico, Pemex's leaders could not say where exactly or even *when* the attacks were taking place. Without accurate data, they were unable to measure the exact scale of the losses, on the one hand, or plug the holes, guard the pipelines, and fix the problem, on the other.

As a senior information executive, Bernal had been asked to lead the overall digital transformation of Pemex. Old systems would be replaced or migrate to the cloud; data analytics would come to the forefront. Harvard Business School–trained CEO Gonzalez Anaya wanted to see numbers for everything and expected his executives to get the data to him and the CEOs of the Pemex sub-businesses. Bernal's charge was to turn Pemex's operations transparent for these heads of all its divisions and for his own boss, creating one shared truth about Pemex's financial, operating, and safety performance. Despite the mandate to lead an overall digital transformation, there was no mistaking the very tactical nature of Bernal's first and most urgent task: find those leaks and stop them.

Bernal knew he had to begin by capturing the data. The pipeline was in his sights. Raw oil ran from the wells through Pemex's six refineries to Pemex's distribution business, at each stage shifting from control by one of three separate Pemex businesses: exploration and production, refineries, and distribution. In the pipelines, the only sensors were pressure gauges that were built for maintenance, not for detection of fuel theft. They showed when pressure rose or fell in unexpected ways. If a change in reading seemed significant, a maintenance crew might go and take a look. But sophisticated criminals quickly disappeared into the night, as thieves drilled and then sealed the hole with a collar to make the gouging appear as nothing more than a momentary blip on system screens that could easily be ignored. Months later, the gangs would return to do the siphoning in bulk, discreetly, with little change in pressure noticed.

Bernal crafted a plan. First, new sensors would track the oil flow across the infrastructure. These sensor signals would be supplemented with motion-sensitive cameras near crucial distribution points to reveal physical activity when alarms were triggered. Second, all incoming data would be pushed to a virtual data room and cleaned of noise that could trigger false alerts. Third, the pressure gauges' historical data would be modeled to determine what meant a legitimate loss in pressure and what seemed to be a criminal assault—and an algorithm-driven alert system would trigger executive and operating unit attention to theft events, and with it a local law enforcement or military response. Finally, all that data would be brought together into the control system and displayed in various dashboards to managers and executives, holding them accountable for results that their data would reveal to all.

In every instance, this plan would involve repurposing infrastructure and workflows originally installed for maintenance purposes but never intended to deliver metrics that precision insight to theft required, or to integrate them to a system that did. The essential work was modeling the combined sensor data, then using those models to predict whether any given alarm was a maintenance issue or a potential crime—and send the alerts.

Pemex had no data scientists, leaving Bernal to recruit talented young economists already on-board at Pemex for classic financial modeling and repurpose *them* with the analytics work. Their models proved 80 percent accurate from the start and, with fine-tuning, into the 1990s. Accurate models were essential because no one could afford to send local police or military units scrambling on false alarms. Although Pemex was a first responder for its own security problems, it had no police powers. In serious matters, it relied entirely on local law enforcement and Mexico's army—all armed—to roll in when called. All of this came at great risk and challenges.

"The effort took us from chasing bad buys to going against the internal bureaucracy, making them accountable to what happens in between," Bernal says. "So you start to close doors on the whole chain, finding and then closing the margin of error of leakage as the oil flowed between the companies. It would rely on the subunit CEOs holding people who reported to them responsible."

Experimenting with different sensors both along the insides and outsides of the pipelines, supplemented by drones capturing imagery from above, Pemex began to build a body of reliable data about pipeline integrity. There were serious challenges to overcome: gaps in old technology, bureaucratic and organizational hesitance, and even corruption and threats of physical violence. But little by little, the truth came to the surface through the data. When the first pipeline sensors showed large discrepancies between what was claimed to have been shipped and received, managers denounced the sensors as inaccurate. "But the numbers were solid," Bernal says. "They showed an off-the-charts difference between how much they put in and how much was lost." All the loss—what retailers suffering employee theft call "shrinkage"—was technically within the legal margins of error. That was the amount of oil Pemex was allowed to lose, by law, in the transfer of "custody of its product" across its businesses.

"Then there was the day we realized," Bernal says, "that Pemex had written its *own* laws around how much tolerance there is to lose product." And its losses fit that tolerated range perfectly.

Early on, Bernal tapped the CEO of Pemex's petrochemical plants and cajoled him to be the first to jump on board the digital transformation. His plants were new and had a relatively large number of sensors already—although they were not built for the new world of the CEO's accountability for financial, operating, and safety performance, of which the antitheft element was a small part. As a first move toward digital transformation, Bernal had picked as supportive a target as he could find. Still, there was mistrust from the start.

"Pemex is run by oil engineers," Bernal observed. "The guys who analyze huge data sets to find where to dig are rock stars, like Silicon Valley developers." But the bosses were different—lifers who had little patience for the reformers from Pemex corporate.

"All you IT guys come here and want to tell me the solution to my problem," Bernal recalled being told. "We spend a lot of money. You become a barrel without a bottom. Just money, money, money and nothing happens. It's gonna take a while before I trust you."

Over the years, data quality had crumbled at Pemex, and confusion, mistrust, and poor decisions all became part of the Pemex culture. Krasadakis would not have been surprised. "I've seen great business intelligence and data warehousing initiatives fail due to low engagement by key users," he wrote—the direct result of a lack of trust in the data. "If users don't trust the data, they will gradually abandon the system impacting its major KPIs [key performance indicators] and success criteria."

Bernal attacked the data quality issue head on. His analysts needed to know, for example, whether some outlier data on leaks constituted a critical discovery or an unknown and poorly handled data issue. There was even confusion *about* the confusion—was it accidentally caused by poor data or intentionally broken by corrupted data fed to the system? This hampered decision making and suited a hardcore group of corrupt Pemex executives, managers, and operators who one report described as "working with transnational criminal organizations in a multi-layered web of corruption, difficult to untangle." As part of the problem, they profited from obscuring the truth and fought transparency, sometimes with violence.

"We actually had people killed," Bernal recalled. "It happened in the Salamanca refinery." Bernal's team bought digital cameras with sophisticated motion detector sensors and installed them at the Salamanca distribution center—having first emptied the refinery with a fire drill. The cameras were activated, but they were destroyed two days later. The camera installation team was murdered.

"Nobody knew about it when we did the operation," Bernal recalled. "Someone leaked where it was."

The digital transformation of Pemex to a sensor-based precision system didn't happen quickly or easily. It could not have happened without the cover of a larger digital transformation going on, or without CEO González Anaya's demands of his chiefs that they know their business exactly as Bernal's data would reveal. By creating an accurate and up-to-date view, powered by data from this extraordinary sensor network, showing precisely where oil was being lost and how much, Bernal's organization would create a true view into the state of operations and the controls needed to act on it.

"You have to have humility," Bernal said. "There are no silver bullets here. But we were able to consolidate all the data from machinery, plants and people into a one-shared-truth picture of operations for leadership, and make it actionable."

Without that insight, Pemex would have continued to be blind to its vulnerabilities and have no better hope of remediation than by instinct and pure luck.

## Diagnosing Problems with Data Availability and Quality

The Pemex story illustrates some of the challenges leaders face in assessing the data they have before using it to launch a precision system. As was true for Bernal and his team, three questions should be top of mind for managers. Start, first, with the basic question, "What is the business problem we must solve?" Keep that statement simple in the extreme: a few words should make the priority clear to every manager. For Pemex, for example, it might be, "Detect every pipeline leak that involves theft." That framing embeds the measure of success—everyone knows what counts as a win. It provides the basis for testable hypotheses, as we discussed in chapter 2. In this instance, data scientists and oil engineers can model the leaks that involve theft, then test current detection systems to see if they can discover those leaks from among all the leaks Pemex experiences.

Next, move to the second question: "What information do we need?" Answering it should be a straightforward, intuitive exercise. If you were to attempt, alone, to evaluate the testable hypothesis you developed earlier, what information would you need to do it? Err on the side of including rather than excluding any data you're unsure about. For Pemex, for example, you'd certainly want data on *all* leak events—the where, what, when and how—together with the unique flow and pressure signatures of each leak. You'd go through case files and figure out which leak events were actually crimes, then build models that flagged incoming leaks as likely crimes, too. But don't stop your hunt for data there. Large-scale theft should show up in data elsewhere—for example, in disparities between the gallons of fuel that arrived for distribution and the gallons actually distributed. The difference could be fuel that was siphoned off and stolen.

Then move to the third question: "Is the data we need accessible?" An easy shortcut to the answer is to ask for a sample of each of the data sets needed or attempt to extract them yourself if you have access.

You'll have plenty to worry about at the highest level—consistency, integrity, accuracy, and completeness. But in our experience three such issues need to be addressed head on:

- *One or more of the data sets needed are not instrumented—or instrumented for another purpose.* Data do not exist by default; active intervention is required to decide what information needs to be captured, to set up the tools and/or software, harvest it, and make it available for analysis. This process is known as *instrumentation*. At Pemex, for example, the pipeline sensor systems had been instrumented for controlling flow and pressure in pipeline operations, not for detecting leaks. Instrumenting for leak detection would require far different physical placement of sensors and use of different sensors built to detect those much smaller pressure changes typical of leaks. These decisions would depend on how *sensitive* the sensors should be (set to avoid both false alarms *and* missed events), how *robust* (able to withstand the rough-and-tumble of pipeline operations), and how *accurate* (getting the location of the leak exactly right). Bernal's first task, then, was to instrument Pemex's pipeline sensor systems for leak detection, not just pipeline operations.
- *The data are stored in mixed and incompatible formats.* A simple example is when data stored in a structured way, such as in columns and rows, cannot be merged easily with unstructured data that take the form of, for example, images unless some way of connecting the two data sets is established. Data

scientists often complain that most of their time is spent cleaning and preparing data for use in a model and less of it doing the modeling. It's a bit like chopping all the ingredients for a meal before cooking it: the chopping can take a lot longer than the cooking!

- *Some or all the data are corrupted, broken, or otherwise unusable.* Data that are incorrect, erroneous, and/or incomplete are worse than no data at all. Incomplete data requires significant effort to correct or fill in. Some data may degrade as they are communicated across systems whose bandwidth is inadequate. At worst, they can lead analysts on a wild goose chase where the insights produced from the data are as flawed as the data themselves. At Pemex, to protect against this, a variety of different leak detection systems were tested. Many were ruled out as costly, and it was impossible to discern if the systems provided data that were any better than the data provided by the incumbent sensor system. If, on the basis of insights confounded by poor data, you choose to make decisions of consequence such as budgeting, investing, hiring, firing and more, those decisions will be misinformed, and you may not even know it.

The importance of data quality for precision projects cannot be overstated. A model fed poor data will always produce poor outputs. The ability to get from starting-point data, which can be rife with the kind of problems described above and more, to usable data for model input (which is still far from the final impact goal) is part of what makes data scientists and engineers so skilled and so in demand. There is a chasm between what someone can learn in college about data management and what a job in the industry requires them to be able to do, between what any manager may believe about the state of the data and its actual issues in use.

Bridging that chasm involves profiling your data and fixing gaps in what you have, what you need, and what you can get. "There is a long, painful and never-ending re-alignment of idealized projects against more real-world work practices," a data scientist wrote in a post titled "Ten Fallacies of Data Science." He might just as well have called it, "Ten Delusions of Data Science."

Fortunately, as you bring the special talents of engineering and business insights to bear on the data challenges you face, you may not be a hostage with little recourse but to accept compromised data with its serious implications. You can go ahead, fix some crucial problems, and get precision going.

## Righting the Wrongs in Bone-Age Assessment

In radiology, physicians use bone-age assessments to determine the true age of a child. "True age" is an important piece of information, but it is often unknown because many children around the world may have no documentation of birth. Doctors who are treating illness or developmental issues need to know their patients' true age. Police officers, lawyers, and judges need "true age" to determine whether to treat a young person as a juvenile or an adult.

Traditionally, the process of doing a bone-age assessment is a methodical, rather mundane task of comparing an X-ray of a child's hand and wrist against textbook examples of bone-age images. But now, scientists are applying artificial intelligence (AI) to make the process far more accurate than ever before.

As chief resident in radiology at Columbia University's Department of Radiology, Dr. Simukayi Mutasa took on the important challenge of building an AI-powered model to determine the correct age of a child based on hand bone structure. It's easy for AI-enabled image recognition to make that call if the textbook examples of hand bone structure are any good. That's where Mutasa quickly bumped into a problem—not just any problem but a profound flaw in the reference data set that has been in widest use around the world for over sixty years.

"We all use this giant book," Mutasa told us—Greulich and Pyle's *Radiographic Atlas of Skeletal Development of the Hand and Wrist*—where we compare an X-ray of both hands. It takes about five minutes of our time and requires no thought. It's just literally looking at pictures and finding the correct one. . . . It is one of those tasks that at high volumes become mind-numbing," Mutasa said. "Nobody would miss doing that exam."

Since its publication in 1959, Greulich and Pyle's *Radiographic Atlas* had become iconic and Stanford University Press's most profitable title in its 120-year history. Yet it has been prone to unnoticed, undiscovered, and uncorrected bias every day of its published life. In essence, for sixty years, physicians who have used the *Radiographic Atlas* unknowingly risked providing potentially inaccurate estimates of true bone age in every courtroom and doctor's office where it was used, whether to classify a child truly as a juvenile under the law or to assess how a child was whose growth appeared stunted.

What was going on in the textbook? The reference data published in the *Radiographic Atlas* comprised hand and wrist X-rays of 300 Caucasian children who grew up in Cleveland in the 1950s. "How can we use their bone age as a standard for the bone age of, say, a group of Hispanic kids or black kids or Chinese kids in 2022?" Mutasa asked.

"Back in the 1950s someone went to a school in Cleveland, Ohio, and took pictures of healthy upper-class white kids' hands, age zero to eighteen years," Dr. Joshua Weintraub recounted. Weintraub is executive vice chairman of the Department of Radiology at Columbia University and a practicing interventional radiologist at New York Presbyterian Hospital. "The medical community decided that was standard bone age. That's how we place your bone age and how we always have. But we're in Washington Heights—a principally Dominican community," Weintraub reflected. "People come to the hospital with developmentally delayed kids—and we're trying to figure out how delayed they are and whether we can get them back on the growth curve. We plug their images into our Gruelich and Pyle *Atlas* of healthy white kids from the 1950s Midwest and the AI automatically brings up the answer, 'You correspond to six years, six months.' That becomes the diagnosis. It couldn't care less if you're a kid from Washington Heights. That's how it places your bone age."

Self-trained in coding and equipped with high-resolution imaging and computers donated for his work—essentially, a physician software engineer—and working under Weintraub's general direction, Mutasa set about fixing the problem. The only solution was to start again.

"We actually redid the entire *Atlas*," he explains. "We took thousands of children from New York City, which represented a diverse population, tons of Hispanic, Black, Jewish and Asian kids—30,000 of those in a publicly available data set." They leveraged another 12,000 bone scans from New York Presbyterian Hospital and developed a bone-age algorithm based on that.

As a high-volume, repetitive, time-consuming, simple-to-do task comparing images, bone scans are something neural networks and deep learning can be good at. "Well-suited to image classification tasks associated with screening, triage, diagnosis, and monitoring," Dr. Eric Topol affirmed. "Actually, significantly better than humans," Mutasa says, citing his own and others' published research. "When humans do it, there is an element of subjectivity to it. The intraobserver variability in humans is high. But with computers, it does not exist."

***

The issue of bias baked into reference data sets is not limited to radiology. The herculean effort to sequence the human genome was finally completed in 2022, taking nineteen years and nearly $3 billion. Yet the human genome reference data also had limiting flaws. The original data set was based on DNA from just eleven individuals—70 percent of it from just one male, 80 percent of it from people of European descent. "All of that information is nearly worthless for anybody other than those of European ancestry," Eric Green, director of the National Human Genome Research Institute, told a reporter.

Whether in radiology or oncology, new projects are underway that, like Mutasa's, reject the concept of a single reference set for all, sequencing instead diverse populations that reflect the human species more completely. The value of such efforts cannot be disputed. The *Cancer Genome Atlas*, for example, is now based on the sequencing of 20,000 patients across thirty-three tumor types. "We can measure every mutation, every genetic change the tumor makes to become cancer," Nathan Siemens recently told an audience. "That has changed the paradigm of cancer therapy. Instead of toxic chemotherapeutic drugs, therapists sequence every patient, find the weak points in tumors, and treat them with the appropriate therapy, including unleashing the body's own immune systems on the tumor. It's called precision medicine and it can work remarkably well."

The same holds true for radiology. With physician software engineers blazing trails, applications will proliferate—and with them will come improved effectiveness and greater equity in diagnosis and treatment. "It used to take four years and a PhD to build one of these systems," Dr. Curtis P. Langlotz of Stanford told an audience. "Now, with the right training data, in a matter of weeks you can build a very accurate model. And we know bone development is different across different ethnic groups. So rather than having one model that gives the right answer for Caucasian kids in Cleveland, in the 1950s, you can build a separate model for each individual demographic group."

Moving to higher volume, more accurate AI-enhanced radiology will no doubt find its way to ubiquitous clinical practice as peer-reviewed studies show AI-enabled precision systems can equal or exceed the performance of even experienced radiologists, pathologists, or cardiologists. "If you're confident in your diagnosis you probably won't need it," Dr. Peter D. Chang said when he reflected on yet another tool he and his students built—this one to read chest radiographs, scanning for the smallest indications of pneumonia (we introduced you to Chang in chapter 2). "But it is useful for the less trained physician." In the early days of COVID, that tool let doctors at the University of California, Irvine (UCI) Medical Center get a jump on treating likely COVID cases even before formal COVID tests came back.

That's good news even now, as around the world radiologists are in short supply but imaging is proliferating. Meanwhile, barriers to uptake and adoption of AI-enabled medicine are cracking. The Food and Drug Administration (FDA) approved the first fully autonomous AI-based diagnostic system in a primary-care setting in 2018. More systems are in the pipeline today. Medicare is beginning to reimburse for adoptions of such systems. The dam is bursting: "Every single common, non-serious non-life-threatening diagnosis, one by one is getting approached by AI," Topol asserted in 2020.

For now, Weintraub said that "the places where we've gotten to true clinical diagnoses are *de minimis* at best." Rather, AI-enabled enterprise systems bring intelligence to workflows, surfacing priority images for reads in the radiologists' queues—both by urgency of the matter and, in some cases, having presorted "negative" diagnoses from "positive" ones that require physicians' review to be certain.

For the future of these precision innovations, leading physicians peering over the horizon see possibilities that amaze. Weintraub described the promise of radiomics, for example, which looks at the pixel relationship of tumors, determines their pathological correlation, then predicts their sensitivities to various chemotherapies.

"By looking at that pixel relationship," Weintraub said, "the computer will do an algorithm and say, 'This tumor is sensitive to gemcitabine [a drug frequently used to treat several cancers].' As the cancer changes, the pixel relationship changes, relating either to the pathologic type or the chemo sensitivity. We will get to the point that my job as an interventional radiologist disappears because I won't have to biopsy it anymore. We'll be able to look at the relationship of those pixels and say, 'This is a leiomyoma sarcoma—just because of the relationship of those pixels, as opposed to tissue biopsy.' It's a whole different field completely."

New kinds of collaborations will be essential to design and develop such systems and then place them into the clinical workflow. On the forefront of radiomics, for example, Dr. Richard Ha at New York–Presbyterian Hospital is changing the future of breast cancer screening by stratifying high-risk versus low-risk patients, not by clinical or family history, which he asserts are poor predictors of future cancer, but by direct inspection of tumor images for features that may predict an individual's precise risk of progression to cancer—sending fewer off for surgical lumpectomies, today's one-size-fits-all solution. For the high-resolution tumor images, Ha collaborated with Dr. Christine Hendon's biomedical engineering group, which developed the necessary optical coherence tomography technology; with surgery and pathology teams to identify the right surgical specimens to image as a reference set; and even with Dr. Carolyn Westhoff's ob/gyn group to learn whether new treatments with fewer side effects than the standard Tamoxifen could be devised for the high-risk women once identified. Ha's final contribution would be to determine whether any new medication was working—again, by direct examination of the tumor images.

Change, no doubt, is ahead—but observers are cautious. "Machines will not replace physicians," Dr. Antonio di Ieva recently wrote, "but physicians using AI will soon replace those not using it." Where it all lands and when will depend on practitioners' confidence in the data, the track record of models in use, new regulatory approaches to continuous learning, and the gravity of decisions at hand. Above all, Chang cautions, one should never underestimate the power of the physician bias against change. "It is not uncommon for physicians to believe the insights they generate can fundamentally never by replaced by an intelligent machine system."

## Doing Data Due Diligence

Simukayi Mutasa is one of many pioneers of precision medicine. Fixing a reference set is but one of many levers of change practitioners can pull who are skilled both in computing and the sciences. But it takes something special to convert that new transparency to value: the passion to right a wrong. Mutasa, for example, didn't stop at simply recognizing the problem. Keenly aware that somewhere in the world an algorithm that was built on old, incomplete data was still misstating a child's true age, and impatient with waiting *another* generation, Mutasa went out and fixed the problem. His on-ramp to the reform of a flawed precision system was the data as he first wondered about its origins and then set about re-creating the data from scratch.

As Mutasa's story suggests, before embarking on a precision project, the state of the data must be known; its quality understood; and, if need be, it must be challenged, tested, and proved. *Profiling the data* can be a valuable exercise because it might yield discoveries as consequential as the bias of race and location in a reference set.

From medicine to marketing, whether profoundly flawed or simply annoying, quality issues can be revealed through summary statistics that allow domain experts and system engineers to sense-check. For a marketing project, for example, they can scan the statistics and ask questions like, "How can the number of new customers within the last thirty days be larger than the total number of customers overall?" It cannot—a sign that something is wrong and needs to be investigated.

Similarly, characteristics of the data, like its completeness and sparsity, can be examined for indicators of quality. If there is a row in a database for every purchase ever made on a website, but only 90 percent of those records contain the price of the purchase, what's wrong with the other 10 percent? If no attempt to verify the quality of this data is made and someone simply goes ahead and uses it to calculate the total dollar sum of all purchases, they will be 10 percent short without knowing it. Any "average price per purchase" or "average dollars spent per customer" values would also be wrong. If a manager acts based on these flawed insights, their actions would be misguided.

Lesson for leaders: data quality assessments can be time consuming and costly. but they are worth the effort. They can increase confidence in both the project and its result, and are extremely effective at reducing risk in a precision project early on. An error in any of these checkpoints—the availability, accessibility, and quality of the data required—can cause a precision project to miss its marks.

## Platforms

If data is the life blood of precision, platforms are its beating heart. The word "platform" has many meanings. For the U.S. Navy, a destroyer, a submarine, or the F-14 *Hornet* fighter plane is each a combat platform designed for specific missions of armed defense or attack. For a hospital, an operating room is a surgical platform. For a school, a K–12 classroom can be thought of as a platform for teaching and learning. Jorge Heraud and Lee Redden turned the everyday John Deere tractor into a precision platform for weed control. For our purposes, platforms are where the work of precision systems gets done.

*Precision platforms* share common basic traits. Borrowing from Thomas Eisenmann's general framework for platforms, precision platforms are highly digitized but have a physical footprint and a human interface. All platforms have *rules* that govern users' access, use, and interactions, and *components* of hardware, software and service modules. Precision platforms are often powered by AI and machine learning; handle vast stores of data; and, if so, have an architecture designed for continuous improvement. Platforms are designed for a specific purpose; they must have well-defined workflows and jobs geared to those tasks. A mission-driven and purpose-built precision platform's success is measured by whether and how the platform permits operators not just to find and engage each other (technical success) but also to change the status quo measurably and precisely as required.

From ships to surgical theaters, to classrooms, precision platforms involve other platforms that are central to their performance. Among these are *data platforms* that serve up the data comprising the precision system's fuel. Data platforms are *purpose-neutral* and flexible because they are designed to enable many varied applications comprising the work of the platform. These include the common types of data applications we described in chapter 2: pattern finding, personalization, value predicting, entity recognition, geotracking, and scenario modeling.

Organizations invest in data platforms to automate these challenging processes. As a result of that prior investment, the users of the data platform don't have to figure those things out every time they want to run a new analysis or train a new model. Today, data platforms are commoditized; in other words, they are purchasable from a variety of major platform vendors and don't usually need to be built in-house. They are durable because they are long-term investments that can support a variety of applications over many years. (Splunk, a data platform vendor, offers this definition: "A data platform is a complete solution for ingesting, processing, analyzing and presenting the data generated by the systems, processes and infrastructures of the modern digital organization." It supports diverse tasks from data ingestion and preparation to model creation, training, and test, to deployment, monitoring, and maintenance, all geared to the incorporation of predictive and prescriptive models into business processes, infrastructure, products, and applications.)

Early in the life of a precision system, data and platforms can be kept lean. In many cases, it's possible to prove complex concepts and achieve remarkable impact without much investment in the way of data and platforms. But as precision projects mature and tackle more real-world complexity, the management of data and platforms almost always becomes more complicated and crucial. The more people who are involved, the bigger the network of machines online and the more complex the management of their output becomes. The broader the scope of the program, the more data need to be managed.

And when working within the context of an established organization with preexisting data platforms, another layer of complexity is added: the challenge of achieving new things with old technology. Existing legacy platforms can sometimes act as a launch pad for precision ventures, but more often they are a drag or even a liability. Heavyweight, expensive technology that requires many teams of engineers to maintain simply isn't fit for new and fundamentally different purposes.

It is possible to transform old systems. Data management teams can sometimes extend legacy systems to meet the ever-evolving demands of sophisticated precision. But the job can be enormously complicated and challenging, especially when a big company with a complex history—like the *New York Times*—commits itself to the new pursuit of precision.

## Replatforming the *New York Times*

Nick Rockwell joined the *New York Times* as its CTO in 2015. Word of change was all around. Since 2011, when the *Times* launched its first pay-for-subscription digital model, the *Times* had been moving toward a new strategy that would replace ad-driven revenues with subscription revenues. With forecasts of doom for the future of the print edition all around, the move to a subscription-based business was a gamble then *Times* CEO Janet Robinson was willing to make.

But the venerable "Grey Lady," as the *Times* was known, was falling down and might not get up. When Mark Thompson succeeded Robinson in 2012 as the *Times*'s CEO, red lights were flashing everywhere, especially for subscriptions, where growth was not just stalled, but collapsing:

> It was something like 74,000 [subscriptions] in my very first quarter, the last quarter of 2012. By the second quarter of 2013, it was 22,000 or 23,000. It looked like the model was plateauing. For a company with four main revenue streams, with print subscriptions essentially stagnant, print advertising in real decline, digital advertising had just turned into decline, to be told that your one hope, digital subscriptions, is plateauing—that's really bad news. So the most urgent thing to do was to figure out how to get the digital-subscription model to work.

If Thompson's doubters were many, few were sure what the best next step should be in the digital future. It was unclear even to Rockwell what his role in the big shift might be. For now, his job was simple enough, standard fare for CTOs, and big: to make sure that "everything worked." To Rockwell, that meant one thing: deliver a computing environment that could "process a lot of data, run a lot of queries, and do it fast," same as he'd done as CTO at Conde Nast and MTV before.

Benchmarking the state of the *Times*'s systems as Rockwell found them was easy, eye-opening, and rattling. *Times*'s rival Bloomberg served up a million news stories on its platforms every day, with a 100-millisecond delay. It had developed robust reader metrics and measures, tracking not just "page views," or "scroll depth," or "time spent on page"—but bundles of metrics in "average revenue per user"—all to know how much money Bloomberg made on each user and how much a user clicking over from Facebook was worth versus one from Google Search. "Overall," Bloomberg's global head of digital claimed, "our metrics tell us that our audience is engaged, educated, affluent, and older."

The *Times*, by comparison, was running on fumes. Its servers, software, and networks were painfully slow. Data analysts stumbled through its data and systems, clawing to gain insight, and petitioned Rockwell to do *something*. "No feedback," Rockwell said, "wasting vast amounts of time, moving incredibly slowly. We were flying blind, with nothing to guide us. It was quick and easy to see that we weren't anywhere we needed to be with respect to data."

Given the state of the *Times*'s infrastructure transformation would hardly be possible—certainly not as CEO Mark Thompson had envisioned it. "Unusable, essentially," Rockwell found, "without a massive investment." Rockwell was clear about *his* next move. "The very first thing we did was a broad replatforming—moving our entire data platform to Google's Cloud Platform infrastructure," he said. He'd gone comparison shopping and found the *Times*'s incumbent cloud provider, Amazon Web Services, lacking. "We recognized and wanted the power of Google's data tools, like 'Big Query.'"

"That was a really good choice," recalled Katerina Iliakopoulou, a lead software engineer. "We needed that very high speed when it comes to querying our user history service—which was in the hot path of every request we get for recommendations. Google Cloud Product turned out to be the right tool for that job."

"Done with literally no effort on our part, creating as low friction as possible for the whole organization, but particularly for the analysts," Rockwell said. "A rare and wonderful case where the impact you have is, like, two orders of magnitude."

The gains in performance came fast. "It literally cut queries that would take a couple of hours to a couple of minutes," Rockwell said. Vast stores of the *Times*'s data were now easily accessible and manipulable—"a single source of truth, all queryable." The import was obvious: system reform was a necessary precursor of strategic transformation as Thompson had laid out. "We could actually start to use data in an operational way, and really start to run the business off of it," Rockwell said. And it confirmed his mantra: "You've got to solve your organization first"—get your platforms right, your data right, your people and teams right. Nothing transformational is possible until then; everything is possible afterward.

The breakthrough in platforms and data triggered a dormant "analytics debt" to pay down—a backlog of analytics jobs to run. Working through it let the *Times* build insight into its customers' journeys as never before—how people started to use the *Times*, for example, and how they discovered the deeper dimensions of its reporting. "It let us gain deeper insight into the business and particularly the subscription business and conversion dynamics and the correlated question of reader engagement," Rockwell said.

Standing at the dawn of enterprise precision, the *Times* was moving past mere digital transformation to *precision systems* that could answer the questions and reform the practice of a full-on move to subscription-based revenue. How much precisely to charge to boost revenues without losing customers? When precisely to drop the paywall—was it after three free articles, or seven, or ten? How best to segment customers precisely, exciting potential new subscribers and engaging existing ones even more?

Josh Arak, then the *Times*'s director of data and insights, would soon unleash a tsunami of testing new products and innovation that would give answers. But there was work to do first, "getting the organization right." Like Rockwell, Arak found the *Times*'s platform too fractured to handle the heavy lifting of a high-functioning enterprise insight engine. There were multiple data fiefdoms with no central testing platform, no centralized test result database, and no unified testing protocols or strategy. Five different channels—desktop web, mobile web, mobile apps, newsroom, and marketing—each ran their own separate testing operations. Each had its own methods of tracking and reporting results. Each was a happy band of users disconnected from the others and each was, potentially, poisoning the others' work. One team could inadvertently taint another's results by running experiments on a cohort of subscribers, for example, while the other was injecting the very same cohort with a different treatment somewhere else. It would skew the data—but no one would know. With this blind interference going on, it was almost impossible for Arak's team to create a single coherent picture of reader behavior, gain insight into *what* to change, or develop a plan for *how*.

The future was on hold until this could get sorted. Arak's best next move was clear: an investment of time, people, and money that would simplify and standardize the organization's entire testing platform—its infrastructure, applications, and workflows. Breaking apart technology fiefdoms that are built up around platforms is a third rail for any technology manager. Going too fast could cost Arak supporters and ruffle the wrong feathers, while going too slow might mask or minimize the problem's true urgency. Losers in the move forward would feel their loss sharply and fast. The winner would be the *Times*, but that was diffuse. Up-and-coming software engineers, data scientists, or product managers eager for new tools should also favor it, but that could take time.

Arak's best move now would be a proof of concept that brought the vague promise of benefits into sharp relief, give winners a taste of the future, and gain a green light from the top. No other move could trump a proof of concept. Arak's proof rolled out two main features. First, it pulled information from each of the disparate testing platforms concerning *which Times* readers had been placed into *whose* test groups, making sure the allocation of readers was truly randomized and controlled across experiments. That would fix the problem of one experiment tainting another.

Second, the proof produced a uniform set of metrics. No matter over which channel or via which tool a test was run, every team would calculate success or failure by the same metrics, making them consistent, measurable, and comparable across all domains. Which article got more engagement (clicks) and which article caused more subscriptions, for example, were important metrics for almost every experiment any digital team could run. On the consolidated platform, the insights could be found in one, clean place and mean the same thing. That made comparison of alternatives easier.

Proof of concept in hand, Thompson backed Arak's plan. And why not? Arak's proposal sat squarely in the high beams of Thompson's strategy to transform the *Times*'s revenue stream. He had hired smart people like Arak to tell him how to do it, and that Arak plainly did: build a unified testing platform and embed it in the workflow and culture of teams across the *Times*. Foster systematic experimentation, supporting many more experiments *and* improving their quality all at once, and thus obviate concerns for interference, normalization, or any such data-quality issues. Achieve new insights into the customer journey from its very start; answer the questions of when, how, and where to price the subscription and drop the paywall. Arak described the decision to build a single, unified platform this way:

> It was an interesting path that took us back to the roots of the decision to rip away the foundations of how we were doing data and build it from the ground up again, and to be more flexible and more nimble and more scalable. We would have never been able to achieve what we achieved if we didn't do that. And the takeaway is, "Cut your losses when you're wrong and stop building upon the problem" because ultimately, you're holding yourself back from building and scaling at the speed that you want.

Arak's team unlocked new breadth, speed, and effectiveness for the *Times*'s transformation from merely digital to enterprise-wide precision. The result was a strategic correction to a subscription-based business model that set the *Times* on a path toward becoming the multibillion-dollar subscription business it is today. It's a story we'll examine further, from other angles, later in this book.

## Getting Out the Democratic Vote in 2020

Let's conclude this chapter with a glimpse of a precision system project with greater social and political implications than the digital transformation of the *New York Times*. In January 2021, members of the Democratic National Committee (DNC) reflected on the 2020 presidential election cycle—unprecedented and chaotic in so many ways—from the vantage point of the winner's side. From this view, it is easy to gloss over how hard-fought the election had been, especially because so much of America was keen to forget the last four years.

But it should be remembered. The traditional battle of voter engagement had to be fought under conditions of a global pandemic, socially distanced campaigning, viral disinformation, cybersecurity threats both foreign and domestic, and a climate of political division that culminated in violence and fatalities. In the end, Joe Biden pulled through, and Donald Trump was denied a second term as president. Hundreds of thousands of volunteers, making millions of calls and texts over 100 days, delivered the Democrats a win. But several factors weaving together behind the scenes helped seal that victory.

One of those little-known factors was the work of Nellwyn Thomas, CTO of the DNC. Thomas had spent the last three years leading her team to develop the party's data stores into an election-winning precision asset, leveraged into a crucial win on November 4. Her team had developed a new and improved data platform to be used by numerous campaign teams for their various strategic purposes. She reflects:

> Democrats succeeded because we, as a party, have a shared foundation made up of high-quality data about voters, reliable data infrastructure, and scalable tools. This foundation, rebuilt over the past three years by the DNC . . . allowed campaigns to be more nimble and pivot more effectively.

Applying precision strategy to politics poses some unique challenges. Presidential election cycles happen once every four years, and each cycle is disconnected from the last, which leads to short-term thinking by campaigns. They don't usually need to invest in strategies and tools that last beyond Election Day, nor do they have the luxury of time nor the capacity to do so. Every minute of every campaign day is an opportunity that must be maximized because the outcome is zero-sum; there's no value gained by achieving second place in a presidential election. These realities create intense pressures for technology strategy and for precision systems, which can involve complex landscapes of databases, transformations, access and application tools, and more, all stitched together to achieve a specific result—in this case, to help the DNC find and persuade the segment of American people most likely to convert to vote for Biden.

To begin, the phone numbers, email addresses, and voter registrations of millions of citizens were scattered across state, local, and federal databases. The DNC tech team had the formidable task of synthesizing and cleaning this data and modeling it into usable insights to influence the deployment of their limited campaign resources. This goal required hundreds of hours of data deduplication. For example, registered Democrats who moved between states could have multiple state registration records, but it's only worth spending the time to contact them in the state they live in now. This was a data quality challenge for the DNC tech team, which invested precious time into carefully eliminating redundant and outdated voter records to ensure that each voter had one clear current registration address.

Thomas's work in developing the data and the platform for an effective Democratic campaign was only part of the story of the 2020 victory. As we'll explore in chapter 8, an equally important factor was the creation, deployment, and management of a nationwide team of professional staffers and passionate volunteers who understood how to use the data and the platform as tools for political persuasion. But none of it would have been possible without the prior work of Nell Thomas and her team.

What's more, Thomas and her team went one giant step further than campaign strategists generally do. They built the DNC data infrastructure to serve as the foundation for future wins in 2024, 2028, and beyond. They cleaned up the entire DNC voter registration database from top to bottom, creating a valuable data asset that would not only help the Biden campaign win but also those of his successors in years to come.

Thomas calls this "breaking the boom-and-bust cycle." It involved a two-pronged strategy: building up the technology platforms required into lasting, sophisticated engines, and attracting and retaining the best talent in the game to make it happen. "This work—building data infrastructure; investing in, acquiring, and cleaning messy data about voters; and systematically monitoring and tracking disinformation—is not sexy," she says. "It is not easy. But it is this work that gives campaigns a strong technology platform and frees them up to experiment and innovate." Rather than spin up one-time analyses and tools for the 2020 election, she built a precision launch pad for multiple Democratic wins. If Thomas's vision comes true, the same launch pad will help create Democratic victories in 2024 and beyond.

## Tools for Data Management: Platforms of Precision

As you've already seen, data work is different from traditional software engineering. Among other challenges, data science requires systems to streamline and maintain the data models that scientists create. Data platforms for models are more complex than data platforms for regular analyses and dashboards because they consume more data at finer levels of granularity and often need to run continuously.

For those building up precision muscles within an organization, understanding that data platforms take various forms—from old-school enterprise resource planning (ERP) systems to cutting-edge smartphone networks—is the first step toward knowing which types of platforms are right for their unique circumstances. There is a wide variety of data platforms on the market. Depending on the size of the endeavor, the financial resources available, the legacy systems in place, and the level of in-house experience with particular platforms, any number of platforms might be appropriate. A search for "data platform landscape comparison" will reveal competitive analyses and countless other guides to the full market. A cursory introduction to the field of data platforms might include the following categories: data storage and compute; data analysis, data science, and machine learning; and data visualization. We'll discuss each one next.

### Data Storage and Compute Platforms

The collection and storage of data is a critical, if comparatively unsexy, precursor to any data-focused project. Just like the physical warehouses that keep billions of goods waiting to be sold warm and dry, so do data warehouses keep your data "warm and dry" until they are ready to be put to work.

Note that data storage and compute systems work best when accompanied by a discipline of data governance. Imagine walking into a warehouse the size of a football field. Hundreds of aisles of racks are neatly stacked with boxes full of items, but the boxes are unlabeled and there is no map to the aisles. Some of the boxes are full of flammable items, and some are full of rotting perishables, but you don't know which boxes hold which. You are asked to find girls' soccer shoes, a pack of sponges, and a copy of Sun Tzu's *The Art of War*. Baffled, you start at aisle one, box one, because you have no better option.

This is a problem of governance, and it exists in data warehouses, too. When data stores are poorly maintained, the results can be disastrous. Missing labels and a lack of descriptive information, known as metadata, make the right data hard to find. Poor data governance practices also lead to data that decays in quality. A lack of data governance can even lead to legal liability. For example, some organizations do not realize that among their stored data exists personally identifiable information (PII) about individual people, which carries with it obligations for higher standards of security and protection.

Data governance is its own microcosm within the broad data universe, and fortunately a few enterprising companies have developed and offer data governance as a service, including plug-and-play data search and cataloguing, metadata management, and PII discovery tools along with guidance for complying with data privacy laws like Europe's General Data Protection Regulation (GDPR) and California's Consumer Privacy Act (CCPA). Whether through a purchased service or an in-house function, organizations find that there's a direct correlation between the strength of their data governance effort and their ability to get value out of their data.

## Data Analysis, Data Science, and Machine-Learning Platforms

Anybody venturing beyond Microsoft Excel's analysis tools is in the realm of data analysis solutions. Here, the volume of companies who can afford to develop and sell solutions is far greater than in the world of storage because it is less stupendously expensive to do so. They offer scientists a wealth of out-of-the box tools to:

- Create and run models with suggested libraries of model templates for different purposes: "For Beginners: Your First Neural Network" to classify images of clothing, or "For Experts: Generative Adversarial Networks" to generate images of handwritten numbers.
- Build, train, launch, and maintain models, including automatic data labeling to increase the quality of data input to a model, and even automated value predictions for specific data such as house prices.
- Carry out statistical techniques like Bayesian analysis, categorical data analysis, causal inference, cluster analysis, descriptive statistics, uni- and multivariate analysis, and more.
- Manage data streaming and perform real-time analytics on the data as they stream, including interfaces for easy creation and scaling of machine-learning models.
- Access open-source, machine-learning libraries suitable for use cases like computer vision and natural language processing.

If you're a nontechnical manager, you won't need to understand to this level of detail what kind of analysis platform your project needs or what kind of models the team will need to build with it. This is the domain of data experts, scientists, and engineers. However, the more you understand about what tools are out there and what they're capable of, the better. Even without technical expertise, you can ask questions and raise issues that will help your team make smart investments and avoid overly short- or long-term thinking.

One key takeaway is that precision can be accessible to a broad range of organizations and people, including nontechnologists. Don't assume that you need to use high-powered tools such as machine learning or AI. For example, Microsoft Excel, first released in 1987, is still the most widely used data analysis program in the world. Having brought concepts like macros and pivot tables into mainstream understanding and offering an entry-level bridge to numeracy, Excel has outlasted countless newfangled competitors and is arguably one of the most impactful software products ever released. Just recently, the *New York Times's* Tiff Fehr and team won a Pulitzer Prize for journalism that revealed the true extent of the COVID-19 pandemic across the world using Excel spreadsheets as its foundational application.

As one well-regarded data scientist recently told an audience, "Simple heuristics are surprisingly hard to beat, and they will allow you to debug the system end-to-end without mysterious machine-learning black boxes with hypertuned hyperparameters in the middle. This is also why my favorite data science algorithm is division."

Still, new developments in data science are happening all the time, and it's useful to be aware of them. For example, a new discipline called model management is emerging, and it is aimed at using both technology and process to make the development and ongoing maintenance of data models easier. Model management is a crucial part of scaling precision systems, allowing data teams to move on to new challenges without being dragged back by continual maintenance obligations for older models.

The ways models are built and deployed is also changing rapidly. Today, complex algorithms that would have been run on large industrial computers decades ago can now be run on the smartphones sitting in our pockets. This development is bringing powerful neural networks—facilitators of a process sometimes called deep learning—within reach of more and more organizations. Neural networks are an example of the kind of data science application that demands a lot of data. They facilitate machine learning by using training data to teach a computer how to perform a task independently. For example, an entity recognition system might use neural networks to learn how to detect cancerous growths from medical scan photos more quickly and accurately than a doctor could manually.

In late 2020, *MIT Technology Review*'s Karen Hao explained the challenges that had limited the use of neural networks:

> Deep learning is an inefficient energy hog. It requires massive amounts of data and abundant computational resources, which explodes its electricity

> consumption. In the last few years, the overall research trend has made the problem worse. Models of gargantuan proportions—trained on billions of data points for several days—are in vogue and likely won't be going away any time soon.

Now that is changing. Hao reported that researchers from IBM had discovered how a small computer—like the one in a smartphone—could be used to train powerful neural networks. It's a breakthrough whose implications will only become clear in the years to come.

## Data Visualization Platforms

Data visualization, often the last step in a project, is when all the effort of collecting, cleaning, and modeling the data comes to fruition and the discovered insights are shared with the customer or stakeholder. Those who work with data daily as analysts know how important storytelling with data is; it's how the answer yielded by the data is revealed and communicated to the nonanalytical masses. Not many of us can look at a spreadsheet full of sales data and automatically glean the summary, but when it's packaged into a neat pie chart or line graph, we can.

If you've ever appreciated an infographic that presents complex and dense information in a compact, visually appealing way, then you've appreciated the power of data visualization. Data visualization is the difference between saying that the sun is 109 times bigger than Earth and showing it by drawing a small circle and a 109-times bigger circle next to it. For communicating an insight powerfully and filtering away extraneous information for maximum impact, data visualization is a critical skill and one that many organizations choose to augment via the purchase and integration of specialized data visualization platforms.

In the world of data visualization tools, the name of the game is flexibility. You want a platform that can connect to any brand of data warehouse, create the exact visual needed and customize it, share it with any other communications platform, and more. This requires speed, interoperability, and an intuitive user experience with powerful query and calculation functionality.

It's a competitive field of offerings, each seeking to dominate the market and to provide features such as powerful data analysis and presentation capabilities, speed and flexibility, sleek graphics, and even machine-learning automated discovery of insights within your data, thus doing some of the lifting for you. All of them play nicely with many different data sources and aim to reduce the friction of importing data to be prepared for storytelling.

The right data visualization platform should be chosen carefully. If your data storytelling tools fall short, presenting information in a form that is too complex, too rigid, or too slow, your key message may never get across. This is a last-mile problem in data that can negate the value of all the preceding work.

One key take away is this: simplicity is always to be preferred. As this tweet from Kareem Carr suggests, it may be within easy reach:

> One of the most fascinating data science lessons I've learned on social media is something like this is often a much more effective way of communicating information to the general public than a line plot.
>
> Monkeypox cases around the world timeline:
> May 6th: 1 case
> May 21st: 100 cases
> June 6: 1,000 cases
> June 14: 2,000 cases
> June 21: 3,000 cases
> June 23: 4,000 cases
> June 28: 5,000 cases
> July 12: 10,000 cases
> July 20: 15,000 cases
> July 27: 20,000 cases
> Doubling time: 2 weeks.

## Making Smart Platform Investments

The range of technologies available today to wrangle the production, collection, storage, computation, analysis, governance, and visualization of data is vast and varied, reflecting the range of different circumstances for each organization and the unique goals and investments of each. The assessment, selection, and integration of data platform technology is a problem native to CTOs and CDOs worldwide, and it continues to be one of the most important investment decisions they will make.

Where you are on your journey from traditional to digital, or from digital to precision, should inform the approach you take to data technology. As with other elements of a precision project, you shouldn't assume that you need to invest in the largest-scale, highest-uptime, most avant-garde data platform options. For some organizations, that will be necessary, but for many challenges, like the one faced by Simukayi Mutasa, simpler and fewer tools can be very effective.

Consider the type of precision project you're embarking upon. Will it involve mainly one-off analyses that aren't frequently updated or more continuous, large-scale machine-learning models? If your first precision efforts are new, small-scale, and ambiguous, avoid big changes to your existing data infrastructure. Instead, find a way to prove a concept with easily available resources.

Using open-source software can keep costs lower, as can the flexibility afforded by trial periods available on most tools.

If you're a little further down the road—if you've already had some precision wins and now are hitting the ceiling on what can be achieved without technology change—it may be the time to take a more methodical approach. Develop a detailed request for information (RFI) process to solicit proposals and bids from qualified platform suppliers. Spend time to clarify your current situation in terms of volume of data, expected volume of daily queries, types of data sources, level of support services needed, uptime guarantees needed, feature functionality needed, budget, and expected future growth. Proper research takes time, but not as much time as a rushed decision takes to unwind.

## Build Versus Buy

Every organization with its own team of engineers at some point wonders whether it wouldn't simply be easier to build its own data platform rather than bother with the mammoth hassle of RFIs, vendors, integrations, and maintenance. The answer is, perhaps unsurprisingly, it depends.

A proliferation of no-code and low-code data management and machine-learning products have entered the market in recent years, enabling nontechnical people to build and launch websites, mobile apps, and more without needing to write a single line of code. This has lowered the barriers to entry for millions of people worldwide to produce and monetize software products. The trend has extended to precision, too. No-code machine learning products have brought the same massive reach to AI. The tech giants and many smaller vendors have several incredibly competitive and useful offerings.

Why then would any organization beginning a precision transformation choose to invest time, money, and irretrievable opportunity cost in building what could be bought? Buying can sometimes be the best option. According to Andy Tsao, executive vice president of data and insights at Amazon's Audible, the rule of thumb is intuitive: companies should build only when they have the capacity to do so, and when the product they seek to offer is inherently differentiating, competitive, or unique to the mission. In other words, there's no point in building a product that's already commoditized, not unique, or not competitive. Put your precision people where they can get you ahead of the rest rather than reinventing a wheel everyone already has. The opportunity cost of building something you could have bought depends on what else you could have built with the same time and resources, but it could be the difference between creating a world-changing product or not.

This is what Josh Arak and the *Times*'s data teams did. Knowing that their data was running in silos and wasting its potential, they invested in building a new product to stitch together all the disparate data sources. This allowed them to standardize testing across all the *Times*'s customer-facing digital properties and develop the more advanced experimentation expertise that contributes to their massively successful digital subscription business today.

* * *

As you can see, getting the data and the platforms right is a big, complex process. It's almost inevitable that, once you decide to begin an organizational journey toward becoming a precision enterprise, you'll need to devote a lot of time, energy, money, and creative thinking to these crucial tasks. Don't try to cut them short, or you may end up regretting it.

## CHAPTER 5

# THE PRECISION VANGUARD

Steven Kelman at Harvard Kennedy School coined the term "change vanguard" to denote a core group of ready-to-go practitioners that a leader can mobilize in support of a change initiative. Where the status quo is formidable, Kelman found, "unleashing change" is key, and the change vanguard plays a critical role. Typically comprising some 20 percent of any workforce and part of a larger "reform coalition" that often totals nearly 40 percent, the change vanguard provides frontline blocking for the manager in her role as quarterback of change.

Kelman drew inspiration from the original models of innovation proposed by Everett C. Rogers and others in their classic studies of innovation on the American farm and factory. They proposed the gospel of "diffusion of innovation" as a framework that segments people into "early adopters," "early" and "late" majorities, and "laggards"—bystanders all, measured by their willingness to adopt *someone else's* innovation. Rarely were they the innovators themselves, who typically numbered about 2.5 percent of the workforce, Rogers claimed.

In the fifteen years since Kelman wrote, a sea change in computing and telecommunications technologies has shifted power from a few innovators to many—turning far more, potentially, into leaders of change than the meager 2.5 percent of Rogers's model. Give this *new* vanguard a well-framed business problem; accessible and available data; the right platforms, tools, and skills, and they will often initiate change themselves, no permission asked, none granted. If they hit their marks, bringing change precisely where, when, and how they wish, they can take home handsome paydays, deliver many-zeroed corporate revenue gains, and sometimes achieve significant new social and political value at scale.

We identify this new vanguard as the *precision vanguard*. It comprises a core group of make-change-happen-now developers of services and products who design and build precision solutions at scale on a foundation of data and analysis. This core group includes software engineers, data scientists, applied mathematicians, marketers, business unit leaders, and product managers. No longer mere passive recipients of innovation, the precision vanguard brims with active creators and disturbers, supercharged and feeling empowered by their talents and systems to take on even the most wicked of problems. They are practically offended by blunt force solutions that kill the crops with the weeds and poison the soil with the water, all in the name of higher yields, or that settle for workplace safety fixes that check the boxes, placate inspectors and regulators, but leave risks all around. All feel a great urgency to get in the game now: suited up with data and insight, they bristle on the sidelines, watching disasters loom, anxious like Keyshawn Johnson, the legendary National Football League (NFL) receiver who seethed, "Just give me the damn ball." Once unleashed, they join the battle for a vision and with a plan: for Keller Rinaudo and Keenan Wyrobek at Zipline, to keep millions healthy with perishable medicines; for Nellwyn Thomas and Nathan Rifkin at the Democratic National Committee (DNC), to take the election for Biden; for Alex Cholas-Wood and Evan Levine at the New York Police Department (NYPD), to avert the next violent crime; for Lee Redden and Jorge Heraud at Blue River, to reduce waste and pollution when farmers weed their croplands.

The magnitude of the precision revolution—and it is a revolution—is directly ascribable to the advent of these technologies and the precision vanguard. In the words of author Cathy O'Neil, armed with "weapons of math destruction" the precision vanguard cracks open the status quo with insurgent, breakthrough innovations, counting "hit or miss" as old-school failure, and overkill as underperformance. Biased to action, the precision vanguard is impatient to get on with it, as Napoleon supposedly admonished: "Engage, then see."

## The Precision Vanguard Today

What sort of people make up the precision vanguard today? What makes the precision vanguard stand out? How can executives or managers spot them, recruit them, and put their energies to work?

Our experience and research reveal the precision vanguard to be *high achievers*, recent graduates with advanced degrees in their fields, often with formal training in software design and engineering, data science, applied mathematics, and testing methods. They speak the languages of multiple professions and are fluent in everything from Python and Scala, the lingua francas of data science and engineering, respectively, to lean and agile project methods. They have the tradecraft and artistry to play in many worlds at once and, at their best, communicate with ease across the boundaries of guilds and professions.

They have *an abiding faith in the power of data* to fix even monumental problems and welcome opportunities to stretch the limits of what conventional wisdom thinks possible. Inspired by the examples of born-digital giants like Facebook, Apple, Amazon, and Google, they know that the very same principles of precision and personalization those firms use are matters of science rather than mystery. Using these principles, the precision vanguard believes they can tame the most unruly messes of the world, bringing about the changes they want where, when, and how they want them, just as science predicts and business requires, testing and proving as they go.

They are *deeply connected* to networks of like-minded and similarly talented individuals, and they embrace *a strong culture of precision* built on widely shared stories of heroes and victories (exemplifying the dynamic of culture building described by corporate culture guru Edgar Schein). A belief in the science, the art, and the power of a precision culture is continually replenished by the flow of graduates from professional schools armed with the latest tools, techniques, and tales of engineering prowess, data science success, and business unit victory against the intractable problems of the day.

The precision vanguard is *given to risk-taking*—"willing to take a chance that there's a new way to think about this," as Josh Arak of the *New York Times* told us. Its brashness and talent empower the precision vanguard to make first starts with the barest of investment or authority, confident they can make things work, as they have before, and eager to tweak and build on the promise of results that no one else can produce.

They are, above all, *called to higher purpose and energized by mission*—the prospect of rescuing a nation or a people, strengthening an organization they love, or delivering health and well-being where poverty and disease needlessly cripple millions. Give the precision vanguard a strategic beacon—a north star—and they are easily activated. When strong leaders manage the members of the precision vanguard correctly, they will unleash the change that the vanguard holds in its hands. But if they are managed too tightly, they will wither, leaving bosses scratching their heads wondering what went wrong.

The fact is that precision solutions do not just happen. Many of the challenges faced by the precision vanguard today are sea anchors lingering from past eras, causing trouble—platforms that slow or horde data, people satisfied with paychecks from good-enough performance, workflows that look like Rube Goldberg creations and probably were. It takes savvy managers to take on those headwinds and find the path forward, frame the right problem, choose the right tools, putt them in the right hands, organize and lead teams to the joy of achievement and the power of results—proving the power of precision, unleashed.

The big challenges for the precision vanguard and its managers are evidenced in the following two accounts of the precision vanguard in action. One is the story of an experienced manager at Cisco Systems struggling to predict who among the firm's many Fortune 500 customers was the most ready, willing, and able to buy its products next—even without direct contact with any of Cisco's customers. The second is the story of a new manager, younger by a generation or more than senior managers at the media firm Axel Springer, who sees clearly the prospects for applying artificial intelligence (AI) and machine learning to the field of programmatic advertising and is ready to lead the firm to its first dip into this soon-to-be-billion-dollar pond.

## Theresa Kushner at Cisco: Building a Precision System to Beat a Billion-Dollar Benchmark

Theresa Kushner at Cisco Systems had absolutely no idea who would be her next customer, which was a problem: Kushner was Cisco's head of strategic marketing, and it was her job to know. It was 2012, the dawn of the age of precision. Data were getting big but weren't yet considered "big data." The cloud was a revolution in progress, and mobile devices were in some pockets but not in everyone's. From consumer products to computing, poor data about customer need and sentiment created great gaps between marketing and product development. Winners spent much time living, working, and breathing the customer experience on the factory floor or in consumers' kitchens and bathrooms, earnestly struggling to get a view to customers' worlds, to get better products and services designed, built, and sold.

Modeling and predicting customer intent was still a crapshoot. The tools used were timeworn and blunt. Little had changed since the advent of the spreadsheet and the word processor—and before that, the pen-and-ink ledger and green eyeshades. "Half the money I spend on advertising is wasted," the department store magnate John Wanamaker once said. "The trouble is I don't know which half."

It was especially frustrating for Kushner that 90 percent of all Cisco's sales went through third-party resellers. That meant Cisco had almost no direct access to its end-use customers. The best Kushner could do was to ask the resellers, "So, how are things going?" Only the gods of data knew where *those* numbers came from. This bugged Kushner. She knew that, with the right data, the right tools, and the right staff, she could easily model who among the millions of actual and possible Cisco users was ready, willing, and able to buy. She believed that a simple *propensity-to-buy* model could help do that, and, if married to the emerging trove of real-time data revealing who among the prone to buy was also *ready to buy* and had *budget to buy*, she could focus Cisco's sales and marketing teams precisely on the highest prospect leads. "You can go after ten clients and make the same amount of money as you could if you're going after a thousand," she told an audience of operations researchers.

Against this were the timeworn job designs, mindsets, and workflows of legacy Cisco. "If marketing exists at all with high-tech companies," Kushner later wrote, "it was in support of the sales teams—the hunter-gatherers of the corporate world. Marketing was relegated to preparing the ammunition for the warriors, sharpening their axes with the right presentations, helping to open doors, the equivalent of Stone Age women who built barriers in canyons so that men could chase game into the ultimate trap."

Kushner set out to blaze a trail and "crack the case" for Cisco CEO John Chambers. Chambers's strategy was abundantly clear: meet the customer at the pass with a new technology offering that meets the customer's emerging needs—"even before the customer knows about them," as fellow CEO Enrique Lores of HP once said. Learn their plans, build to them, and get there first with the offering. The strategy would drive revenue, customer satisfaction, and employee bonuses. "When you are able to merge these two things," Lores said, "where technology is going, with needs that our customers still don't know they have, is when real innovation—magic—happens."

As this case unfolds, we will see Theresa Kushner exemplify a number of key traits of the precision vanguard. She will *challenge the status quo* that accepts low performance as the best anyone can do. Unsatisfied with old-school guesswork and a do-today-pretty-much-what-we-did-yesterday mindset, Kushner will fuel new high performance with a mix of structured and unstructured data available as never before, earning it that new name: *Big* data.

She will *use new tools, techniques, and talents to work around the limits of the old methods and unleash higher performance.* Rather than bemoan the lack of direct customer contact, Kushner trains her radar (as she told one audience) on "the great big wide universe, hoping beyond all hope, that there is some signal from someone out there that says, 'Yes, I want your product and I love what you do.'"

She will focus on a *clear business problem*—boost sales by targeting leads of highest value—the customer, new or old, who is ready, willing, and able to buy. Those gems will align perfectly with Chambers's strategy—and become Kushner's north star.

She will *launch a discovery project* that melds all the data into a coherent whole, ready for probing and insight. Kushner has no answers in hand—just the right questions—who is ready, willing, and able to buy? She will focus on building repeatable models that predict those next best customers.

She will *rework her teams* to position talent and tools for the critical tasks at hand, from engineering the data to building the models and gaining the insight. That takes collaboration across the boundaries of disciplines and units, including marketing and sales.

She will focus on *delivering end-to-end* on the value of precision, never stopping until her data lead to new sales that Cisco would never have otherwise had. It's not enough to get the math right—that's just the start. The math must be converted to new sales. That's the test of Kushner's precision system in action and its key metric. At Cisco's scale, it will need to be in the *billions* of dollars.

She is *humble yet optimistic*. Even not knowing the answers, Kushner is confident about her process and her path, trusting in both science and intuition that she can deliver on the promise.

So armed, Kushner set out to discover and predict for Chambers who among all current and potential Cisco users would soon be ready, willing and able to buy the newest products Cisco had on the drawing board. She already had billions of bytes of customer data coming from all sorts of places—digital traces of visits to chat rooms, help desks, industry surveys and conferences, the Cisco website, blog and social media posts, support listening centers, specific industry watering holes, and user comments. Some of the data was soft and qualitative, some hard and structured in spreadsheet columns.

About 14 million people visited the Cisco website each month. Scanning the network addresses of the visitors, Kushner's team could tell which visitors had just started their conversation with Cisco. Cisco worked up detailed profiles on some. It reached out by email and telemarketing to a subset, combined that with archived and purchased business intelligence data on the firms, and together computed how much of Cisco's wares each firm had already bought, how fat their wallets were now, and how much they would probably spend on product next year.

That all meant they had budget and propensity to buy. But when? Who among these firms was also *ready to buy*? The most intriguing data had still to be built. Kushner's team scanned social media for chatter, not about Cisco but about the target firms. What was the whisper on the street about what Corporation X or Company Z were getting ready to buy, and when? Linking the hard, structured, propensity-to-buy data of past sales with the soft, squishy ready-to-buy data of social media produced nine strong sales leads—all predictably ready, willing, and able to buy *now*.

None of this came cheap. Kushner would have to prove that the hours and days her software engineers and data scientists spent on data preparation and model building was worth every penny. By Kushner's account, as much as 80 percent of their time had been spent cleaning the data, ascribing it to particular firms, and building and testing the models. All the preparation work in the world delivered nothing of value to Chambers. Only new sales would do that.

Kushner offered the nine strong sales leads to Marketing. "Not interested," came the reply. Amazing—after all that! It turns out that the Marketing team's bonuses were based on activity levels, not results. Kushner was promising precision targeting and *reduced* activity. It would be more efficient and more effective to use her nine, but Marketing's incentives were aligned with being busy, not necessarily being efficient or effective.

She'd send those nine leads to Sales instead. They got it: Sales was given bonuses on results, so the faster and bigger the results, the better. The best sales leads would now be targeted by the top sales teams in the world, a perfect match.

"They thought that was the most miraculous thing they had ever seen," Kushner quipped, adding with a touch of humility, "They're easily swayed, I must admit."

There was even more value here that was once hidden and was now in plain view. Sales learned news about its customers. One was getting ready to install an online marketing system, completely redoing its entire operation. Sales had had no idea—they'd had their sales apparatus geared up for networking. Kushner's nine opened Sales's eyes to a different view of that customer, one they could make moves on.

This was Cisco's precision strategy in action: Be there when the market or the customer transitioned, with the precise products and services the customer was ready to buy. Finally, what Kushner gave Cisco was precision that transformed: *reusable* models that could be continuously improved, updated, and customized with machine learning—for fourteen technologies, twenty countries, and every Cisco sales team. Soon Cisco was running 75,000 models each quarter answering the question, "Who is ready, willing, and able to buy now?" By Kushner's estimate, the models produced a $4.2 billion sales uplift—a figure well in excess of the billion-dollar benchmark Kushner had discovered mattered most at Cisco. The lift reflected sales that, but for Kushner's work, would have eluded Cisco.

Kushner was no newcomer to corporate life. As a seasoned professional, she ran her precision project and team with deep instincts for the pace and direction of change that she could mobilize and that her bosses would support. Kushner was not hired to head strategic marketing because she fit the bill of a highly entrepreneurial manager who would invent new systems and methods and bring in billion-dollar paydays for Cisco. Yet here she was, with the curiosity of a detective and the methods of a scientist, dauntless as an explorer, and leading like a quarterback. As is often the case with the precision vanguard, she invented her own position as the opportunity broke for her and exactly as she would have it.

Above all, as a seasoned manager, Kushner was guided by Chambers's strategy as her north star. She trained a laserlike focus on the prize: those very few customers, lurking like needles in a haystack, who were prone to buy, who budgeted for it, and who were ready to buy. For Kushner, that meant drilling the data for signals of propensity, budget wherewithal, and intent, and connecting all those dots and more to see if they predicted readiness. Then test, fix, retest, and prove they did, and repeat until nine emerged.

This was definitely a team sport—with position players highly expert in engineering, analysis, data science, marketing, and business running the plays Kushner called, no doubt with very good advice on the field and from the sidelines. But it was her call and her score. Notice that this work crossed boundaries; Kushner had to collaborate with many units and managers from across Cisco, few of whom were her direct reports. In fact, Cisco was famously struggling at the time with a wild experiment in creating a highly matrixed organization. With all the conflict and the confusion, it's a wonder she prevailed.

Remember, too, that one last-mile challenge nearly collapsed the effort. When it came to the essential task of converting Kushner's nine leads into business value, delivering on the promise, Marketing balked. It was people, *wrongly* incentivized, who nearly brought the project to its knees. But people also made it happen in the end. Kushner pivoted smartly, pitching her findings in the currency the Sales department valued—fewer calls, high-probability hits. It paid handsomely for Sales to buy what Kushner was selling.

Kushner's work was the essence of precision strategy in action. She used forethought and planning; clarity of purpose; great data across multiple platforms; the right charter and authorization; and a disciplined team made up of the right people, asking, probing, modeling, and proving. In the end, she stuck to the big plan while shifting her tactics as her circumstances required to sell the win, locking in the powerful incentives for Sales and sidestepping the blowback from Marketing. It took all that to convert some big data out there into operations of astounding value.

## Maddy Want and the Advent of Programmatic Advertising at Axel Springer

Today's tools of AI and machine learning open new vistas, invite new challenges—and doubts—and offer new wins for the precision vanguard, often on an astounding scale. Axel Springer's journey to a powerhouse of AI-driven programmatic advertising takes us to Berlin in 2015, where the 21st century was poised to deliver a revenue stream to the European publishing giant that it could barely have conceived. Change would arrive via a partnership with Samsung, the largest producer of mobile phones in the world, and a venturesome, courageous twenty-five-year-old product manager who saw something no one else on either side of the deal did: the value of new technologies for mass, precision personalization.

The deal was set. Axel Springer agreed to make an app for Samsung that would be the preinstalled European news app on Samsung's new S7 flagship mobile phone. It had always been the agreement that Samsung would not pay for the app, but Axel was free to monetize it. That meant ads. And that, it turned out, was a tremendous amount of money.

The app would be named *upday*, and the task of developing it fell to Madeleine ("Maddy") Want, one of the coauthors of this book. Want was product manager for the Axel Springer engineering team responsible for ads. Their task was to design and build the user interface for showing the user premium ads between articles and the software engine that would place them. Axel Springer certainly had a stable of premium clients ready to purchase ad space on the phone—the BMWs, Mercedes, and Audis of the world. Those advertisers and the Samsung client were the totality of what Axel cared about.

What no one saw was that, because of the scale of the S7 phone rollout (10 million new users in the first thirty days), the opportunities for ad placements on *upday* would quickly dwarf Axel Springer's premium clients' appetite for buying ad space. Want's back-of-the-envelope calculation showed 200 million daily chances to show an ad. The premium ads would take 5 percent of those, at best.

"We had a surplus of inventory of ad placements so massive that no manual process could fill it," she remembers. "It was all very *Mad Men*. Digital was just a new place to publish the old-school ad deals the sales guys made with their premium clients." With no robust strategy to monetize the app, Want knew the lost opportunities would soon pile up.

With little permission asked but meeting no objection, Want and her team brainstormed a precision strategy for programmatic advertising built on a software engine that would sell the surplus ad space via automated high-speed auction, then push ads to the phones, targeted precisely to users who might like the ads best. The more users clicked through ads, the better machine learning could make the targeting even smarter, increasing click-throughs and raising *upday*'s fees and revenue. That was the Jim Collins "flywheel effect" in action—the same one, in principle, that powered the monster platforms of digital natives Amazon and Netflix, but adapted by a talented product manager and product team to the emerging platform of an old-school media business making its first moves into the precision world.

Want had enough experience to make these moves. "Since I was already building the user interface and the engine," she said, "stretching to incorporate programmatic was still in my court." But smart as it was and feasible as Want knew it to be, this was no done deal. She would need to pitch the executive staff for authorization, staff, and budget. Being a twenty-five-year old Australian woman in a German organization of "tall men in gray suits," Want knew she'd need a bigger, badder voice with her at the table.

Want briefed the vision and plan to *upday*'s head of sales, a direct report to the CEO. They framed their proposal to the executive team as all-gain, no-threat. "Your premium placements get first preference, and we fill the rest," they promised. The *Mad Men* in sales would cheerily carry on, making commissions the old-fashioned way, while Want's app would pull in millions more in per-click revenues.

With the project getting the green light, Want and her team went to work. Scoping the engine build, Want knew she'd need engineers. "This was definitely in-house work," Want says. "You don't want to outsource a proprietary moneymaker that is core to your business. And I knew enough to know that what we were building is not super-complicated and can be done incrementally. You can get a minimum viable product out there quickly and build on it;"

First stop was the CTO's office for engineers. Two volunteered, looking to pick up some new experience. Other experts chipped in; data scientists and product managers in the adjoining cubicles were already personalizing news stories. "I was there for the ads," Want recalls, "but they already knew how to target. They were a godsend."

Want's team then launched a two-step attack. First, they built, tested, and proved the ad placement system on the phones, using pure and simple sales automation with no personalized targeting. They used a commercially available engine to run the automated high-speed auctions and to place the ads. A few weeks in, they tested and proved the engine's integration. All systems were go for the proof of concept.

But the second step was the crucial one: adding intelligence to the ad placement. One simple rule guided the engine: If the customer clicked on a news story in a particular category like football and there was an ad available for football, the app would show the football ad before any other.

Click, bang: that did it. That was personalization, and that worked. Even with the most rudimentary targeting, the number of click-throughs started rising. Want strolled to the head of sales' office. She'd already seen the data for herself.

"Is this doing what I think it's doing?" Maddy recalls her asking. "I was in the happy position of being able to say, 'Unbelievably, yes!' "

Revenue from programmatic advertising at Axel Springer soon overtook the *Mad Men*'s premium placements, growing 20 percent month after month. Want's team continued to roll out upgrades, fine-tuning the system, chasing higher and higher click rates, and providing a more satisfying user experience and growing revenue. Precision programmatic advertising is now the primary revenue stream for *upday*, and it is now the single largest ad inventory supplier in Europe, with a reach of 25 million monthly users across thirty-four countries. Axel Springer's partnership with Samsung has expanded to new device types and new marketplaces.

As for Maddy Want, she moved on to Amazon, taking on product management for the data team at the Audible audio book publisher—a team that was triple the size of the one at Axel Springer and with exponentially greater reach. In 2021 Want was named Vice President, Data at Fanatics Gaming and Betting. "I got what I earned," she now muses. "But I had to go get it myself." Addressing her challenge with humility for what she did not know, eager to learn and acknowledging her gaps, and equipped with her energy and vision, Want motivated and mobilized the professional resources she needed at each step of the way, engaging senior management incrementally, at the right time and place; getting people above and beyond her pay grade to do things with her and for her; and supplying her with talent, buffer, and cover. In the end, she and her team enjoyed a well-earned victory lap.

"The truth is marketing is being marginalized today," Theresa Kushner later wrote. "Gone are the days of *Mad Men* and large advertising budgets. Welcome instead to the world of messaging, microsegments, and metrics. We no longer wonder which half of our advertising budget is working; we wonder instead if our messaging is hitting the right number of prospects with the right content at the right time. These capabilities can come from anywhere within the organization, and do."

## The Lessons of the Precision Vanguard

What lessons can managers take away from these examples of the precision vanguard in action? We discuss a few in this section.

*Exploit whatever technology you have—but seek out promising new precision systems when you can.* Theresa Kushner faced a cumbersome process of data preparation and crunch. She prevailed, but her overhead just to make the data ready for analysis was a time and money sink. Maddy Want had more powerful tools available to help, especially in off-the-shelf software that could drive the new high-speed auctions. That gave her new capacity fast.

*Use corporate strategy as your beacon, and partnerships for your power.* It's quite likely that in any formal statement of corporate strategy, there is nary a mention of "precision." Nonetheless, you can mine corporate strategy to put a precision initiative squarely in the high beams. Kushner drew power for her project from its clear alignment with Cisco's strategy to offer new technology that met customers' emerging needs. Want's project was a radical departure from established *Mad Men* practice: by forming powerful partnerships, Want made common cause with a senior member of the firm—and then with her peers. Each contributed uniquely, with her senior partner navigating the boardroom issues while she managed the nuts and bolts from the middle. Together they brought Axel Springer forward to its future.

*Turn the world transparent.* Precision is impossible without data and platforms—and then science and art to reveal the essential truths of the world. In 2012, about 80 percent of the Kushner team's time was spent readying the data—before the team even began to mine the data for its secrets. The essential work of science followed: understanding the secrets the data held and the truth of the world by framing, testing, and proving hypotheses and models that revealed which customers were ready, willing, and about to buy Cisco's products.

Want's work was the same exploration, facing the same demand that she turn the world transparent and reveal its secrets: who among *upday*'s millions of users was ready to buy muffins or jogging shoes or sports cars? Want let the software do the learning and run the high-speed auctions that converted all that insight into revenues. In the time-honored business of chasing leads, new technology transformed the game and the gamer: it made the play smarter, faster, and cheaper.

*Precisely engineer the change you want—test it, prove it, and get your metrics and measures right.* Whether at Cisco or at Axel Springer, a well-turned business problem framed the changes that Kushner and Want, respectively, were after: create new customers by prying loose the essential truth of customer preference from the mass of customers before them, then capturing those wallets with compelling offers. Using traditional tools and techniques of statistical analysis, Kushner discovered the nine highest value targets for sales from within the mass of possible targets, ultimately generating global sales worth more than $4 billion. Her metric was easy: money Cisco never had before, guided by Kushner's models, sealed with Sales's skills.

Want also recognized a new tantalizing vision of the possible—not to find "the nine" but to find the *millions* from an addressable universe of 200 million ad placements. That became the new metric—new money Axel never had before, easily captured in high-speed auction. That left the *Mad Men* way in the dust and powered the new future.

*Be ready to lead across boundaries of units and disciplines.* As an executive, Kushner pulled together a multidisciplinary team of experts from software engineering, data analysis, data science, and business systems. These were the early days of precision where such teams were first cutting their teeth. Soon, the Wants of the world would emerge, providing similar leadership but tapping new skill sets, energized by new challenges and incentives. Addressing her challenge with humility and curiosity, eager to learn and acknowledging her gaps, Want motivated and mobilized professional resources from across her networks.

*Treat the process as a practical, tactical journey of discovery.* Both Kushner and Want planned and positioned their teams for discovery, learning, and improvement. Kushner took a large problem, tightly framed, and broke it down in to a series of solvable mini-mysteries rather than trying to solve the whole problem at once. Generating increasingly refined hypotheses, Kushner moved steadily forward, persisting until she had the answers to her question: Who is ready, willing, and able to buy?

Want used the discovery process to expand her base of support, and she used an incremental test-and-learn strategy to demonstrate value from programmatic moves. As she progressed, she gained legitimacy and converted her incremental wins to greater commitment to her team and to the required technology.

*Manage through the headwinds of status quo bias and resistance.* Kushner was resilient in the face of setback, even when, on the last mile, she ran up against the status quo mindsets and practices embodied in marketing's performance incentives. Kushner managed the setback deftly, turning rejection into success and delivering a win for her team and Cisco.

Want combined an ask-no-permission, do-it-yourself mindset with a sensitive understanding of the need to maintain support from her corporate sponsors. When she realized how her project might threaten the *Mad Men*'s premium ad business, she designed the programmatic advertising workflow to leave their revenues and relationships untouched.

*Be prepared to find new business opportunities by swimming "out of your lane."* Just as no one at Cisco hired Kushner to find the path to sales increases of $4 billion by data science and other means, no one at Axel Springer asked Want in her role as a product manager to develop programmatic advertising. The company's marketing and sales executives were happy making money the old-fashioned way. Only a few knew programmatic advertising was coming, Want among them.

Managers need not look to their formal job descriptions for specific authorization to launch a precision initiative because they likely won't find it there. Those who hire corporate managers and write position descriptions rarely understand what precision systems entail. It's up to those in the precision vanguard to take the initiative and pull together the talents and resources needed to take advantage of the precision opportunities they perceive. In the process, they'll likely move beyond the limits of their job descriptions.

*Be comfortable in the several roles of precision leadership.* As Kushner and Want both found, there will be days when precision system managers must wear the admiral's bicorn—keeping one eye on the strategic goal and the other on the work of the fleet, assuring it is fed and resourced, engaged and happy, moving at the right pace, on time and on target. On other days, the manager is more like an orchestra leader, using a baton with a light touch to ensure that all the orchestra members are reading from the same score, on the same page, and harmonizing their many efforts. Still other days, the manager can find herself in the role of pit boss, managing the day-to-day slog with hammer, wrench, and screwdriver, working through the nuts and bolts of the precision work, keeping people on task and on plan, and filling in gaps left by vacancy or nonperformance.

* * *

Here are two final takeaways from this chapter. First, if you are an executive leading your organization on its first steps toward precision, your challenge is to energize and sustain the precision vanguard, unleashing the change they represent. Understanding what motivates them and how they operate is an essential first step. Use this understanding to encourage their dreams and passions while keeping them focused on the big goals set by corporate strategy.

Second, if you are a member of the precision vanguard or aspire to be one, you'll need the support of one or more executive sponsors who can bridge the two worlds of the status quo and the future you see. Their job is to help you make hard, first progress; supply you with sufficient money and authority to keep going; and help you interpret and manage worlds in conflict, whether on your team or all around it. Above all, they should help you build a coalition for precision that engages friends and holds off foes while you navigate your journey to success. *Purpose-built* coalitions, Kanter wrote in 2022, "connect otherwise disparate spheres of activity that bear on big problems by aligning powerful actors behind a purpose-driven mission."

We'll delve even more deeply into how this works in the chapters that follow.

# CHAPTER 6

# THE POLITICAL MANAGEMENT OF PRECISION

In the cases and examples we have presented in previous chapters, we have seen managers navigating bureaucracies and networks to bring about change they want precisely where, when, and how they need it. In each case, it took an entrepreneurial manager—the hallmark of the precision vanguard—to navigate the bureaucracy and right the ship. All spotted opportunities for vast improvements using precision engineering and performance. Each was passionate to seize the moment. None had any interest in standing still. Each was positive of a vision of technology making change happen even in the unlikeliest of places—between strangers on a street in Singapore, on a farm in Salinas, or at a 160-year-old newspaper in New York. And to do so on an astonishing scale, at warp speed, and with massive benefit.

And in each we have seen a special kind of entrepreneurial manager—one who is adept at technical and business matters of precision systems, certainly—but reliably, as well, with people and politics, one who can navigate bureaucracies and networks large and small; rally support; develop, test, and prove the change; and repeat. That's a manager's *political intelligence* at work. It's like emotional intelligence, which helps a manager lead with empathy. Political intelligence helps managers lead with *savvy*: being keenly aware of the resources they will need, the deals they must make, the buttons they can push to get them, the bargains they can make to get what they need.

In this chapter, we'll consider these political practices that the would-be builder of a precision organization must master if they hope to get the job done.

## The Map and the Compass

As you've already begun to see, building a precision system is profoundly cross-boundary and cross-discipline, touching interests far and wide, gathering support but sparking opposition as well. The politically savvy manager starts her journey to precision with a map of the terrain ahead. It might be a mental image, a back-of-the-envelope drawing, or the printout of a planning tool. She sees a path marked by the milestone events coming up—"business problem framed," for example, then "team built," "data gathered," and beyond.

The savvy manager also has in her mind's eye or on paper the image of a *bureaucratic compass*. Her compass includes everyone who might help or hinder her moves forward. To the "north" she places her bosses; subordinates to the "south," colleagues to the "west;" external supporters and providers to the "east." Each controls or influences some essential resource that the manager may need to start up or stay alive—people, budgets, authorizations, access to data and platforms, and the like.

Map and compass in hand, the manager will soon forge alliances, negotiate deals, pitch prospects, make promises—and deliver. Her compass will begin to look less like a four-pointed star and more like a network map—because a network of supporters is exactly what she will have built and what she will need to crack the status quo. That network is crucial.

A savvy manager knows that sooner or later she will need help dealing with a person or matter in her path, blocking her, over whom she has no formal power or control. That, in essence, is a political problem. The road to precision is filled with them, requiring that the manager "borrow" someone else's power and influence to stay on track, keep on target. She will negotiate terms and bargain with her ally, trading for whatever her ally needs, striking the best deal for whatever she needs. It's no sign of weakness that she is negotiating and bargaining. Negotiating, bargaining, and persuading are part of every manager's job and are tests of a manager's political savvy. "Managers are not self-sufficient," Robert Kaplan wrote. "They must engage in 'foreign' trade to get what they need. They will get what they want provided they have what their fellow managers need."

## What You Will Be Shopping For

Your tactics to clear the path range from using hard power threats and coercion to soft power persuasion and attraction to your cause. You will likely have on your shopping list crucial "asks" for precision—the power resources you will need to make precision work. These power resources include the following:

- ***Political power:*** An *executive sponsor* who likes what you promise, authorizes it, and can champion it to his or her peers *confers legitimacy and authority* for the mission, which is essential. His or her task is help you get a *charter or brief* that gives you license to move across units, *broker deals* for you with other bosses to gather people and data for the precision job, *keep you on target* and on track so you don't drift perilously away from your brief, and *build moats* around your operation and *give you cover* from attack by stakeholders who may see in your initiatives mortal threat to their own paydays.
- ***People power:*** *Team members* who bring needed skill sets to the table include engineers, data analysts, operations researchers, and marketing teams. *End users, both internal and external,* will be involved with you from the start, designing, testing, and proving the system. Colleagues will contribute time, staff, and their own voice in support. *Managerial authority and influence* will be needed for team members over whom you have no formal power or control.
- ***Technical power:*** There are two kinds of technical power. *Hard systems,* like platforms and data, are what you will need to work the precision levers for change. *Soft systems,* such as insights into formal and informal workflows and procedures, will let you know where your system must link to the way people are getting their work done today. All their routines, workarounds, and hand-offs must be either replaced and disrupted or left untouched.

## Bargaining for Support

The savvy manager knows that she is not the only one with a compass and a plan—practically *everyone* at any organization has their own compass and, if they are navigating change, a plan they are bargaining for. In the upcoming case about Teach to One, for example, principals, teachers, and school superintendents all have *different* compasses—and each is bargaining constantly with *their* networks to negotiate for assets of power and influence.

That bargaining could include conversations about you and the precision system you are stewarding. At the highest levels, executives may be debating whether the organization needs the system you are touting *at all*—is there a strategic fit? What problem will it solve? And what problems will it cause? Mid-level executives may be debating whether and how to fund or otherwise support you—where will the resources come from? At the lowest level, technical professionals, legal shops, and human resources may chime in about the design and details of execution—its legal sufficiency, costs, and staffing.

Laurence Lynn described these conversations as elements of "games" played at all levels of an organization and "in the manner of a chess master who simultaneously engages several opponents." The lowest-level games, Lynn wrote, are especially consequential because decisions there "have profound effects on the costs and effectiveness of [corporate] activity. No political executive who purports to be concerned about . . . performance can afford to remain aloof from low games."

Among the most consequential of these conversations are those occurring on *value networks*, described by Clayton Christensen as the organization's web of suppliers and producers whose exchanges create value for the organization. No one goes about establishing a value network; you cannot apply to join. They arise spontaneously over time and keep needed services flowing, steady, and dependable. Efficient and effective value networks embed the status quo and anchor it in workflows, procedures, job designs, and infrastructure.

In a value network supporting the provision of computers to schools, for example, vendors provide hardware; information technology (IT) departments design and run networks and hire technicians to keep everything humming; software developers create applications that teachers use; teachers develop lesson plans and teach from them, earning rewards of status, pay and other benefits; students head to the colleges of their choice. Although they cannot see each other, the computer vendor, the teacher, and everyone in between are locked into their value network by contracts, procurement rules, training and certifications, all in service to the status quo, all making change very difficult. Just try shifting a school system from Apple computers to Dell (or vice versa) to see how the value network built up around the incumbent systems resists. Everyone knows their roles in a value network, has tooled up and staffed up to provide it, knows what is expected of them, how to make their part of the "deal" work, what others need to do, and what to expect from performance and pay.

Indeed, behavioral scientists have won Nobel Prizes trying to figure out just how *much* it takes to pry a beloved technology out of a user's hands. People (and firms and agencies) are *really* attached to what they have, *even if it's objectively much worse than what's being offered*. A predilection for the status quo and a mindset against the new go by several names—"risk aversion" and "status quo bias" among them.

Just to illustrate: if we were to propose to swap out your email system for, say, collaboration software like Slack or Teams, many corporate users would (and *did*) feel the acute sting of loss as they were psychologically *"endowed"* with email as a perk of the job, were *averse* to trying its substitute, and were therefore *biased* for the status quo and against the swap. That's the *endowment effect* at work—a potent component of the status quo bias.

The status quo bias and components like the endowment effect, as explicated by Richard Thaler, Amos Twersky, Daniel Kahneman, Richard Zeckhauser, John Gourville, and others, help explain why so many innovations die in the hands of managers on a mission to crack the status quo. The preference for what we already use is so strong that Intel CEO Andy Grove once guesstimated that, to be successful, any new tech product or service entering the marketplace had to be *ten times* better than what consumers were already using. Handheld calculators compared to slide rules did that handily, for example—and created a huge global industry (you can still find slide rules as relics of a bygone era on eBay). But electric cars took forever to get a toehold in the marketplace because they forced car owners to give up something they felt entitled to and that they were *endowed* with—virtually limitless driving range powered by gas stations from coast to coast. Electric cars couldn't promise that for a very long time. Even through 2022 shortages of charging stations persisted, with news sites running headlines like the *Wall Street Journal*'s "While Electric Vehicles Proliferate, Charging Stations Lag Behind," or *MotorBiscuit.com*'s "Lack of EV Charging Stations Creates Range Anxiety in the West."

The status quo bias and the endowment effect can be bad news for, say, a school superintendent who wants to switch computer vendors for the district's classrooms. Her proposed replacement can't just be a little bit better—it has to be *stellar* to stand a chance with users. And she'll need a precision vanguard clamoring for it and designing, testing and proving it in operation.

The difficulties are even greater for revolutionary change, not just in technology but in the methodology of teaching itself when based on a different set of principles, different engineering, and different science. Writing of such "radical innovation," Rebecca Henderson of the Massachusetts Institute of Technology (MIT) and Kim B. Clark of Harvard Business School cautioned that such change "forces established organizations to ask a new set of questions, to draw on new technical and commercial skills, and to employ new problem-solving approaches." As we will see later in this chapter, it will take great political power and savvy to navigate such shifts through the thicket of incumbents standing guard at the gates to the status quo.

## Your Best Moves: Clearing the Path to Yes

The political challenges facing the vanguard manager working to innovate a precision system into an existing value network is considerable. The status quo is not just an annoyance managers can scoff at. It is an emotional predisposition of players on every manager's compass—a *bias* against change, embedded in a Maginot Line–like series of fortresses comprising today's value networks.

Sooner or later, the manager will need help dealing with a person standing at the gate or a problem blocking her progress. Helping to overcome institutional resistance is where executives and sponsors of precision systems play a huge role. Even so, threats to value networks can alarm a CEO who is worried about disruptions to a time-tested way of doing business and concerned to hold support from *her* compass—whether from proven high performers among *Mad Men*, for example, or physicians who are influential with peers and hospital board members alike. The CEO also has a compass and could have a serious case of the status quo bias.

Here are some of the specific moves savvy members of the precision vanguard make to boost their chances of winning support from the top and from people around their compass:

- They *design for user acceptance* and minimize resistance. At Axel Springer, Madeleine ("Maddy") Want designed in rules for her high-speed ad-placement auctions that left marketers and their accounts untouched. This kept them on the bench and out of the fray, leaving Want free to plunder the wealth of new programmatic advertising.
- They *design to eliminate technical complexity* from new platforms, as Blue River Technologies did by embracing the John Deere brand and making the See and Spray towed array easy to use, thus minimizing the need for training or for changing work routines. Although it was complex, it was in essence like any other tractor attachment a farmer might use.
- They place the new precision system *in the flow* of existing procedures and incentives, streamlining them, making their use easer. At the *New York Times*, for example, Josh Arak engineered a new testing platform that created uniform precision metrics and methods across all products and services. While it shifted control from multiple owners to a shared platform, testing itself was part of data science and product management. Arak's work streamlined it.
- They *exploit crisis or urgency* and reframe the move to precision as a rescue line. The Zipline team brought its precision approach to drone deliveries to one of the world's most urgent human problems—providing lifesaving medical supplies to people in remote villages of Africa. The circumstances enabled them to get permission and buy-in from people and organizations that might otherwise have been dubious about such a radical new innovation.
- They *work around the "endowed."* Marketers at Axel Springer, for example, felt heavily "endowed" by the relationship networks they built and the compensation they earned from selling direct ads. Want's plan worked around the endowed, leaving those relationship networks untouched while creating tremendous corporate benefits elsewhere.
- They position the precision initiative squarely *in the sweet spot of corporate strategy*. People may disagree about the how but not about the what. By aligning on the CEO strategy and measured by those same metrics, the precision manager makes it easier for any CEO to get to yes in her negotiations with senior executives.

## Swapping Value for Value

Anyone from whom you seek help will always ask, "What's in it for me?" Your answer must be, "I have something you want." Above we listed power resources you may have to bargain *for*. Here's what you can bargain *with*—the "currencies of exchange" that you can offer in trade for the resource you need for your journey to precision. Always check against the following in your value proposition:

- At an *organizational level*, you can promise new high-impact results exactly where, when, and how your partners want them, without the usual downsides—impacts they didn't intend, costs they didn't have to pay. Precision systems can have high investment costs, but they also boast compelling benefits. That all adds up to compelling returns. And you have a way to measure this, thus proving your case.
- At a *unit level*, maybe someone's already at work on the problem: there's a pilot or a project already in progress, but it is different. Perhaps they have to complete it anyway, and you promise getting it done faster, better, cheaper, and with precision. Maybe bonuses or rewards are tied to outcomes you can deliver more easily. If you can pinpoint high-value sales leads, for example—customers *ready* and *able* to buy—you make life easy for sales teams.
- At an *individual level*, maybe your project gives engineers opportunities to work with a new programming language they want some experience with. Data scientists will like being in the forefront of solving an important business problem. Team members may relish working closely with C-level sponsors on a high-profile project. Maybe as CIO, you want to see what it takes to pull a precision operation together.

What's emerging here is an image of the precision vanguard operating like those Rosabeth Moss Kanter has called "corporate entrepreneurs"—"people who test limits and create new possibilities for organization action by pushing and directing the innovation process." Among their traits, Mitchell Weiss writes, entrepreneurial managers are "more driven by perception of opportunity than by resources at hand . . . they commit to those opportunities very quickly and for short periods . . . they stage their commitment with minimal exposure at each stage . . . they use resources only episodically, often 'renting' instead of buying . . . they tend to organize with minimal hierarchy."

In every such instance, the central task of managers to *move the precision disruption forward* takes political intelligence to understand context and terrain, foresight to *eliminate* likely conflict, and alliances and friends to manage resistance. Whether change bumps up against a mindset only or a value network in open revolt, the threat can quickly become existential, powerful enough to defeat even a CEO's best efforts at innovation—and even take down an administration. That means a manager seeking to build a precision system where an incumbent system is already delivering reasonable results will need to muster all of his skills of political management to prevail.

## Joel Rose of Teach to One: Navigating a Complex Value Network

Joel Rose, CEO of Newclassrooms.org, thinks of "education debt" as the most impactful and debilitating drag on children's K–12 schooling in the United States. On his account, the industrial model of schooling—one teacher, twenty-five kids all learning the same thing at the same time—was fine when public schools were just starting out as "common schools" in the nineteenth century and unifying curriculum for wave after wave of immigrants was important.

"And then we just kept doing it," Rose says. For decades, school systems across the country were moving millions of kids in lockstep from childhood to citizenship. Except schools are not assembly lines, and kids are not cars. Not every child masters the grade-level curriculum right on time. If you keep the assembly line of schooling moving, some children fall behind. They build education debt—deeper and more profound each year. The child who didn't master fifth-grade skills isn't ready for sixth grade when it's served up; that child then stumbles through sixth and falls even further behind in seventh.

Schooling so conceived leaks failing children all along the way, crushing them, Rose believes. The experience crushes teachers, too. "As a fifth-grade math teacher," Rose recalls, "I had a classroom full of kids with second-grade math skills and eighth-grade math skills. I was given a stack of fifth-grade books and told, 'Good luck.'" Rose leans forward to make sure we understand:

> This is not about optimizing the existing model. The K–12 teacher's job is just not designed for success. You can train them, hold them accountable, give them better materials. But if you just graft that on top of the same delivery model, schools will continue to leak failed kids. You actually have to change the delivery model.

If you were going to invent a school system today, Rose doubts you would ever design it like this. With his cofounder and CTO Chris Rush, Rose has pioneered Teach to One, a new model of K–12 education focusing on math skills. It embodies *tailored acceleration*, which Rose describes as a method for "matching precisely the skill that the kid is trying to learn with where she happens to be in her own progress and giving her just enough stretch that the kid is challenged—not so much that she feels overwhelmed."

In effect, Teach to One creates individualized "assembly lines," personalized for each student in a classroom, each powered by machine learning enabled by artificial intelligence (AI) and Teach to One's models, which test, prove, and prescribe a unique path forward for each child following a daily diagnostic. Every day, the student proves her progress. If she stumbles and falls, she goes back and enhances the learning selectively, where she is weak. Working through the exercises, she reduces her debt enough to handle the challenges coming up next. Then, and only then, does she advance.

It all gels around a series of classroom sessions prescribed and modified during the day, depending on the last session's exit quiz. Chris Rush explains:

> There are 300 mathematical skills and concepts that a kid needs to know in order to go from basic numeracy to college readiness. We know each of those 300 skills, the relationships they have to each other, and we're using data to figure out, based on where you're starting, the path you should take to get where you need to be.
>
> Let's say you need to learn trinomials. Our skills map tells us there are four predecessor skills. What if the kid knows none of the four—does that mean he has to learn all four? What is the best skill to give the student right now? What is the best set of students to be working together? Who is the best teacher to work next with the student?
>
> These questions make the system smart. It's widely assumed that when learning fractions, you should learn adding fractions first, then subtraction, then multiplication, and dividing, and learn that all as a unit. But what machine learning helps identify is that, oddly enough, after learning adding fractions, for some kids, you should break from fractions and do something with place value—then come back to multiply fractions, and then to subtracting.
>
> These are things that humans don't naturally assume when they're trying to design good progressions. Especially for outlier situations, machine learning identifies more efficient pathways for kids to move forward.

Teach to One accelerates the rate of growth so that kids who appear stuck can move faster to catch up and stay there. Teach to One's models, perfected by its data scientists over time, now predict with 91 percent confidence whether, at any given point, a student's education debt will hold him back or let him keep up with the crowd, given his level of readiness for the things *they* are working on.

"If I'm generally functioning on the fifth-grade level," Rush explains, "and I start working on something on the seventh-grade level and I am not prepped for it, I'm not going to be successful, ever. This is where kids start getting stuck. Teachers say, 'Don't work off grade-level stuff. That's too advanced.' Yes, but it turns out there are select predecessors for seventh-grade work that if the student masters can get him back up to grade-level speed."

That's where Teach to One's personalization comes in for the low performers: work for success on the predecessor and catch up on the debt that matters.

Once a student can keep up, getting to that level will accelerate his development faster than if he remained outside it. In technical terms, it's called the "zone of proximal development"—and that's where every student needs to be working.

"It's like the movie *Finding Nemo*," Rush says. "They go on this whole journey. They only make up so much ground—and then, all of sudden, they tap the East Australian current, and they just go. We see that happening in K–12 schooling. And it shows us we can close gaps faster by focusing on prep and stretch—hitting the essential grade-level predecessors and getting the kid into the flow where the current can carry him faster."

Moving stuck kids faster is crucial. One of the unbudging metrics in K–12 education is grade-level tests that come around every year. Everyone must be ready: those scores matter. Superintendents' futures and teachers' tenure decisions are pegged to them. Parents watch scores, eyes on both college admissions and the Zillow factor: real estate prices rise and fall depending on the rankings of local schools. Entire school systems can be seen as one huge value network locked in on test scores. Everyone knows their role, their expectations, and their rewards—professional promotions, job security, college acceptances, and vendor contracts.

But failing kids drag down overall test scores. As teachers in traditional, industrial-model classrooms push forward through the grade-level curriculum and teach to the test, the casualties of education debt fall behind. For Rose and Rush, harnessing the classroom's East Australian current in the form of tailored acceleration may be the key to engaging teachers and helping students at every skill level achieve solid performance on those all-important tests.

## The Imperatives for Managers

Winning acceptance for Teach to One from the complex, entrenched value network of American education—the network of schoolteachers, principals, school board members, district superintendents, union leaders, textbook suppliers, state and local legislators, and others who help to shape and run K–12 schools in this country—is a huge challenge. Rose and Rush have been engaged in a ten-year effort to overcome resistance and gain adoptions. Here are some of the strategies they've employed (you'll recognize some of them from our earlier discussion of the methods used by politically savvy members of the precision vanguard):

- ***Hold the mirror to the organization.*** Use your data to make underperformance visible. Teach to One's value proposition is its promise to close the education debt for underperforming students. For Teach to One to get its foot in the door, the very notion of learning loss and education debt must be understood and accepted. Acknowledgment that there is a problem starts with transparency of the data.

- ***Shrink the ask for precision.*** Reduce how much change in business routines, workflows, and technology is really necessary to achieve the gains in precision you are touting. Large changes in behavior—teaching method, for example—coupled with big changes in technology—the move to precision systems—can be complex, prone to failure, and highly resisted. Over the years, Rose and Rush designed technical complexity out of their precision platform and restructured operations to require less behavior change. They built easy-to-use apps and other tools. They shifted from unorthodox thirty-five-minute classes to standard sixty-minute periods. They simplified workflow with an AI-enabled scheduling engine that prescribed where every kid should move next in the school day, with what lesson and with which teacher.
- ***Customize the offering to context.*** The Teach to One classroom sometimes held 100 children at a time in a gymnasium-like setting, causing resentment among some teachers, parents, and students. In search of a better way, Teach to One customized their modules so the customer could configure for small, midsized, or large classrooms. Rush and Rose were hopeful that customizing the modules, when combined with personalization for the student, would make their product a winner.
- ***Shorten the time to a win.*** Benefits take time to develop. Managers may not have much time. Managers can shift their approach to deliver wins faster. With Teach to One, *lower-debt students can be targeted*—for example, younger students who may be *one-grade level* behind for whom debt is easier to retire faster.
- ***Know what counts as a win, and to whom.*** A tenfold improvement in *cost* from a new precision system could be highly appealing to superintendents and finance administrators. A tenfold improvement in *effectiveness* for a target segment of low-performing, "debt-ridden" students could be highly appealing to students, their parents, and their teachers. Savvy managers can tailor program design to build support from these important constituencies.
- ***Design for the unendowed; customize for the endowed.*** Seek out stakeholders *unendowed* by the status quo for whom the current system is *not* working—students, for example, who are already deeply in education debt and for whom major change could be most welcome. For the highly endowed—parents and their high-performing students, all college-bound—Teach to One planned to introduce accelerated learning of their own so that high-performing students would be ready for high school–level calculus sooner
- ***Aim for the decision maker's sweet spot of strategy.*** The key was the test. Teach to One marketed itself as offering to bring all children to grade-level performance within a defined period of time, accommodating the superintendents'

needs and controlling the pace, timing, and breadth of the initiative. Having executive support was an essential to their authority to move.

- ***Shore up legitimacy of the team as innovators, and of the innovation as effective.*** Managers at Teach to One used testimonials, independent evaluations, and research as they battled for support. With the status quo often in opposition, they faced a constant challenge to show why a school system should change to the Teach to One system. No progress was assured merely by research, but none was even possible without proof that accelerated learning worked as a system for the organization and for the parents, students, and teachers.
- ***Be part of the conversations going on at all three levels of the "game."*** At the highest level, this means the rationale for the program itself; at the middle levels, how it will be resourced; and at the low levels, how it will operate. Not all managers have the ability or interest to play at each level. While each is essential, the low game is especially so, involving what one analyst has called the "fine-resolution, small motor process of government" that reflects the concerns of teachers and principals, as well as administrators in procurement, personnel, and budget.
- ***Be prepared to bargain for what you need.*** "Managers are not self-sufficient," Kaplan wrote. "They must engage in 'foreign' trade to get what they need." Managers at Teach to One activated resources across their networks for political support, financing, and authorizations.
- ***Diversify away from status quo metrics; loosen their stranglehold on program shape and design.*** The current system may operate with a powerful set of status quo metrics. Standardized test scores are a perfect example in schools. But precision systems can and must be measured by so much more. Teach to One can be measured by percentage reduction in education debt, for example, as measured by multiple metrics. Managers may prefer to be measured by the existing business unit metrics of performance, but they should add metrics specific to the intent and design of the precision system. As an alternative, managers may seek some "neutral" innovation zone for proving and testing their systems where they are excused from the status quo metrics.
- ***Create purpose-built coalitions.*** Make common cause with those who share your policy outlook. Who can support valuable investments in your research and development (R&D)? Who feels a sense of urgency from an immediate opportunity? Rose and Rush fostered three coalitions—for example, a policy coalition that helped pave the path to personalization; a donor coalition to support R&D for data teams; and a coalition of actual buyers and users of the program, including teachers as key influencers.

***

The jury is still out on large-scale uptake and adoption of Teach to One. Competition has emerged from other providers of accelerated learning techniques, including the Zearn system. But the evidence is clear: the COVID-19 pandemic created new education debt for all, new federal funds became widely available for procurements, and research increasingly showed that accelerated learning outperforms old-school approaches that hold back children who are underperforming rather than accelerates them with precision.

Yet there is no environment in the United States today more politically volatile than public education. Every dimension of the culture wars—from critical race theory to vaccine mandates and masks, from sex education to health, transgender, and LGBTQ issues—plays itself out loudly at meetings and elections for 15,000 state and local boards of education.

In that environment, third rails abound, frying school superintendents with their jolts, leaving these leaders among the shortest-serving executives in public life, easily bounced when a majority on a school board shifts from election to election. There is no innovation that could *not* run afoul of linkage to some broader political strain, no matter how good the idea.

This all puts innovation for precision on the clock, with time always running out. No one can doubt that moving to grade level fast is a good thing—everyone can agree on the what. But the how can be highly contentious. It puts the onus on innovators like Rose and Rush who, in making the great promise of precision, must concentrate value, deliver fast, activate the vanguard, move the undecided, engage the unendowed, skip the endowed, move children into the zone of proximal development, and let the current do its job.

In this context, building and implementing a successful precision system is much more than a technical project. It is a deeply human challenge that requires high levels of managerial savvy, strategic insight, communication skill, political intelligence—in short, the entire array of skills that make up political management and leadership for precision.

## CHAPTER 7

# BOOTSTRAPPING PRECISION

In making the move to tech-enabled precision, no one has all the power needed to get the job done. It's the nature of our work and our world today that risk and opportunities can start halfway around the world and land on our doorstep fully massed in an hour, in minutes, or even seconds. Whether chief executive or senior manager, supervisor or team lead, few among us can go it alone. We often need to go "out of our lane," even out of our immediate network, and get help.

Our networked world is replete with these moments. Volatile weather makes mincemeat of global airline operations. Cyberattacks take months to develop—but they are delivered in seconds. Terrorists poke their heads up in strange places but fade away just as quickly into the surrounding city.

On such days, you are called to action, and you must make fast pivots away from disaster or toward rising opportunity, from a future you don't want toward goodness—even greatness—and create change you *do* want. Maybe you don't have the full complement of resources you'd assemble if you had a year to think things through and pursue them, but it's on you to block the threat or seize the day. The status quo will not hold. Nothing offers managers greater impact at lower cost to deal with today's opportunities and crises than precision systems. Well-designed, well-developed, and well-managed precision systems let managers find the needle in the haystack of threat and extract it, bend it to their will, and change the path of history.

It's a conundrum of the networked world: in the face of light-speed challenge and threat, precision is best, but neither nature nor markets wait for us to figure out the full-bore collaboration of technology, people, and power we'd like, confident we could crush any threat, any opportunity thrown our way, with precision. The great skill of the precision vanguard manager comes from *cobbling together* the precision solution that works even in that most precarious moment. She draws some platforms and data from here; some people and expertise from there; duct-tapes them together in good-enough bonds, weak and strong ties alike; combining resources on loan, from executive attention and to physical, political, and logical assets, all shared, all waning or claimed back as soon as the crisis seems to pass.

In this chapter, we'll look at the essential elements of the art of bootstrapping the precision solution:

- *Bootstrapping a coalition for precision*, forging the network of potential allies you need right now, each ally having something to offer a precision system that will run at its most basic.
- *Bootstrapping the operational capacity for precision*, which means piecing together people, platforms, and data into a working operation that proves the concept, the point, and your leadership for the move to scale ahead.
- *Bootstrapping authority and legitimacy for precision*, which is about claiming the stature and standing you need to make your precision-oriented moves, even when nothing in your title or your job description says you can.

## Jason Bay and Contact Tracing in Singapore: Bootstrapping a Precision Response to COVID-19

In the winter of 2020, COVID-19 was just a whisper in Wuhan, China. It would soon pummel the world. In Singapore, China's island-nation neighbor to the south, leadership recognized the early warning signs of this new plague. Having battled multiple epidemics in the past, Singapore knew the drill: everything had to be done right to stop it, and fast—especially in the holiday season, with thousands of citizens traveling between China and Singapore for business and leisure, each and every one a potential carrier.

By mid-January, as Washington, Moscow, and London dozed, Singapore was rolling. Jason Bay, a senior executive in Singapore's "smart nation" digital governance initiative, saw an opening he wanted to take. Trained as an engineer, Bay felt certain that the smartphone could be a powerful tool that was now ready for its place in Singapore's pandemic-fighting toolchest.

In many ways, Singapore was the perfect place to test Bay's hunch. It is a digitally savvy nation—80 percent smartphone-enabled, accustomed to bold talk of digital solutions. In surveys, 75 percent of Singaporeans said that they trusted the government to handle their data. Just as critically, Singapore had a battle-tested contact-tracing team ready to track down all the contacts of anyone diagnosed with COVID—moving fast to nip the spread in the bud.

Contact tracing was recognized globally as an essential tool for effective epidemic response. But Bay knew that, if COVID-19 were truly to break as a *pandemic*, there was no way even Singapore's vaunted contact-tracing team could handle it. He'd heard Singaporean health colleagues complain about how manual, tedious, time-intensive, and labor-intensive contact tracing was—and that was with a mere thirty cases of Middle East respiratory syndrome (MERS) in 2015. How would they deal with hundreds of thousands of cases of COVID-19?

If the contact-tracing process could be made easier, faster, and more precise, there was a chance that Singapore could slow the spread of COVID-19 dramatically. Years earlier, Bay had toyed with the idea of a smartphone-enabled contact-tracing app—but at the time, he was lower down in the bureaucracy, the smartphone technology was not yet ripe nor in everyone's pocket, and the last contagion had been contained. Now that same vision of contact tracing would come racing back, this time centered on a smartphone app.

The idea was this: Bay's engineering team would build the app. They would call it TraceTogether. Anyone could download it. With the app loaded, smartphones in a crowd would continuously ping each other, leaving a coded, digital log entry on any phone they passed saying, in effect, "I was here." If and when an app user fell ill, he would notify the health authorities; the health authorities would notify anyone whose phone had left its digital footprint on the sick person's phone when they had brushed by each other and advise them to get tested.

The possibility for transformation here was obvious. If mobile app adoptions went as Bay hoped, *every* instance of an infected person passing another would be picked up by the Bluetooth phone-to-phone pinging. That would create a large new pool of traceable citizens who could be contacted if one became ill. Only a big data-powered, precision system enabled by artificial intelligence (AI) could work against a total actual universe of possibilities like that. In similar fashion, at the Democratic National Committee (DNC), Nellwyn Thomas worked with data embracing the *entire universe of possible Biden voters*; at Blue River, Lee Redden worked with *the entire universe of plants in a field.*

There were big ifs to consider. Would everyone download the app? Would the Bluetooth-powered ping-a-phone system, fully automated, work? Would in fact Bay's system detect every instance of an infected person passing another? Would health authorities be able to keep up with the load—and could privacy be ensured to build trust in the system right away? The idea was simple to state but complicated to carry out. Bay contemplated the requirements, and they were staggering:

- *Precision*. Exactitude was mission critical. If the app cast its tracing net too wide, the notifications would become constant, meaningless, and then ignored. Cast the net too narrowly, and lives would be needlessly lost—pings that should have been made would not be. To hit the sweet spot, Bay's team would leverage Bluetooth—which had an active pinging range of about thirty feet. With Bluetooth always on, User A's smartphone app would ping User B's smartphone only if they came within thirty feet of one another. That "handshake" would take less than a second, and a log of all these pings would be kept on each phone. If days later, User B contracted COVID-19, anyone who had brushed by User B and was logged in would be notified to seek care.
- *Multiplicity*. Bay's contact-tracing method would have to complement, not replace, the existing manual system—adding capacity, not substituting for it. If 60 to 70 percent of smartphone users could be convinced to download the app, then millions of Singaporeans would become contact points for the system as the pandemic burst on the nation. Bay and his team would have to build the app for all devices and operating systems, including Google's Android, Apple's iPhone, and any other manufacturer's smartphone.
- *Lightness*: Bay's proposed app was, essentially, a software solution to a real-world, real-life problem. There would be no huge infrastructure build. No training required. Those 80 percent of Singaporeans with smartphones had all the gear needed. Each would download the app and start using it immediately. Everyone in the all-digital nation of Singapore would know how to do this.
- *Privacy:* Privacy loomed large as an issue to be addressed. TraceTogether could be considered just a "back door" for government to track users from one location to the next. Safeguards would have to be built into the software design right from the start such that TraceTogether could reveal providing the *proximity* of two users to each other without giving away their *location*; rules for records retention of the proximity logs, specifying who would keep them and for how long; and governance of the whole enterprise. Trust was essential and would flow from these decisions.
- *Quickness*. The pandemic was happening right now. Bay needed to field something fast to beat it. Once in the field, the app would need to shorten times to notification. That in turn would take massive coordination with Singaporean hospitals lest the addition of thousands of new cases overwhelm the existing procedures, infrastructure, and workflows.
- *Consistency*. The app would have to work each time. It could not buckle under the load, work sometimes but not always, or work differently for some people than others. Consistency would build the trust essential for widespread adoption.

These were big asks. Taking stock, Bay knew the problem to address, his goal, and the broad outlines of a plan, but he could not see all the way through to the end or be sure of the path ahead. Time was of the essence. The "north, east, south, and west" of his bureaucratic compass teamed with individuals and groups whose powers, skills, and connections made them either must-haves for the work ahead, nice-to-haves, or never-haves. Sooner rather than later, Bay knew he would need help—he would have to summon his network to action—call a *coalition for precision contact tracing* into existence. But for now, his vision and plan were sound. His authority and legitimacy for action were well within his charter. The capacity for building the app surely lay within his organization or close by. Cobbled together, Bay had enough to power up and jump into the fray.

On January 24, Bay texted a friend in the Ministry of Health, which owned the official Singapore response to COVID-19. "So, here's the idea. What if . . ." and Bay sketched the use case for smartphone-enabled contact tracing. "In your view, is this doable, technically?"

"Yes," came the reply from Bay's friend.

"Is anyone at the Ministry of Health already working on a contact-tracing app?"

"Not that I know of," came the reply. "But let me check."

This initial testing of the waters was critical to Bay's hopes. By sending his idea up the flagpole, Bay ensured that if someone else was already at work on it or if the concept raised a problem for someone powerful, he would hear about it quickly. Bay did all this in a single flurry of text messages—a fast, well-targeted, and impactful step, the first on the road to building a coalition for precision contact tracing.

Bay now began using his well-developed networking skills. To the north, Bay brought his bosses into the loop. At this early date, he kept his messaging light, couching his plans with Ping Soon Kok, GovTech's chief executive, as just, "Something we're toying around with," and getting, "Sounds interesting, keep me informed" kind of reply. He got the same kind of noncommittal answer from GovTech's chairman Chee Khern Ng. This was enough of a green light for Bay to get going. He would keep these overseers in the loop, making sure they were never surprised by what he was doing.

To his south, Bay would source his technical team initially from a trusted inner circle of collaborators, then expand. His first recruit was Joel Kek, a software engineer on his staff, to whom he would turn repeatedly over the coming months. Via his Slack channel, Bay invited other interested software engineers from his team to participate. He held out the promise of taking deep dives into the workings of Bluetooth, which would be novel and interesting to many.

"I had a few people come forward over the next couple of days to hack away on Android and iOS," Bay recalls. "They created a prototype app on the iOS device over that long holiday weekend." This fast and light proof of concept was good enough to test the idea and send it up the flagpole.

To his west, Bay further reduced the chances for conflict by sharing his moves with the internal units with whom he would need to collaborate. He pinged the Defense Science and Technology Agency and the information technology (IT) arm of the Ministry of Health. After confirming with these units that nobody else was doing anything close to what he was suggesting, he would ask, "Then let's talk about what you need from the system." He consulted with staff from Singapore's Personal Data Protection Commission, who would help him frame the privacy architecture for the app. To learn how to integrate his app into the existing contact tracing workflow while adding more value, members of Bay's team sat with Singapore's contact tracing teams and observed their process.

Knowing he would need marketing to boost app downloads, Bay partnered with the Ministry of Culture, Community, and Youth. They could build grassroots and nongovernmental organization (NGO) support and reach out to the entire population using languages that included Tamil, Bengali, and Burmese. And he partnered with the Ministry of Health for marketing on WhatsApp (used by 86 percent of Singapore's internet users) and in print, online, and broadcast news. To broaden his base of support further, Bay secured the endorsement of Singapore's foreign minister, who announced that the TraceTogether app—which by now had been given its name officially—would be open-sourced and available to the international community. "Together, we can make the world safer for everyone," was the message.

Finally, to his east, Bay surveyed potential end users for feedback on the user interface, modifying his plans as needed. He leveraged his Singapore government contacts to engage Apple and Google for permission to post the app to their app stores. (This was denied by Apple at first because the TraceTogether app ran continuously in the background, drained batteries, and violated Apple's terms of service. Eventually, however, Apple relented.) Bay connected with hardware engineers from external institutions like Nanyang Polytechnic, who provided him with much-needed testing capacity in specialized environments. He consulted university researchers via his own network at Stanford who had recently published their own Bluetooth experiments to ask about any pitfalls they had encountered and any code snippets they had found especially useful. And he engaged data protection advocates for their views, further modifying his plans based on their input.

On March 20, 2020, just eight weeks after the launch of the project, Singapore rolled out the TraceTogether app. In its first days, nearly 1 million Singaporeans downloaded it—16 percent of the target group. Ultimately, 95 percent of Singaporeans signed on to TraceTogether, but only after Bay further adapted by providing older Singaporeans and those who did not want the app on their phones with a smaller Bluetooth-enabled token that would do the same work of pinging when close.

Bay had done it. He had swiftly assembled a coalition for precision, calling it into existence, mobilized it around his vision and goal, boosted his operational capacity as required, gained the authority and political support he needed, and delivered the TraceTogether precision app just as planned—right where Singapore needed it to be.

## The Bureaucratic Compass and the Strategic Triangle

Everything about bootstrapping a precision solution is light, fast, and impactful—and must be. Even if it doesn't completely solve a problem, the cobbled-together solution can and should still go *against* a problem and count as a win: light enough to get the job done fast and big enough to make a difference. The fix—call it a *hack*—buys you time to breathe, watch the solution in action, and fix it to last. For now, it blocks a threat or grabs an opportunity, gets you a clearer view of the terrain ahead, and proves your approach. For the day, and maybe even *forever*, it works, works, and works—for the problem, the customer or citizen, and the team.

Bay's move forward had two universal components that managers find quite helpful in navigating the shoals of precision innovations. The first—his *bureaucratic compass*—gave Bay a way to visualize quickly the many actors and forces in his world, some to avoid, many to call on. The second—his *strategic triangle*—gave Bay a three-pronged test of readiness to act that he could use at the outset and at various milestones on the TraceTogether journey.

Introduced by Mark H. Moore, the concept of the strategic triangle includes three key components: a vision of value, operational capacity, and authority and legitimacy. All three are essential to the manager undertaking a big new initiative. Let's consider them one by one.

1. Did Bay have a *vision of value* around which he could mobilize support for his change? Bay's vision of value was simple: detect more cases faster, better, cheaper than a manual system could *at the scale needed*, ensuring privacy all the while. It contained a moral imperative, essential for coalition building, what Kanter calls "a significant, inspiring mission [that] attracts followers"—protect the nation. His hypothesis was that early discovery, matched with early intervention, could bring the pandemic to a halt. His key assumptions were that the technologists could build it, the public would use it, politicians would support it, and the health system could carry the load.

Above all, Bay's vision of value did not entail replacing the manual system but rather complementing it, extending the range and reach of Singapore's total contact-tracing effort. The two systems, working together, would yield greater results when combined than either system could alone.

This was classic precision system placing: fill gaps left by existing efforts, doing it at least as accurately, perhaps at much higher volume, faster, and at lower total cost. Run the precision system concurrently with status quo systems rather than replace them, at least to start. In fact, *leverage* the existing system for the capacities that Bay needed but could not provide, such as making the actual contacts when notified of an ill user, thus yielding improved overall performance.

2. Did Bay have the *operational capacity* to turn vision into action, deliver on the promise of value, quiet naysayers and satisfy his supporters with the results? The issue of operational capacity for precision systems always turns on whether a manager has the right data, software, and hardware; the right analysts, product managers and software engineers to design and develop the tools she wants; and the right users positioned to deliver value on the "last mile" of service. Ultimately, for Bay, the test will be whether he can discover and identify the entire universe of possibilities—then touch each with his solution in time, check its ultimate impact, then fix to perfection. Does he have people with the skills to capture and analyze the data, and to design and test the proof of concept? His precision "game" depended on his answers to these questions.

3. Did Bay have the *authority and legitimacy* he needed to move on the problem? Bay's direct line of reports included 500 or so engineers. They knew the mobile world well and, if it came to it, could hammer out solutions and test and prove them fast. Clearly, from a technical viewpoint, TraceTogether was in Bay's wheelhouse. Although Bay could check that box, he still needed license to move and support to operate. His charter would come from his bosses, who would give him the license to find and deliver a solution. Support would come from his east and west compass points—his colleagues, and external collaborators. He needed to make them all believers—or at least to suspend their doubts. Move too fast, and he'd surely outrun the headlights of his support. Run too slow and it would all be too late. First steps—quickness, managed with care—would be decisive. The need for some, more modest initial claims and proofs was paramount, and Bay delivered in the early proof of concept.

Neither the bureaucratic compass nor the strategic triangle are set in stone. They are dynamic, just as the world around is, and require checking, and maybe resetting, as you move forward with precision innovation. Your progress changes your terrain. As Bay did, you'll reassess your context. Should you update your value proposition to engage new support and tune up product or service designs? Have new prospective winners and losers emerged whom you must engage? Is the problem still "live" and worthy? Are staff members, platforms, and funds right for the next stage of work? Do you have the authorities you need and legitimacy to make the next best moves?

At the same time, you'll need to communicate your progress around your bureaucratic compass—to your bosses, your teams, your colleagues, and your external collaborators. Each needs to hear different messages. Your team needs to understand how their contribution matters to the big picture and they are clear what they need to do next. Your bosses need to hear that their support is on track for results and any tremors of discontent can be managed. Your colleagues need to know that you are winning still and that they should stand ready to assist or, if need be, stay on the bench and out of your way. Your external collaborators need to know you are on track and delivering, and that you'll soon be ready with a next round of asks. Your goal throughout is to keep your friends engaged, your opposition back on its heels, and your next round of support teed up. All of it should be guided by your strategic triangle. Keep that in your back pocket and top of mind.

## Bootstrapping Predator, the Precision Drone

It was August 20, 1998, thirteen days after al-Qaeda had destroyed U.S. embassies in Nairobi and Dar es Salaam. By all rights, Osama bin Laden, the organization's already notorious leader, should have been a dead man. He'd apparently been spotted by the Central Intelligence Agency's (CIA) overhead drones at an al-Qaeda encampment in Khost, Afghanistan. Hours later, on President Bill Clinton's order, four U.S. Navy ships waiting in the Arabian Sea off Pakistan sent sixty-six Tomahawk cruise missiles screaming into Afghanistan to obliterate the encampment. But bin Laden wasn't there. Perhaps he'd been alerted and fled, or perhaps he'd never been there at all. In the aftermath, no one was quite sure.

That near miss, just one of many, would cost the United States and the world dearly. For the next fourteen years, bin Laden wreaked global havoc, attacking the USS *Cole* in Yemen; Khobar Towers in Riyadh; targets in Mombasa, Bali, Casablanca, and Istanbul; the Jakarta Marriott; trains in Madrid and London; Northwest Flight 253; the Islamabad Marriott; and, of course, the World Trade Center and the Pentagon in the United States. Dozens of attacks, thousands dead, followed bin Laden's escape at Khost.

If there was fault here it lay at least as much in the design of the drone as in the entire precision system around it. It was 1998, early yet in drone warfare. The drone that spotted bin Laden was a simple surveillance aircraft, unarmed, propeller-driven, not much more than a glider powered by an Austrian racing snowmobile engine and equipped with a fat underbelly camera ball. It couldn't shoot, couldn't target, couldn't do much of anything except stay aloft for hours at a time—what was called "persistent stare"–flying in areas considered too dangerous for piloted aircraft and transmitting images of the battlespace below.

Fast-forward to March 1999. The U.S. Air Force was now in a shooting war over Kosovo. The Predator, as the drone had been named, had been perfected as an intelligence gatherer, but U.S. Air Force General John Jumper, commander of all forces in Europe, was deeply dissatisfied with its capabilities as a weapon. The problem was a simple one: just as at Khost, Predator's kill chain was broken. The lag between when the drone spotted a target and the moment when a lethal weapon could be delivered against that target represented "a big furball of delay, handoff, and decision," as Jumper called it. That lag made the Predator ineffective in engaging battlefield targets that might pop up for mere minutes or even seconds—then vanish.

"The realities in the battlefield made it evident as to what needed to be done," Jumper recalls. "Shorten that kill chain. Scrunch the time lines up to near real-time." Or else, as some proposed, just leave it in the hanger. Unless fixed, Predator was unfit for combat.

On March 24, 1999, Lieutenant Colonel Đorđe Aničić, a Serb missile battery commander who later boasted that no one had told him that the US F-117 Nighthawk stealth bomber was invisible, shot one out of the sky using Soviet era surface-to-air missiles ("SAMs"). That pushed American fighters above the SAMs' 15,000 foot ceiling. It also drove relentless development of laser targeting for the Predator. If Predator could "lase" the ground target—illuminate it so that high-flying fighters could lock their munitions on to it—that could prove highly effective, and keep the fighters out of SAM range.

Jumper saw possibilities for addressing the "Khost problem." He lit a fire under a network of like-minded partisans to get Predator combat-ready. They included General Mike Ryan, the Air Force chief of staff; Colonel James G. ("Snake") Clark, who knew the military bureaucracies well; and Colonel Bill Grimes, who ran Big Safari, a program that had been doing rapid prototyping of highly experimental Air Force craft for years.

Big Safari took on the challenge of fixing Predator. Weeks later, the program shipped a new Predator equipped with an underbelly ball that added a global positioning system (GPS) and a laser designator to the camera. Now, when Predator saw a likely target, it could mark the object with its laser. Overhead jet fighters, high up and out of ground-based missile range, could lock on the GPS coordinates, launch a strike, and destroy the target. The sensor was now also the pointer. Days later, a Predator equipped with the new combination ball lased a Serbian tank for an A-10 *Warthog* flying overhead which destroyed it, giving Predator its first "real time" kill.

Seemingly overnight, the war in Kosovo ended. Jumper packed his bags and returned to the United States and his office as head of U.S. Air Force Combat Command. That's where Jumper learned that the new underbelly balls were being *stripped off* the Predators. Why? The war was over, the exigency passed, and this innovation was not part of the original Predator design—the program of record, in Pentagon jargon. Off they came.

"I didn't react well to that," Jumper said. He immediately sought to revise the program-of-record specifications for Predator and ordered the balls reinstalled. Predator would continue to be both sensor and pointer. Round one in the bureaucratic war over drone innovation went to Jumper. But this was a heavyweight match with a few rounds to go.

In the spring of 2000, the pressure to find and kill bin Laden was ramping up. Killing bin Laden had become a top priority of the U.S. counterterrorism operation. But after multiple tries and misses, there was nothing but frustration to show for it. Jumper sensed opportunity. This could be the right moment to upgrade Predator even further by redefining its mission.

"Let's go the distance," Jumper told his staff. "We're going to weaponize Predator. We'll put Hellfire missiles on it—the lightest we have that can still kill an armored vehicle. When Predator finds a high-value target, we'll do something about it right then and there. If we can see it, we can shoot it."

The Hellfire missiles seemed perfect for the job. They were available in quantity immediately from the U.S. Army, and they were light, easily mounted to Predator's underwing assembly, and low-priced by Pentagon standards at $485,000 each.

"Putting a weapon on the Predator wasn't just a good idea," Jumper later told author Richard Whittle. "It was a *great* idea." But bureaucratic opposition quickly surfaced.

"The acquisition community took hold of this," Jumper says. "'Woah! We're incorporating weapons? Now, this is a big deal,' they said. They still viewed Predator as a disposable sort of almost model airplane. This made it a lethal warfighting machine."

It would take an act of Congress to approve such a change, coming off-cycle between annual appropriation bills. Till then, Air Force lawyers proscribed any "touch labor" (i.e., hands-on manufacturing activities) aimed at arming Predator. Jumper worked with his precision coalition to address the new challenges. The team now included Chief of Staff Mike Ryan; his vice chief General William Begert, who would manage congressional approvals; and Richard Clarke, then counterterrorism chief to Bush, who was hot on bin Laden's trail and high on Predator. They realized that, before they could put a fleet of weaponized Predator drones into service, they would have to tackle several emerging risks—some typical of many precision systems projects, some unique to the military.

There was *regulatory risk*. The shift from surveillance tool to weapon had already run afoul of the Pentagon's acquisitions procedures. The Big Safari team took the wings off Predator so they could develop and test the underwing Hellfire launcher separately. This meant that, technically, Predator was still not armed. But once Predator was weaponized with a Hellfire missile, Jumper told us that the "biggest challenge came when the policy people said that meant Predator had to be defined as a cruise missile, in violation of the Missile Technology Control Regime. That would mean renegotiating an existing international treaty that included the Russians. The idea that a 100-pound missile with a twenty-pound warhead carried on a nonstealthy platform that flew at seventy miles per hour posed a leap in missile technology was ridiculous."

Clarke stepped in. He'd been personally involved in the treaty negotiations. He interceded with Navy lawyers, pointing out that Predator was no missile—it *carried* missiles, but was itself a mere platform designed for a return to base and reuse. It took months, but the lawyers bought that argument and released Predator from its hold.

There was *budget risk*. Who would pay for the project? As in any large organization, that was a fundamental question that needed an answer. Jumper's innovation fell into a bureaucratic no-man's-land: the intelligence community was losing a surveillance asset and was not particularly interested in its kill-chain benefit. The operator community was gaining an asset but was not really sure what to make of Predator or how to coordinate with it. Even the Air Force was reluctant; it bought planes, but Predator was—well—a drone. "Nobody had something like this to deal with before," Jumper recalls.

It was ultimately Ryan who ponied up the $3 million needed to keep Big Safari in the game. To smooth the administrative path, Clark sent letters to ten members of Congress stating and explaining the intention to transfer the funds.

There was *operating risk*. How would the Air Force adapt to a see it/shoot it, real-time kill chain after years of experience with a slower process in which the action was parceled out in stages, first to observers and only later to shooters? And who would be empowered to authorize a drone attack: the Air Force? The CIA? The president?

The issue made it all the way to the top. President George W. Bush decided that authorization would be in the hands of the CIA director (then George Tenet) or his designee, coordinating with the four-star general in of U.S. Central Command (then Tommy Franks), or his designee. The Big Safari team built that command structure into the Predator's software and procedures.

There was *basing risk*. Predators needed to be close to the battle zone. But no country in the region wanted to host American drone "pilots" sitting at terminals launching missiles from Predators into Muslim Afghanistan. To solve this dilemma, the Big Safari technical team engineered a revolutionary split operation that positioned operators in the United States linked by satellite communication to drones physically based near Afghanistan. With just a 250-millisecond delay, stateside pilots would be able to fly, see, and shoot Hellfires from drones over Afghanistan.

Finally, there was new *technical* risk. Would the weight of the missile bring down the drone? Could the Hellfire software be modified from the Army's design built for helicopters firing missiles at tanks from 2,000 feet to fit the design of Predator shooting from 15,000 feet? Would Hellfire's exhaust shoot the rear-mounted propeller off Predator? If the missile got hung up on the launch rail, would it cause the Predator to spin out?

Answering those questions and more would normally take elaborate testing on a giant range. The acquisition bosses told Jumper to expect a process that would cost $15 million and take several years. Clark suggested an alternative. The new Predator could be tested by propping it on top of a hill on a static stand, starting the engine, putting the Hellfire on the wing, and firing it at a static target. It worked. Just sixty-one days and $3 million later, Predator was qualified to use the Hellfire.

On October 7, 2001, four weeks after the attack on the World Trade Center and the Pentagon, U.S. Air Force Captain Scott Swanson sat in a control room in a complex of containers built in the woods behind the CIA campus at Langley, Virginia. Swanson had the authorization to fire at a moving image from a drone over Kandahar on the screen in front of him. A bare quarter of a second after the pressing of a trigger at Langley, a Predator missile 6,900 miles away found its first target, thought to be a team of bodyguards to Mullah Omar, supreme commander of the Taliban.

Over the winter of 2002, a new fleet of Predators would be armed with makeshift Hellfire tank missiles prepared not just to hunt but to kill. The sensor had become the shooter, collapsing the kill chain from hours to minutes and then seconds. Jumper's dog was finally in the fight.

It had happened through a masterful job of bootstrapping: *Predator's weaponry, propulsion, authorities, governance, communications, imaging, positioning systems, talent and budgets* had all once been anchored somewhere else—designed for some other purpose, owned by sometimes bitter service rivals. Little by little, each had been pried loose by threat or promise, modified and repurposed, and *cobbled together* in Predator, creating an astonishingly lethal new platform.

The results quickly altered calculations on the ground and transformed the face of battle. Rather than choose between safety for pilots and aircraft operating from a distance, or risk all with close-in targeting, generals could have both safety *and* precision targeting. Rather than choose between preserving a village *or* destroying an adversary in its midst, Army captains could spare the village, destroy the adversary, keep their own troops safe and losses low, holding political support for the war on Main Street, USA. That massive shift from "either/or" to a radical new "and" sidelined opposition, emboldened leaders, and empowered warfighters. It would transform the very nature of the combat space, and bend it to U.S. forces' advantages, for years. That was smart. That was advantage. That was value. It was precision strategy in action.

## Afterword

In the twenty years since Predator, drones have steadily and completely altered the face of battle. Proved with precision strikes against insurgents by American forces in Afghanistan and Iraq, the drone emerged full force in 2020 as a punishing and determining factor in set-piece, open battlefield warfare. During the conflict in Nagorno-Karabakh, Turkish Bayraktar TB2 drones and Israeli Harop "suicide" drones under Azerbaijani command took out Armenian tanks, armored vehicles, and artillery pieces, and scores of air defense systems, served up in what the *Economist*'s Shashank Joshi called "a bonfire of armor," bringing Armenia to its knees and the conflict to an end.

That same Turkish drone—the Bayraktar TB2—is now sold as a platform with portable command stations and communications equipment, with months of training as part of the package. Israeli advanced mission-planning software, replicated by Turkey, has found its way into these systems, where it steers drones and missiles through gaps in radar coverage onto defenseless air defense systems below. The Bayraktar TB2 precision drone system now dominates the skies in many conflicts, making Turkey both an industrial powerhouse and a player at the diplomatic table.

Yet sophisticated as the aircraft, software, and munitions now are, drone technologies are particularly amenable, still, to the ingenuity of soldiers cobbling together precision battlefield solutions on the fly. In late April 2022, a video surfaced on Twitter. It was eye-popping. The camera tracked a commotion beneath a Ukrainian drone among a squad of Russian troops as they ran for cover. Two made it into a car. In a few seconds, the drone released some sort of improvised bomb that plunged straight down through the car's sunroof and blew it up.

The accuracy of the dropped-bomb attack was rated by some on Twitter as "a hole-in-one level of luck," but that wasn't quite true. If the first drop missed, the drone operator could tweak the next to perfection. For it turned out this latest battlefield drone—a version of one used by Isis over Iraq—was likely an off-the-shelf octocopter like one you'd buy at Walmart, but modified for an extended flight time of forty minutes and capable of carrying three VOG-17 Russian-made fragmentation grenades (plenty of them around Ukraine those days.) Add a thermal imager for spotting warmed Russian trucks and tanks at night or hidden in woods by day, and a Canon camera to record and broadcast every attack, hover at 100 meters where it was invisible and noiseless, and nothing would prove quite so true as the boast of the Ukrainian drone units' uniform patch: "You Shall Not Sleep."

But it also reflected what one observer called "a start-up" mentality among Ukrainian troops. "For 8 years," @nolanpeterson tweeted, "Ukraine's military has transformed, allowing front-line personnel to operate creatively & as autonomously as possible. That is a big change from the strict, top-down Soviet chain of command—which Russia still employs." With terrifying consequence, on May 9, 2022, Ukrainian drones, missiles, and artillery destroyed seventy Russian tanks, armored vehicles, and even a tug boat spotted forging the Siverskyi Donets River, killing hundreds of soldiers—essentially a battalion's worth. Days later intelligence analysts reported Russian forces appeared to be trying to forge the river *again*, tweeting, "It went about as well as their previous attempts." Down the road, yet another improvised Ukrainian drone dropped two $475 bombs on a Russian $3.5 million T-90 tank, the most advanced in the Russian arsenal, destroying it.

In August 2022 the *New York Times* reported that along the front lines Ukraine's experimentation has produced an array of inexpensive, plastic aircraft, jerry-rigged to drop grenades or other munitions. Off -the-shelf products or hand-built in makeshift workshops around Ukraine they now comprise the bulk of Ukraine's drone fleet.

Whether high-end military drones or nimble off-the-shelf concoctions with improvised munitions, on both sides of the Ukraine battlefield drones have taken the place of human eyes in the sky for reconnaissance—and compressed the kill chain to a moment or two between spotting troops on the move, or the flash of artillery, and pivoting to shoot, forcing adaptation and innovation on the ground. Using advanced howitzers that are built to launch a ferocious salvo in little or no time, for example, artillery teams have adapted their old "lock and load" routines to "shoot and scoot"—an end-of-cycle mad dash for cover before the retaliatory drone strike hits. In the duel between two battlefield precision systems, it is the speed of humans and machines working in tandem, one attacking, the other escaping, both learning and adapting to the other that determines who wins. Above all, for the drone, there is no precision and no outcomes that matter without the platforms and data, people and politics that bring it all together in the battlespace. "This war offers a lesson," Peterson tweeted. "The power of platforms depends on the character and competency of the personnel who operate them."

Finally, we opened this brief history with an account of the historic miss of Osama bin Laden at Khost in 1998, blaming it on immature drone technology and a "furball" of indecision in the kill chain of a not-yet-ready-for-prime time precision system. Two decades later American military activity in Afghanistan had all but vanished—at least until early July 2022, when Ayman al-Zawahri, bin Laden's number 2 and heir to the al-Qaida organization was tracked to a Taliban safe house in Kabul. On July 31, a drone had him in its cross hairs—the drones now "tip of the spear" of a fully evolved precision system—which now meant months of planning, simulations, and tailoring, waiting for exactly the right moment.

> WASHINGTON (AP)—As the sun was rising in Kabul on Sunday, two Hellfire missiles fired by a U.S. drone ended Ayman al-Zawahri's decade-long reign as the leader of al-Qaida. The seeds of the audacious counterterrorism operation had been planted over many months.
>
> U.S. officials had built a scale model of the safe house where al-Zawahri had been located, and brought it into the White House Situation Room to

show President Joe Biden. They knew al-Zawahri was partial to sitting on the home's balcony.

[Having] painstakingly constructed "a pattern of life," as one official put it, [they readied] a "tailored airstrike," designed so that the two missiles would destroy only the balcony of the safe house where the terrorist leader was holed up, sparing occupants elsewhere in the building. They were confident he was on the balcony when the missiles flew, officials said.

"Clear and convincing," Biden called the evidence. "I authorized the precision strike that would remove him from the battlefield once and for all. Justice has been delivered."

## Lessons for Leaders: The Art and Science of Bootstrapping Precision

Bay's TraceTogether system and Jumper's Predator illustrate the unique value of precision systems in addressing the kinds of challenges that arise in today's risk-laden, fast-changing, networked world. Whether the enemy is a quickly spreading pandemic or elusive terrorist targets, moving fast and light with precision is the order of the day. Yet the very nature of these challenges also means that the complex job of creating precision systems to respond to them is particularly daunting. When time, money, official authority, and other resources are all in short supply, members of the precision vanguard must apply the art of bootstrapping to create the precision system that desperate times demand.

What lessons can managers learn from the ways Jason Bay and John Jumper cobbled together their precision systems? First are five criteria by which any precision system can be judged, whether cobbled together or otherwise:

- *Will the cobbled-together precision system do what is intended, and only that?* For TraceTogether, will it alert millions of individual users to their exposure to COVID-19—and only them? For Predator, will it kill what it sees and shoots—and nothing else?
- *Will the cobbled-together precision system draw support for uptake and adoption in operations?* Can TraceTogether work in conjunction with other tracing systems to quickly tie off the spread of COVID-19? For Predator, will the U.S. Air Force adapt and adopt unmanned drones in its warfighting plans?
- *Can the cobbled-together system's performance be measured with precision-specific metrics in operations?* Will TraceTogether correctly alert its user, and no one but its user, of COVID-19 risk, every time? Will Predator accurately destroy all targets, for example, and nothing but its targets, every time?
- *Can the cobbled system give managers valuable new choices that they will use?* Will TraceTogether allow health agencies to locate accurately all those

exposed to COVID-19 without intrusively violating citizens' privacy? Will Predator allow generals to choose *both* operator safety *and* high accuracy in targeting rather than force generals to choose between keeping pilots safe at high altitudes but bombing indiscriminately, *or* improving accuracy but losing pilot safety at lower altitudes?

- *Making those choices, will the precision system, transform the theater of operations exactly as managers intend?* Will TraceTogether keep COVID-19 minimal and manageable in Singapore, and will it keep privacy intact? Will Predator bend the combat space to American advantage for years—and broaden the latitude of decision makers knowing their choices no longer have as a consequence the blunt-force trauma of indiscriminate targeting?

Second are four lessons that are specific to leaders who bootstrap precision systems under conditions of imperfect insight and means, great uncertainty and risk, and overarching necessity:

- *Networks are essential in the provisioning of cobbled-together elements needed for precision.* Neither Bay nor Jumper had all the assets they needed to move—not authority, data, and platforms, nor skilled personnel or budget. For that, they turned to their networks and those of colleagues. Yet each node of a network represents both a potential asset and a possible source of resistance. The greater the networking, the greater the power of each to abet or disrupt.
- *Bootstrapping is inherently cross-boundary, requiring collaboration, persuasion, and agreement.* Both Jumper and Bay had significant levels of command-and-control power within their domains. Again and again, however, their best moves were only possible in someone else's domain. Both had to rely heavily on network brokers—colleagues who spanned several domains through which they could exert influence.
- *Engaging the support of existing, trusted institutions is an essential element in building the legitimacy of a cobbled-together system.* The success of the TraceTogether project derived largely from the fact that Bay could draw on his own network of skilled engineers to work on its design and implementation, and on his contacts in the Department of Health and other key government ministries. The trust these institutions commanded in Singaporean society went a long way toward giving TraceTogether legitimacy in the eyes of the community. Similarly, sending the Predator to the existing Big Safari program for its development gave the initiative a supportive home in the Air Force system, conferring sustainability. The leaders of Big Safari knew how to navigate every convention of procurement, development, and operations, bending all but breaking none, fielding systems fast, helping to establish Predator's viability as a weapon.

- *Bootstrapping precision requires adaptive leadership skilled at political management, effective communication, and technical agility.* Unexpected circumstances prompted the development of both TraceTogether and Predator. In the case of TraceTogether, the cause was the sudden emergency of COVID-19; for Predator, it was the repeated failures to capture and kill bin Laden. The lack of any central planning or control for these two improvised projects meant that, at each stage in the process, those in charge had to check and recheck their strategic triangle and its constantly shifting requirements. When authority was lacking, the precision vanguard borrowed or begged it. When capabilities fell short, the vanguard invented them. When the technology was unavailable, the vanguard brainstormed alternatives.

You may never be called on to create a precision system under the kind of intense pressure that Jason Bay and John Jumper faced. Your work as a member of the precision vanguard may never involve consequences that are literally matters of life and death. But you may well find yourself saddled with a precision project that must be completed fast and effectively under circumstances that are far from ideal. When that happens, the stories of TraceTogether and Predator will offer lessons you're likely to find valuable.

# CHAPTER 8

## CREATING AND MANAGING TEAMS OF PRECISION

Throughout this book, we have seen people engaged in the hard, consequential work of moving change precisely where, when, and how they want. Whether the project has been formally launched from the top of the organization or bootstrapped under the radar, it must align people with processes, and platforms with performance to deliver on the promise of precision. In the quest to make that happen, the most important decisions may be those regarding people—often comprising cross-discipline teams of individuals whose talents, energies, commitments, and insights will determine the success or failure of their precision systems.

The challenge of building and managing a team of precision is one of the most difficult a manager can face. Given the diversity of professional talents involved and the range of personal and professional interests that team members bring to the project, it's not surprising that conflict and confusion can bring teams low. Teams that are incorrectly staffed, chartered, and managed stumble on fixing the right problem. When a team's decision making is gummed up, teams suffer miscues, misfires, and delay. All contribute to performance that falls short of promise and potential. But correctly staffed, organized, and programmed teams can function as powerful drivers of change—force multipliers for continuous learning and institutional reform, even (or perhaps *especially*) when battling against the stiff headwinds of the status quo.

No team is perfect, but even an imperfect team can win. Classic studies of team dynamics enumerate the conditions that foster team effectiveness, including having what is seen as "a real team," a compelling direction, an enabling team structure, a supportive organization context, and expert team coaching. In this chapter, we'll see how these and other factors affect precision teams in particular. What is required for teams of precision to succeed? What makes precision system teams different from all others?

We turn first to the work of the distributed organizing team of the Biden-Harris campaign—a team of twenty-eight managing a network of 160,000 all-volunteer partisans aligned on a single purpose, for a single day, in the race for the presidency of the United States, the most consequential political contest on the planet.

## Distributed Organizing for the Biden-Harris Campaign: Crafting Victory through a Team of Precision

As we saw in chapter 4, the foundation for Joe Biden's victory in the presidential campaign of 2020 was the work done by Nellwyn Thomas of the Democratic National Committee (DNC) and her team of data specialists. They invested countless hours in cleaning, organizing, and instrumenting vast troves of information about voters in thousands of districts in every state across the country. In the process, they turned otherwise useless raw data into a priceless flow of content that the Democratic Party could use to connect with people who might be persuaded to vote Democratic—not just in 2020 but in many future elections as well.

The Thomas team's overarching goal was to make Joe Biden president. That called for a full-on collaboration with other teams, as well as a new model of coordinated team effort for the digital world known as *distributed organizing.* This part of the story begins in July 2020, when Nate Rifkin was tapped to head up the national distributed organizing team for the Biden-Harris campaign. Early voting for the general election would start as early as September—two short months away—and culminate two months later on November 3. With the Democratic primaries over by June, the campaign had expanded rapidly from 200 staffers to 6,000, including 28 campaign veterans who comprised Rifkin's leadership team. That group and its massively scaling force of 160,000 volunteers would play a crucial role in the campaign's success or failure—and comprise the greatest challenge of Rifkin's career.

Rifkin's top staffers were all handpicked all-stars from several now-folded Democratic primary campaigns, or on loan from the DNC. They included Roohi Rustum, relational team director, from the Pete Buttigieg campaign; Claire Wyatt, call team director, from the Bernie Sanders campaign; Catherine Vegis, events team director, from the Biden campaign; Clarice Criss, text team director, from the Tom Steyer campaign; Deputy Director Spencer Neiman, from the Elizabeth Warren campaign; and Rifkin himself, also from the Warren campaign. People from DNC included Distributed Analytics Director Nina Wornhoff and Deputy National Training Director Lucia Nunez.

Skilled and savvy as team members were, never before had a campaign team attempted distributed organizing on the scale Rifkin contemplated. The data platform, cobbled together from many preexisting platforms, was new. The processes were new. The team's ability to collaborate and solve problems together was untested. And within weeks, as the campaign roared into action, Rifkin's team would be sending about 10 million texts per day and making millions of calls to voters in seventeen battleground states using some of the most advanced technology tools available for a rapidly scaling service and team.

These big challenges broke down into a host of small, specific challenges, each potentially crucial, all turning on whether Rifkin's team could mesh the technologies and people together for precision results at an unheard-of scale. Would the "sweepers" function as needed? Would the "assigners" be able to keep up? Would the "captains" be able to answer all the questions that would come up in chat on the campaign's Slack channels? Would the ROB Bot tool they'd built succeed in moving texters from their starting channel into the ready-to-go channel? These and other questions would be answered only under fire.

The basics of distributed organizing for political campaigns have been known for a while. Distributed organizing most often works as an element of traditional place-based organizing, part of the local campaign managers tool kit. It's all managed top down, with staff, money, and resources sent wherever campaign headquarters believes they are needed most urgently—whether to knock on doors, raise money by telephone, or get out the vote.

Digital campaigns came into their own in trailblazing forays on the internet by Democrats Bill Bradley in 1999, Howard Dean in 2003, and John Kerry in 2004, all of whom raised big pots of money online in small-dollar donations with eyebrow-raising ease. Digital campaigns exploded in the hands of skilled Barack Obama campaign staff in 2008, when they caught Hillary Clinton napping with too little social media presence until it was too late, and again in 2012, when they crushed John McCain, who insisted on holding onto the top-down, command-and-control campaign style he cherished.

With the advent of twenty-first-century digital platforms, distributed organizing evolved as a forked offspring of the main digital campaign, operating across networks, principally handling get-out-the-vote initiatives. Wikipedia-like, it operates more like a decentralized web of partisans, with a core of professional staff activating impromptu groups of volunteers. Central staff members design and develop the decentralized platforms, keep the fuel of data in the tank, probe for the opposition's weaknesses, and shape messaging and targeting with analytics, providing front-line volunteers with tools and scripts for specific voter outreach and mobilization.

Digital distributed organizing's secret weapon is the same "flywheel" principle that powers Amazon's or Netflix's sales in e-commerce, where AI-powered computers learn from every consumer choice how to make their recommendations even better, boosting sales to the stratosphere. Distributing organizing counts on the flywheel effect for volunteer recruitment: volunteers beget volunteers, who beget even more, turning distributed organizing into a massive campaign power resource. Managers concentrate the volunteer fire to create the change that matters most—getting prospective voters off the couch and into the fray, ultimately pulling the lever for their candidate.

As the 2020 campaign neared, no one on the Democratic side had big plans for distributed organizing; it was nice to have, but it was less favored than campaigning the old-fashioned way—in person, on doorsteps, raising money, changing minds. Then, in March 2020, the COVID-19 pandemic struck. By July, it was clear that the fall's voter mobilization would, of necessity, fall *entirely* to digital teams. That effort would be led by Jose Nunez, director of digital organizing for Biden-Harris. Distributed organizing was led by Rifkin, one of Nunez's seven direct reports. The fast pivot to *massively* distributed organizing was on.

"We had no choice," Jose Nunez said. "The pandemic hit. We're not going anywhere. I told Nate, 'I want you to go big and don't stop. Bring in everybody from California to Idaho to Alaska, point them in a direction of a state that we need to win,'" Nunez said. "My job was to navigate any way that I could to allow him to be efficient in growing our volunteer base."

Through July, Rifkin hunted for his own leadership team. He describes his ideal candidate as the "three-blade-tool distributed organizer," proven in the big leagues, a "big structural thinker who could really conceive of building this process," and having a "use-your-tools/tech savviness mentality," someone who would be undaunted by setbacks and prepared for challenges that would be "pretty darned technical."

Little wonder, as platforms, software, and data would be central to the work of the distributed organizing team. It was fueled by targeting data from Wornhoff's analytics shop (itself made possible by Thomas's massive overhaul of the DNC data systems), a massive research base of tested and proved messaging scripts, and a constant stream of polling and surveys. Rifkin's volunteers could log onto the distributing organizing team's Slack channel to do voter contact from wherever they were, whenever they had a couple of hours, equipped simply with Wi-Fi, a headset, and a cellphone or laptop. The second a volunteer showed up, they were assigned to a team and given texting or calling tasks into high-value voter segments in any of the seventeen battleground states—Kansas in the morning, for example, Colorado in the afternoon, and Minnesota in the evening. Experienced volunteers hovered over newcomers virtually, sometimes watching the whole team texting or calling from wherever they were, tiled on

Zoom. There were no in-person phonebanks, no boiler rooms, no doorstep visits—all the traditional team tools of in-person organizing. The volunteers were anywhere they needed to be, turning on a dime, thrown into the breach, whether from their favorite Starbucks or a farmhouse kitchen. When finished they could log off, stroll home, or head downstairs to the dinner table.

The method transformed the campaign. At its peak, for example, Criss's texting team would launch a new 100,000-person texting campaigns every sixty seconds, sending 10 million texts a day and 330 million texts over the course of the campaign. To handle this massive job, team members designed and rolled out new tech tools and, with them, new workflows, processes, metrics, and training. So many new volunteers were streaming into Biden's digital ranks that training sessions might need scheduling seven times daily.

Wyatt's calling enterprise, organized via Slack into teams and supported by paid staff moderators, made 88 million volunteer-to-voter calls over four months, using scripts in English and Spanish, all newly drafted for each battleground state. Every call was person to person—not one robocall was used. By Election Day, the call crews surged to over 103,000 volunteers, helped by several new dialing tools to scale their operation nationally. Thrutalk allowed volunteers to call voters without actually dialing, for example. Obama organizers in 2008 and 2012 had made forty-five to sixty calls per hour. Wyatt's volunteers could make about 300 in the same time. One Democratic dialer recalls:

> Calling swing states is relatively easy. All you need is a computer and a phone, and you can do it at your convenience in the comfort of your home. One hour of virtual training, and then you're good to go, logging into the party's "Dem Dialer" technology platform . . .
>
> You follow a script that displays on your computer screen, moving from one section to another based on the answers you get from the person you've contacted. Calls are dished up to you automatically, and you start talking as soon as you hear a tone generated by the system, indicating that someone has answered. That person's name appears in the first line of your script.

It was all "easy peasy"—or made to seem so. With the rate of newbies coming and going onto the system, it *had* to be.

None of it would have been possible without Thomas's powerful new voter file, all cleaned and deduped—unleashing the cellphone as another power resource. "We went from 30 percent of people having a cellphone in our database to 70 percent," Thomas recalled. The platform and people, data and workflow all came together for managed precision, Thomas told us:

> Where before you had a list of people to call randomly chosen, now we can give you the hundred that are most likely to be convinced by that phone call. That

leads to huge efficiency. Instead of targeting "Nell Thomas" in two states, you're talking to her in only one, and only the more recent one. You spend that other time and money on someone else.

Using the tools provided by Thomas, Rustum's relational team focused on making connections with voters through friends, family, and community acquaintances—a method of changing people's hearts and minds that research demonstrated could swing results by as much as 8 percent. They were especially primed and trained to call hard-to-reach supporters: first-time voters, voters who were unregistered, who were unregistered at their current address, or whose registrations had lapsed. Even Republicans thought to be on the fence were included. In the press of time and resources, traditional campaigns often had to pass on these difficult cases. "Not enough juice in that lemon," it was said. "Too hit-and-miss." But the evidence of research was plain: so-called relational conversations recurring over time with such voters could produce double-digit gains in conversion and turnout. With thousands of volunteers streaming to the distributed team's digital ranks, great voter data, and effective platforms, Rustum's relational team could go after them all.

Catherine Vegis's events group helped stitch all the campaign teams together. Everyone knew the presidential race would be tight. With a massive digital organizing effort, even adding 1 percent or 2 percent to the Biden ledger could make all the difference. Vegis's job was to provide that spark with over 11,000 virtual events for the volunteers—from parties to phonebanks, all to foster team identity and solidarity even in a digital-only world of isolation and pandemic.

It should come as no surprise that Jose Nunez and Rifkin faced pushback—principally from state party organizations who felt hamstrung by their own lack of technology and deflated by Rifkin's consumption of volunteer power—and usurpation of their turf. "Why are you texting so-and-so in Missouri to ask them to make calls in to Wisconsin?" Rifkin would hear. "Rifkin's operation is big—but it doesn't work," Jose Nunez would hear. He noticed that the traditionalists rarely questioned whether all the local, in-person canvassing and phonebanks *they* did was working.

Pressed, Rifkin and Nunez responded with what evidence they had. Penetration testing proved that the distributed team's volunteers were measurably locating more people on the lists, contacting more, and having longer conversations than anyone else. New digital tools were showing *their* impact—certainly compared to *no* tools. Wornhoff's postelection analysis for Higher Ground Labs showed that organizing software like Mobilize, analytics and planning software like Deck, and friend-to-friend software like Impactive all helped Democratic campaigns that used them to better manage the surge of millions of out-of-state volunteers onto digital platforms and boosted local vote share by solid percentage gains compared to campaigns that did not. Now that most presidential campaigns have become squeakers, even small gains loom large: they can easily be the margin of victory or defeat.

"This is how we're going to help your state and your campaign win," Rifkin promised. And that was enough for most. But the fact was that the distributed campaign was flying somewhat blind. In the press of a campaign, there was little time to test, prove, and improve targeted messaging. Instead, Rifkin's team imported proven and tested get-out-the-vote scripts from past efforts and from Democratic libraries of resources like the Analyst Institute. The intuition and experience of the distributed team's executives counted heavily, especially in the absence of actual data from live testing.

"The conventional wisdom on this is extremely well-tested and extremely data-driven," Wornhoff asserted. "I've been on a lot of campaigns that have spent time message testing. All come to this exact same conclusion."

Traditional polling could do little to verify scientifically the distributed team's impact. That's the nature of politics. No one knows, finally, how anyone else *actually* voted. When a company runs a Facebook ad, they can see whether it results in a sale, then tweak to improve and retest it. But the impact of campaign activities on individual behavior in the voting booth is unknowable.

Something clicked, however, for Rifkin and the Democratic team. Biden won. And the distributed organizing team could claim some credit, if only by virtue of its massive numbers, a tried-and-true metric of campaigns that no one could contest. At the very least, it offered a powerful new engine for recruiting, training, and mobilizing volunteers, all passionate for the cause. By November 3, 2020, the DNC distributed organizing team had engaged 160,000 volunteers to text, email, or speak with millions of other volunteers and voters. The twenty-eight members of the leadership staff and their 160,000 volunteers had been responsible for one-third of all voter contacts by the Biden campaign; for 330 million texts; 88 million calls; 11,000 events; and 200,000 relational conversations by voice, text, and email. Jose Nunez and Rifkin are now convinced that a nationally distributed organizing campaign will prove invaluable when elections return to the age-old knock-on-doors, chat-on-the-front-stoop relational organizing.

"I don't think that's really clicked with some people I've talked to," Jose Nunez reflected. "Here we have a dedicated team that does all the volunteer recruitment for mobilization efforts for an entire field operation. What if a field organizer on his local turf never had to worry about hopping on call-time for four hours every night? We're trying to think about that holistically and at that scale for 2022, at least."

On election night, Rifkin had dinner with his partner in the Philadelphia apartment where he had spent most of the past year, shut in by the pandemic like everyone else. His one computer monitor tracked the results, as it had the entire campaign for eighteen hours each day. The voting, which had started months earlier, was now down to a trickle. Every voter list and nearly every voter on them had been engaged. There was really nothing left to do. And then Fox News called the state of Arizona for Biden, and Rifkin knew that the election was won.

Relaxing in Florida months later, reflecting on the scale, speed, and power unleashed by the distributed organizing team, a small smile creeps across Rifkin's face, visible even over Zoom. "We were definitely punching above our weight," he says.

Their work didn't end on November 3. The distributed organizing team had its final say in the following days, when Georgia's secretary of state provided a list of voters whose mail-in ballots had problems and would be uncounted if not "cured"—problems like a signature that didn't match a signature on file or a box checked incorrectly. Having made many passes over the Georgia voters, Rifkin had highly accurate contact data on hand. He texted his volunteers to reach out with one-on-one calls alerting voters there might be a problem in their ballot and informing them how to fix it.

"Because we were able to do this," Rifkin said, "there was a much lower rejection rate for Democratic ballots that needed to be cured than for the Republicans, who didn't run a program like this. That's what a large-scale volunteer apparatus lets you do. Whereas the GOP program really kind of stopped on Election Day." As a result, one more state swung from the red column to the blue column—yet another bit of evidence of the unexpected power of precision systems in the hands of capable managers to help change the course of history.

## Lessons for Executives and Managers from the Biden Precision Team

In a presidential political campaign, with a five-month runway and a goal of victory on one day in November, the need for precision was unforgiving. Whether by one vote or 1 million, it was win or go home. Rifkin's precision team helped deliver the win for Biden. What lessons does it offer for other managers of precision teams?

First, *precision flows from precision data, platforms, and organization architectures.* The data distributed by Thomas's operation was much more accurate than any campaign data before. It "stretched the playing field" to 50 million addressable voters in seventeen battleground states. Data scientists embedded with Rifkin's team coordinated targeting with the central organization and helped the distributed teams pivot to new targets morning, noon, and night. A portfolio of tools and platforms connected the campaign with millions of people who might not have been contacted otherwise. Precision-guided texts, calls, and emails to voters who were high in value wasted a minimum of volunteer time, energy, and spirit.

Second, *precision systems are capable of unleashing and channeling phenomenal human energy that can be properly distributed.* The 160,000 Biden volunteers recruited one another using the so-called flywheel effect. Their energy was great. But the secret of Rifkin's distributed precision team was that it was available anywhere, any time, and it required no special technical skills thanks to a distributed organizing platform using Slack or Zoom.

Third, *a precision distributed organization operates best with a balanced team structure*—one that is loose enough to welcome newcomers by the thousands every day, train them, and target their energies, and one that is also structured enough to achieve one shared goal: winning the election. Rifkin's vast nationwide team benefited from clear expectations of nuts-and-bolts results (number of calls made, time spent on calls); transparency to every team and effort; and constant onboarding and training, with virtual support just moments away. The relentless coordination made possible by a powerful infrastructure of platforms and data allowed the central organization to direct the distributed teams toward targets—people, districts, and battleground states—deemed essential.

Fourth, *a leadership team of three-blade-tool executives is essential*—one that shares culture, capabilities, and outlook, and exemplifies the same idea of coming together for a purpose that they espouse for the volunteers. Jose Nunez, Rifkin, and others were veteran campaigners and organizers who truly believed in the power of groups to change the world. Now, armed with data and platforms, they were unstoppable in a contest where millions of messages bolstered by proven content could deliver even small but crucial change.

Fifth, *the ability to communicate the "lift" from distributed organizing helps managers prove precision's value.* Although direct evidence may be lacking, proxy metrics of engagement like "time on calls" can be measured. As a recognized metric for call teams, proxy metrics can help prove precision when scaled across all 50 million addressable voters, as precision systems permit.

Sixth, *team spirit matters.* Rifkin's precision team flourished thanks in part to continual efforts to build and maintain solidarity, to strengthen the ties among network members and overcome the disconnectedness of the virtual world. As a result, perfect strangers began welcoming each other to the campaign, providing reserves of energy and morale that helped all the team members maintain their flat-out efforts through Election Day and beyond.

## Mobilizing Teams of Precision at the *New York Times*

One-hundred miles up the New Jersey Turnpike from Rikfin in Philadelphia, *New York Times* Chief Technology Officer Nick Rockwell had his "eyes on glass," too. At the *Time's* offices in midtown Manhattan, Rockwell watched system metrics as pressure on the *Times*'s servers built over Election Day. It would be one of the year's biggest days for news media. For months, twenty teams of *Times* engineers had prepared for this moment. Like Rifkin, they faced a series of unknowns. Would the digital newspaper's infrastructure and software hold through days and nights ahead as millions of readers checked in for the news updates they craved? Would months of stress testing and proving systems prevail in the heat of battle, in the crunch of Election Day and later that night?

As discussed in chapter 4, the *Times*'s broad replatforming onto Google Cloud Platform had spawned numerous teams on diverse missions. Cross-disciplinary teams had, in fact, become the beating heart of enterprise precision at the *Times*—circulating its data, testing and proving its precision solutions, continuously learning, sharing its discoveries, and improving the *Times*'s performance.

Over the preceding two years, the collaboration between data scientists, product specialists, reporters and editors, and Rockwell's engineers blossomed. Dozens of cross-disciplinary teams flourished, adding people working in other disciplines as needed, from software engineers and data scientists, to user interface designers and marketers. Home-page redesign, meter rule and paywall testing, and ad pattern changes were the three big-ticket transformational items on the *Times*'s precision agenda, all squarely in the spotlight for the move from an ad-based business to subscription. Each day, newsroom editors tested different headlines on the same article, and marketing teams tested the best ways to convert casual visitors to subscribers, probing for the optimal placement of subscribe-now buttons to drive revenue, even testing the placement of individual stories on an individual reader's home page based on their prior reading behavior.

Staffers from the growth mission team, for example, comprised experts in audience research, brand, data, design, marketing, product, and technology. Each focused their skills on a narrow slice of precision, a problem they could crack, all drilling down on CEO Mark Thompson's goal of increasing subscriptions.

As a product designer, Nina Marin welcomed the change from serving as a team jack-of-all-trades. She said, "I am laser focused on just two sections of the customer journey: optimizing the registration processes, and subscription flow. I find it really challenging and exciting to be able to have such clearly defined objectives." Her colleague Asiya Yakhina, another product designer, was focused on onboarding users after they subscribed or created an account. "We think about how to help readers build a meaningful habit of reading the *Times*," she explains.

Conventional wisdom held that cross-functional teams were prone to bogging down. "Where product teams, engineering, infrastructure, marketing and operations teams all have different points of view, metrics, and OKRs [objectives and key results], that's where you really start seeing friction in the seams," a seasoned team manager told a crowd eager to learn the secrets of team success.

But at the *Times*, cross-functional teams *accelerated* project development. "Before I jump into design exploration," Yakhina said, "I want to be aware of technical constraints, and testing requirements, and have the copywriter from brands weigh in. It's pretty great to see early collaboration pay off—avoiding last-minute workarounds because some constraint wasn't accounted for."

Results were in plain view, fast. "Basically, you talk to a member of the qualitative testing team and you kick off, say, on Monday," Marin recalled. "Your prototype is ready Tuesday. A discussion guide is written Wednesday. Then Thursday, you're in user testing all day, and then Friday to Monday you start getting results."

Benefits could surprise—and go far broader than planned. An intern on one of Chris Wiggins's data science teams figured out how to predict the emotions that readers would feel when reading different stories—from "inspired" and "happy" to "saddened" or "amused."

"Sweet," the marketing people told Wiggins. "We can monetize that. Suddenly," Wiggins recalled, "we had a full-on backend team that was going to go sell premium ads based on how an article makes you feel."

The shift to a culture of testing was underway, all built on new infrastructure and standardized practice that made it easy to use, and so easy to require. The team run by Josh Arak, executive director of data and insights, figured out how to design a new workflow so analysts on teams like Yakhina's and Marin's could query and communicate findings from user data faster and easier—"the fine-grained details of user behavior," his team wrote, all "to understand how our readers are interacting with the *Times* online."

Shane Murray's data analytics and platform group reported the discovery of two new metrics that would supplant classic industry benchmarks like "page views" and "time spent on articles" as the *Times*'s new key performance indicators (KPIs). The best predictors of future subscription behavior, Murray reported, were instead *depth of engagement*, "measured as the number of stories and screens viewed," and *breadth of engagement*, "measured as the distinct number of topics read." These two metrics seemed to measure precisely the value of the *Times*'s experience to any reader, all with an ultimate, countable outcome of new and renewed subscriptions.

The Murray team's findings gave other teams new, sharper focus. Using the new measures of "depth" and "breadth" of engagement, for example, Said Ketchman's meter and marketing technology teams tested and proved new meter parameters for the paywall—how many free articles an unsubscribed customer could read before she hit a registration wall.

When combined, experiments in pricing across numerous teams helped the *Times* solve a long-standing optimization problem: when to drop the registration wall and how to price the subscription. The result might be fewer readers per article and lower ad revenues. But it meant, in fact, more subscribers and more revenues from that source—just as CEO Mark Thompson had hoped for—even foreseen. Most important, these models changed as the systems learned via machine learning, reflecting new market conditions, new temperaments among customers, and new pricing that continued to work.

Through this whole process, Rockwell had played an unusual role. CTOs often focus on making big infrastructure moves in support of corporate functions and systems. Instead, Rockwell established himself squarely in the role of executive sponsor and super-manager of precision initiatives geared to the CEO's strategy and mission anywhere his engineers worked—and that was frequently on teams of data scientists, software engineers, and marketing pros. Each team had a set of OKRs and varied leadership, depending on whether their task was principally one of product development or data analytics and engineering. But when rolled up to the KPIs, the result would always be centered on subscriptions—in other words, on outcomes from precision for specific corporate goals.

"It's super-straightforward how this works," Rockwell explained. "But tricky, nonetheless. Take funnel conversion through the purchase process. Once someone's clicked on a *subscribe* button, how many people come out the other end—completing the sign up? That was the teams' broader goal context. They would set specific OKRs—one team focusing on the OKR of 'improved performance of Step 2 by three-tenths of a second,' for example. That would be its goal for the quarter."

The role of precision team leader, Rockwell offered, is to assign these components, then stitch the disparate OKRs into KPIs, "laddering" individual tasks to the overall goals, distant as they might seem, so that every team member can clearly see how even the lowliest task contributed to the loftiest corporate goal.

"It was hard," Rockwell admits. "I probably spent more time on defining, aligning, and measuring goals than anything else at the *Times*. But if you want to know what's happening in the organization, just look at everybody's OKRs. They are the lingua franca across the different functions."

The collaborations among teams sometimes achieved powerful results even without fancy software. Tina Fehr, an engineer embedded in the newsroom with the interactive news team, helped the *Times* win a Pulitzer Prize for its COVID-19 coverage, moving reporters from their simple spreadsheets to fully developed data applications that pulled more than 10 million requests for pandemic data from around the world. "Screen scrapers" ran software over websites, collecting the data in widely disparate formats, all requiring cleaning and preparation to be stored, searched, and analyzed by other teams—all finally published for readers in a clear, understandable form. But it began and ended with Excel spreadsheets customized for the precision purpose.

"It's never an engineering team that ships value," Rockwell said of the many team endeavors he sponsored and lead. "It's always a cross-functional team, small, typically 'two-pizza size,' and technically diverse." With the problems of news production, marketing, and platforms so intertwined, without a cross-functional team, "the best result you might get is good work," Rockwell observed. "But you're going to do it slowly. Without a cross-functional team, you just can't innovate technology at the pace that we need to."

Perhaps no tactical success was as important, if unheralded, as the work of Rockwell's twenty election night teams. Each was small and operated independently, monitoring its own systems and performance autonomously after the *Times*'s move to the cloud. But on election night, all would need to move together as one, whether battling breaks in data systems or supporting deft product changes such as lifting the *Times*'s paywall for the night to attract even more readers. They needed a general—a "directly responsible individual" (a term that Steve Jobs made famous for Apple's products). For the *Times*, it would be Rockwell. Coordination was key, and the planning began months in advance.

Rockwell and his deputies had the teams prioritize the systems that would keep three essential workflows going on election night: constant home-page updates, publishing articles, and pushing out alerts. They'd stress-test them, simulating ten times normal traffic, then fix them as needed to handle the loads. They built playbooks for recovery they might need and ran timed drills for every failure scenario they could think of, all forty-three of them.

Come election night, none of the disasters happened, even with the *Times* handling 237 million readers during election week and 120 million on Election Day alone. The replatforming for precision paid off, coupled with team management across diverse silos and domains, and Rockwell's leadership.

The close of the 2020 election cycle found the *Times* in an enormously strengthened business condition. Headlines from May 2020 ran, "*The New York Times* Tops 6 Million Subscribers as Ad Revenue Plummets;" then on November 5, 2020, "*New York Times* Hits 7 Million Subscribers as Digital Revenue Rises." A "Trump bump," COVID concerns, and data-driven control over its paywalls and offers added 2.3 million digital subscriptions in 2020 alone, bringing total subscriptions to its digital products and print newspaper finally to more than more 7.5 million, with revenues of $1 billion. All this built up from the *Times*'s first paywall of 2011, an altogether daring move taken in the face of a paltry half million subscribers, when CEO Mark Thompson first saw that red light flashing of collapsing ad revenues and plateaued subscriptions.

"A tech org should be evaluated according to the business metrics by which the company itself is assessed," Rockwell argued in his farewell post on *Medium*. "A tech team that performs masterfully against misaligned goals is just a terrible situation." By that standard, Rockwell and the precision teams he helped to create produced vast amounts of value both for the *Times* and for the millions of readers it serves.

## Secrets of Managing Teams of Precision at the *Times*

For the manager of enterprise precision systems, Rockwell's travails offer keys to success to all managers of precision systems. First, *get your organization right*. For a precision system to work, you've got to fix your hard technologies—tools, data, and platform—and your soft technologies—job design, workflows, and collaborations. Rockwell made major moves to fix the hardware first—the teams' data and platforms. The *Times* had fantastic data locked away in silos and cylinders, and bungled up in slow-moving analytics. For their core business to work, they would have to free it. Combined with reforms of its soft technologies, the move to Google Cloud Products enabled the *Times*'s data science, engineering, and product teams to flourish.

Second, *guide the journey to precision with a single strategic beacon*. There will be plenty of conflict and confusion ahead—fiefdoms challenged, platforms consolidated, teams in coordination. For the *Times*, CEO Thompson's declared "north star" of building subscriptions gave everyone a clear idea of purpose against which every project, initiative, and next best move would be tested.

Third, *manage cross-functional precision teams for speed and accuracy*. Engage staff with new job designs, work that sparks employee energies, autonomy focused on big company goals, and team solidarity. Assign the teams narrow-band, precision OKRs and KPIs that trigger the best on-task work of specialists, together. Ladder all on the shared, single corporate purpose. Deliver tangible wins that gets results in the hands of professionals fast.

Fourth, *build in infrastructure and process for design, test, and prove*. Precision value counts on precision results, whether as OKRs or KPIs.

Fifth, *go out of your lane if necessary to manage cross-functional teams of precision*. No one at the *Times* embraced precision systems per se, and they might not have understood their work through this lens. But they would demand solutions that could be achieved only with a level of precision made possible and built on reform of enterprise soft and hard technologies with precision in mind if not the goal.

Sixth, *measure precision systems by the same business goals as those used by all your business units*. But then go beyond to explore the wealth of possible unanticipated outcomes.

Seventh, *use precision systems to eviscerate and break through the limits of old trade-offs*. Well-managed precision systems can produce new and better optimizations—for example, lifting both subscription number *and* overall corporate revenues even when traditional metrics like page views decline and ad revenues sag.

Eighth, *cultivate a culture of precision as a power resource*. Unified under the single, powerful strategic beacon and shaped by the daily work of numerous teams employing new infrastructures and workflows, an emergent culture of precision became a potent resource in the hands of the *Times's* managers. A culture of precision built on the truths of science and on the possibilities for shaping a better world helped Rockwell and his team of teams achieve that balance and create a *New York Times* ready to thrive in the twenty-first century.

## CHAPTER 9

# BUILDING THE PRECISION ENTERPRISE

Executives can use precision systems in simple, relatively modest ways—to solve specific problems involving individual departments, for example, or particular activities within an organization, such as sales, marketing, or customer service. In this limited role, precision can be highly valuable. But precision systems are at their most powerful when they are adopted across the breadth of an entire enterprise, transforming its business model and enabling a range of new value-creating activities.

Of course, building a precision enterprise is a big job. Whether at the *New York Times* or the Biden-Harris campaign, across an enterprise, at scale, there's no flying under the radar. From start to finish success calls for a full-court press, with in-depth awareness and active support at the most senior level as well as execution carefully planned and monitored at every level, from middle management to the front lines, all roles laddered and aligned on a clear guiding purpose, managed through to delivery as promised.

This means that enterprise-wide precision initiatives bring with them both enormous potential value and significant risk. Stakeholders both inside and outside the organization may push back against the change, either through passive resistance or active sabotage. Models that work well in the pristine lab can falter in the rough-and-tumble real world. Learning loops can break down so that the system becomes outdated and when relied upon even dangerous. Seemingly flawless plans can be derailed by both "known unknowns"—things we anticipated as problems—and "unknown unknowns"—problems we never even thought about.

In this chapter, we will explore four enterprise-level precision systems, explicating their essentials, from conception to execution and sustained success. As you'll see, every enterprise move to precision has at least three requirements.

First, *nothing is possible without first gaining insight to the true state of the world.* The great precursor to precision is transparency, whether behavioral on the street, physical in the office or plant, logical in the cloud, or social on the digital platform. Transparency requires data from many sources, cleaned and ready to be joined up. In the hands of data scientists, software engineers, business users and ultimately smart analysts it must reveal and describe to managers the true state of the world they face.

Second, *the results you want don't just happen; they must be managed for.* That means resolving any conflicts about *what* to do and any confusion about *how* to do it. That takes a clear beacon—a north star goal that all can see—clarifying CEO intent and breaking work down into component parts that roll up into the goals of precision. It requires if/then predictions and good models that reveal options, expected impacts, and the path forward.

Third, *every stage in the work requires constant adjustments for impacts good and bad*—aligning and realigning people, platforms, data, and performance for precisely the change managers want, where, when, and how they want it, and no other. As managers bring change forward, experimenting will reveal what works—teams make discoveries, models get smarter, and the path that was once hazy becomes clear—until it becomes hazy again. Precision requires that managers incorporate those insights continuously with thoughtful experimentation, and frequent adjustments.

Following these resolves, executives and managers can create impressive precision systems from a few relatively simple ingredients known to all managers—some new transparency, new accountabilities, new expectations clearly communicated, a continuous stream of accurate data, and concerted leadership attention. Advanced technological tools like artificial intelligence (AI) and machine learning may or may not be required—most leaders of change through history have had no such tools available, yet they have still succeeded. In the real world now, with data so important, technologies like AI and machine learning can be crucial to marketplace advantage or battlefield success, and to changing the tradeoffs managers face. But even without that volume of data, leaders can sometimes do what they do best: play hunches backed up by only rudimentary data and systems. With single-minded purpose, a clear sense of what should happen next and how, and some luck, they may prevail.

That's what happened with Bill Bratton at the New York Police Department (NYPD)—not once but twice.

## Bill Bratton Brings Rough Precision to the NYPD

In 1994, New York's new police commissioner, Bill Bratton, confronted crises in crime, disorder, and sinking views of police and policing. Confident he could get better performance out of police officers and urged on by a new mayor named Rudy Giuliani who, as a former prosecutor, was used to telling cops what to do, Bratton embraced a new approach to crime prevention under the rubric of broken windows theory. Originally articulated in a 1982 article by social scientists James Q. Wilson and George L. Kelling, this theory held that small signs of disorder—broken windows that go unfixed, for example—produce a sense of social breakdown that contributes measurably to the rate of crime in a community.

But perhaps the most important innovation Bratton brought to the NYPD was a new commitment to precision in the use of crime statistics. Acting with Bratton's full authority, two Bratton deputies, Deputy Commissioner Jack Maple and Chief of Department Louis Anemone, both tough street cops, relentlessly demanded transparency from commanders to the true state of crime in their precincts, with data broken down street by street, block by block. Maple, in particular, pried open the spigot of NYPD data, forcing precincts to use a standard reporting scheme, schedule, and platform. The result was week-to-week data and analysis that, while not much more finely grained than the numbers you'd see on the back of a baseball card, utterly transformed the game.

Bratton and his top lieutenants processed the data and its use in the hands of police commanders through a world-famous accountability system they invented called Compstat. The weekly Compstat hearings became the basis for regular excoriations (and occasional salutes) of commanders in front of all their peers, with Maple and Anemone literally pounding the broad table in the cavernous meeting room until they delivered better results. Then the commanders returned to their field offices and did the same with *their* subordinates, the supervisors, uniformed cops, and detectives.

Sure enough, New York crime rates began to move in the right direction, falling 10 percent during Bratton's first year. It was that easy, or that hard, depending on how you looked at it. But with those results, the NYPD had little reason to experiment further because it knew what worked: the rough precision of Compstat- and data-driven policing. All that mattered was that Bratton got it done, even making the cover of *Time* magazine. For that press coverage, two years into his term, in 1995, Bratton was promptly let go by a mayor who felt upstaged by his police commissioner for taking credit for reducing crime in New York, as he did.

During this downshift in reported crime, there was something else going on that would help Bratton manage to a new kind of precision. Unknown to many, in 1994, the same year Compstat was born, a second data revolution began under Bratton, this one comprising surveys and focus groups convened *by the NYPD* to plumb the basis of resistance or support for changes Bratton might make among both citizens and rank-and-file cops. These surveys guided Bratton's incoming agenda and let him take daring chances on the path to reform and transformation, knowing exactly how to stay in the headlights of his support among New York citizens and police officers, the better to win over both.

Precise as they were, the surveys were expensive to build and deploy, and they were never adapted for everyday use or married into Compstat as performance measures. Once the initial probes were completed, no one built models to test further and prove suggested changes or learn whether some options worked better than others to move the needle on citizen sentiment and workforce engagement. Nor was any commander called to account for such measured progress. Compstat stayed uniquely focused on serious crime. Although Bratton arguably held in his hands precision solutions at once scalable, repeatable, sustainable, and enterprise-ready, without building these metrics into Compstat the opportunity was never securely anchored but rather subject to change and disuse by subsequent police commissioners.

In 2014, Bratton returned for a second stint as NYPD commissioner, this time under "reform" Mayor Bill de Blasio. Bratton inherited a department that, in the intervening decade, had taken Compstat and its single, crime-busting metric of police success to extremes. For ten years under the previous mayor, Michael Bloomberg, NYPD officers had ruthlessly expanded an age-old police tactic called stop and frisk, roughing up nearly 1 million principally Black and Latino New Yorkers nearly 5 million times. If ever there was an attempt at blunt force change with all its consequences, stop and frisk was it.

After a decade, stop and frisk had turned many New Yorkers against the NYPD. Yet untethered to measures of citizen sentiment or workforce engagement, only the corrosive drip and then a deluge of mutual ill-will and contempt finally caused change in police practice, brought on not by introspection or reconsideration of data but by election of a new mayor on a plank of police reform and his appointment of Bratton as New York City's new police commissioner.

As before, Bratton was resolved to achieve his mayor's wishes. Bratton's 2014 surveys of citizen sentiment and workforce engagement showed the damage done. Middle-aged Black and Latino women, for example, were especially hostile to the NYPD—not because they themselves had been stopped but because their male children, friends, and relatives had been. Police officers, too, were rich with grievance. Even with enviable pay, benefits, and pensions, many felt unappreciated on the job, even unsafe. When polled, they said (in alarmingly large numbers) that the department was on the wrong track, and they would never take the job today if offered it.

Even with these invaluable insights the two surveys were impossible to use in daily operations. They were a one-time only affair, describing the general state of sentiment rather than a dynamic taking of the pulse of citizen or workforce satisfaction not even block-by-block, or month-by-month, let alone incident-by-incident. Without that fine-grained, real-time data, it would be impossible to see what went "wrong" or "right" from patrols the night before, for example, and fix it or replicate it. The citizen survey was completed within five months, but the workforce survey took upwards of a year—Bratton's first on the job—much to his chagrin and impatience. All the while, Compstat remained the only performance management game in town, unchanged to incorporate impacts either on citizens or cops, to be modified only by Bratton's order that Compstat track the number of stop-and-frisk incidents so that they, too, would soon decline.

In Bratton's second term, Bratton again turned to two long-time cops to help bring his vision to a reality in his second term. (In the intervening years Maple had passed away and Anemone had retired.) Dermot Shea, Bratton's Deputy Commissioner for Crime Control Strategies (and who in 2019 would himself become New York City's police commissioner) ran the numbers side of Compstat each week with a keen eye for results, pressing for follow-through on missed opportunities. Philip Banks, NYPD's chief of department, would keep his eye on police street conduct. Most importantly, Banks had a theory of why Bratton's plan would work, and he regaled officers with it each week. If officers would ease back on hard-edged tactics, Banks said, more Black and Latino residents would report crimes, testify in court, and share their intelligence with police. That would start a virtuous cycle: make officers more effective, build more trust between officers and citizens, improve outcomes in court, create safer streets, build more trust . . . all the while manifesting the rule of law in every encounter.

The theory made sense. But in reality, everyone was playing hunches. NYPD never introduced Compstat-ready *noncrime* metrics that commanders could report and discuss—how citizens felt about every and any encounter with officers, for example, and whether they trusted police and policing more or less, and how officers felt about encounters with citizens and whether they felt more or less engaged on the job. The department never systematically tracked witness failures or intelligence gaps, whether to learn from success or call commanders to account for failures. The NYPD could have engineered some tools and techniques to measure all these dynamics at a fine grain to see whether Banks's "virtuous cycle" was kicking in, as predicted, but it never did. Without those measures, it would be impossible to show commanders the way forward to improvement.

Compstat remained a dull-edged, blunt force instrument of change, "precision" only to the extent that it used some data—any data—to move deployments and tactics. True precision solutions would only evolve in the closing days of Bratton's tenure. Bratton was well aware of innovations in the corporate world, as he had always trained his eye outward for lessons to be learned. There, with the cloud, mobile and social technologies and applications in hand, precision solutions had come into widespread (if early) use, addressing issues central to Bratton's own vision—for example, giving fast insight into customer or citizen sentiment, critical issues in workforce engagement such as "shock events" that could trigger workplace behaviors from withdrawal to aggression. In the right hands, they could deliver high-performance, high-quality services, just as Bratton wanted for citizens and cops.

None of these tools made it into the NYPD toolkit in time. They were essentially unknown to Bratton's inner circle in either of his terms, where staff and consultants fell back on the proven playbooks of years gone by, and resisted changes they neither understood nor wanted. Twitter was regarded by key aides as a frightening security risk—but Bratton liked the prospect of communicating directly with his cops and the public rather than mediated by the city's tabloids. Mobile phones seemed to key aides to put too much power in the hands of individual cops—but Bratton liked this, too as a symbol of his trust and his insistence on "arming" his cops with the latest tools. The cloud was unknown and suspect as somehow more insecure and less trustworthy than the NYPD's own systems—even though the intervening years had left the NYPD's information infrastructure with literally no redundancy and highly vulnerable to attack. From advances in precision polling of workforces and citizens from political campaigns, to social media tools for recruitment and hiring practices, Bratton alone appreciated the power of information and communications technologies to do exactly what Bratton wanted: engage citizens and cops in meaningful work together and the reform of their relationships.

What Bratton did possess was a keen political intelligence, a belief in the virtues of building citizen trust and police engagement, a conviction that a shift in platform was essential and possible for Bratton and the NYPD to engage its officers and citizens, and a driving sense—optimism, really—that more could be had than simply dusting off the playbooks of twenty years earlier.

Overruling doubts of his core team of advisers, for example, Bratton achieved a revolution in policing by, in fact, moving the NYPD onto social media—not just the one central Twitter account that he had inherited from the Bloomberg years, but over 100 accounts that Bratton established for every command. Against the practice of central control of communications and contrary to his CIO's counsel, Bratton approved a policy of smartphones for every officer—even though he was thwarted at first by a questionable acquisition of Windows-based mobile phones just weeks before Microsoft announced it was getting out of the business. But Bratton's move to innovative technology solutions took him as well to proving and testing web-based platforms like Ideascale, where citizens posted and engaged officers about specific neighborhood issues. Modified for the NYPD's peculiar workflows, supervisors could track inbound issues as "jobs" acting upon these matters and closing them similar to traditional units of police work.

If Ideascale could tell officers *what* citizens wanted, Bratton still needed a measure of *how well* officers met the needs of citizens—something trackable and actionable, something that could be managed within the Compstat process. He commissioned a proof of concept with New York City's 311 executives to robocall citizens for satisfaction with 311-dispatched police services, the same day as service was delivered. With 90 percent or more of police workload being service-oriented rather than requiring a rapid response to crimes in progress, Bratton saw service requests as potent, unmined terrain for improving citizens' views of police.

In the borough of Queens, for example, the NYPD under Bratton's successor James O'Neill collaborated with 311 executives to survey by robocall citizens who had called 311 to report their driveways blocked—a frequent complaint by homeowners where street parking was tight, and a personal driveway essential. Using randomized control trials, commanders were shocked to discover 50 percent of citizens were dissatisfied with the NYPD response they had received that day—in part triggered by shortages of commercial tow trucks simply unable to keep up with demand. That pointed the way to prospects for gain from new towing arrangements.

Throughout Bratton's tenure, this same tug of war played itself out time and again—old forms and ways constraining a leader eager for progress on old problems that persisted, begrudging use of the test-and-prove techniques among commanders who simply wanted to "check the boxes" off their to-do lists, new forms of precision made possible by now off-the-shelf technologies that had emerged but which an old guard of consultants and managers mistrusted, oblivious to the possibilities for change that precision systems and technologies offered, precisely as Bratton wanted.

All the same, New York City's bottom-line crime results continued to improve. By 2018, even two years *after* Bratton left the department, the NYPD made only 11,208 stop and frisks per year, a tiny fraction of what it had been a few years earlier under Mayor Bloomberg and his police commissioner. Crime was steadily decreasing, as it was nationwide. The fact was that no one could say why in a way that would provide guidance that others could follow. Even the underlying broken windows theory of crime had withered after years of academic doubt and political attack.

Perhaps the best that can be said is this: a modest recipe of some new transparency, some rudimentary data, new platforms (like Twitter and Ideascale) that softened the boundaries between cops, citizens and commanders, and a proven accountability process worked—led by a senior executive with the vision, legitimacy, and capacity to pull it off, driven by his own optimism and political savvy. Yet as it had in 1994–1995, the possibilities for reform that Bratton grasped and truly believed in passed as he retired from the NYPD in 2016. Pilots faded, lessons learned evanesced, staffs left, leaving the NYPD to miss yet another opportunity to institute durable, sustainable, and effective change, now based on precision systems, for years to come.

## Melissa Gee Kee and the Precision Transformation of Human Resources

As Unilever's global head of college recruiting, Melissa Gee Kee kept Unilever's talent pipeline rich with high-quality prospects; 160,000 employees fueled Unilever's engine, keeping 2.5 billion consumers in 190 countries coming back for more of its 400 brands. But by 2015 Gee Kee had become concerned that she could not guarantee future results without major change. The future was just too uncertain; disruption was everywhere. Millennials connected by social media in the marketplace for products and the workplace for jobs had more power to choose—and move—than members of any other demographic before. In a global world changing fast, Unilever had to be prepared for anything.

Gee Kee believed that the tools she and other human resources (HR) professionals relied on in making hiring decisions were no longer adequate. The typical résumé was backward-looking. But with the future uncertain, it was unlikely that anything on the résumé of a college graduate predicted success in an unknown future world. Gee Kee couldn't afford to keep priming the talent pump with guesswork. By contrast, the interview was forward-looking—but rife with potential for bias of all kinds.

The status quo process was locked in on collecting résumés—but then most were tossed in the circular file. Recruiters were locked in to visiting only a dozen "surefire" schools and having passing contact with another 200. Half of the job applicants abandoned the 100-question application or bailed later in the six-month hiring pipeline. Just one in six final stage survivors was offered a position. This was a brutal hiring funnel, serving no one's purposes well.

Gee Kee needed forward-looking insight that provided precision in assessing every candidate's potential for Unilever, matching their attributes today to qualities that would make them invaluable tomorrow, no matter what the future brought. She also needed a way to deliver to new employees a Unilever rich with rewarding, purposeful work. Survey after survey showed that's the promise millennials required.

"I was given the great opportunity to refresh our college recruitment process," she told an audience recently. "Instead of refreshing, we ripped it out and decided to start all over again. We were looking for a very different type of talent to grow our business. To be purposeful."

Gee Kee was confident in her support—led by two executive sponsors, one from HR, one from the business units. Both were dissatisfied with HR and willing to back Gee Kee's new way. "They flew cover, gave us support, while we developed in stealth mode," she recalls.

Gee Kee set to work to create a precision hiring system for Unilever. She would use some of the new diagnostic tools on the market to discover the attributes that mattered, cast the net wide to scoop up candidates who might have them, validate them through a battery of AI-driven assessments (a Harvard Business School study had found six methods more effective than résumés), present them for hire, and on-board the chosen few fast.

She was confident in her discovery process. It fleshed out traits that mattered, pulling data from focus groups, surveys, and performance reviews, building model personas to search for. One she called Sandy, the other Sam. And she was confident she'd find and target her Sandys and Sams in the recruiting pool, using Facebook ads that targeted particular segments. "Unilever marketing helped us learn how to better segment our recruits," she says. "Understanding the different types of people we're trying to get, and the different channels to get those people."

She was also confident in her new platforms—all AI-driven; all self-service; all easy on, easy off. All were nicely built for the channel of choice—the smartphone—and the user of choice—millennials. She was confident in her new applicant process. First, the interested applicant clicked on the Facebook ad, headed to the Unilever online application, and spent five minutes importing his or her LinkedIn profile. Gee Kee's first automated scan looked there for obvious gaps and disqualifiers.

Survivors went on to the next stage—twelve online minigames built by the firm Pymetrics, testing for traits like logic, reasoning under pressure, and appetite for risk. Again, this stage was fully automated.

The survivors of the minigames—the top third of the applicants—clicked on Unilever's HireView platform where they provided a video of themselves. That video would reveal how candidates respond to business challenges while displaying vocabulary, facial expression, and question response speed—all clues about the candidate's talents.

So far, zero human contact.

"It was definitely a weird feeling to know that robots are sitting in judgment of you," says Jordan Vesey, twenty-one, a Penn State student who was one of the early users of the new system. But it filtered out bias; helped Unilever go wide into the marketplace, hitting tens of thousands of candidates with highly targeted ads; do it in the channel of choice; and then narrow down all to a small cohort of seemingly high-value/high-probability candidates. And do it fast.

Survivors of the video selfie stage made their way to the assessment center—a daylong session with executives and recruiters. This was the applicant's first in-person contact with Unilever. At that stage, the process, platforms, and precision paid off handsomely. Where formerly one in six had survived the assessment center, about four in six were now getting passed on for offers—and 80 percent of those were accepting. They hailed from 2,600 colleges, not 200, and offered an incredible diversity of background and perspectives.

"That's what we want," Gee Kee says. "To be representative of our consumers and really help us grow in places that we haven't been before. "

As for the long-term results, Andy McAllister, a Unilever director of supply chain, was skeptical at first, but soon became a believer, reporting that the candidates he reviewed were as strong as, or stronger than, the candidates he had hand-selected the prior year.

That's as good a start for an AI-assisted precision system in HR as you could hope for. The new system produced results equal to those generated by humans while offering, simultaneously, tremendous gains in efficiency, effectiveness, and equity, saving 75 percent of recruiters' time and freeing them up for more complex, interesting tasks. A hiring process that had taken six months now took two weeks, with far less frustration and abandonment. And Unilever rose to be the number one employer of choice in over forty countries—up from number ten.

Precision vanquishes old trade-offs. You no longer have to choose between speed and accuracy, between scale and quality. With well-managed precision systems, you get it all—speed and accuracy, scale and quality. That's value. And it's transformational.

## The Sensored Enterprise: Burlington Northern Santa Fe's Sixty-Year Precision Journey

From the world of smart manufacturing, sensors and big data drive breakthrough business performance:

- At a Siemens electronics plant in Amberg, Germany, for example, machines and computers handle 75 percent of the value chain autonomously—with 1,000 automation controllers in operation from one end of the production line to the other.
- At General Electric's (GE) Durathon battery plants, 10,000 sensors measure temperature, humidity, air pressure, and machine-operating data.
- At Coca Cola's orange juice bottling plant in Orlando, Florida, sensors and computers juggle a quintillion decision variables, mix 600 flavors, and produce the same crowd-pleasing orange juice in every bottle, no matter the grove where the oranges originated or the fact that new oranges are harvested only three months each year.

But as we saw in chapter 4, in the case of fuel theft at Pemex, the industrial world of sensors is no mere "wave of the wand." In railroading, too, sensors are not just nice to have—they are crucial yet difficult to get right for safety and efficiency in some of the most difficult operating environments on the planet.

In 2013, just outside Casselton, North Dakota, an eastbound Burlington Northern Santa Fe (BNSF) train loaded up with grain derailed as a westbound train carrying crude oil approached on the opposite track. Derailed cars from the grain train blocked both tracks. The locomotives of the crude oil train collided with them only seconds later. The locomotives derailed, along with multiple tank cars. Tank cars ruptured, oil spilled, and a large fire broke out. An explosion rose ten stories into the air. The crash scene was horrific in every way, tanker cars exploding, dozens more derailed, all seared into memories of employees, responders, and residents. Once extinguished, cleanup and an investigation to find the cause began. Shortly, a discovery was made within the wreckage. One of the grain cars' axles had fractured, setting off the entire chain of events. It had been forged badly in manufacture.

"I am a retired technologist from the aerospace industry," a citizen told a hearing. "I retired here and now I'm an environmentalist. I live along your [BNSF's] railroad tracks going through Flathead River. I hear them going all night long. And my concern is derailments. You're carrying highly toxic, flammable tar sands all night long, hundreds of trains coming through while everybody sleeps. It really worries me that you can get a derailment into the waterways that's gonna impact thousands of people's wells and go all the way down to Flathead Lake. How can you reassure me that you're going to prevent a derailment into the waterways?"

BNSF has been laying in sensors since it first patented some in 1956. They tell managers the locations of all its 6,000 locomotives somewhere along 32,000 miles of track—and more. If you look today, you'll see one of the most remarkable array of sensor technologies imaginable—in the sky, at trackside, on the rolling stock, and underfoot. They are part of a precision, sensor-enabled system that converts detectable features of BNSF infrastructure, equipment, and operations to signal, awareness, and action.

For the trains, trackside acoustic sensors interpret noises from the vibrations of roller bearings in train wheels. Thermal sensors placed every thirty miles track changes in bearing temperatures from point to point. Force-sensing systems detect when wheels begin to flatten and strike the rail with every revolution rather than rolling with minimal friction. Trackside cameras image passing trains, checking for cracked wheels, capturing 1.5-million wheel images per day, 3,000 per train.

For the rails, crewed and uncrewed "geo-cars" travel the lines, using ultrasound, radar, and machine vision systems that search rails for microscopic breaks while taking precise measurements of rail geometries. Drones may fly out 200 miles at a time with microimaging, looking for broken ties and bolts wrongly angled or missing altogether. A single drone flight captures a terabyte of imaging data. Ground-penetrating radar looks for subsurface saturation caused by moisture, poor runoff, and vegetation that can destabilize the foundations of rails and ties.

For operations, sensors keep track of the location, speed, and destinations of thousands of cars and locomotives. As trains move toward their destinations and enter the yards, computers optimize the release and placement of cars into newly reformed trains, keeping the number of switches required over the journey of a car to a minimum. They help managers decide whether to move some cars and locomotives off to maintenance or bring other cars and engines back into service.

As trains roar past at seventy miles per hour on the open prairies or creep through congested railyards, all this sensing generates 35 million readings each day. Data travels over a massive private telecommunications network, feeding a central data enterprise in Fort Worth, Texas. There, engineers, data scientists, and operations managers oversee the ingest of today's data from every acoustic, thermal, visual, and imaging sensor in the system, adding it to years of historical data, perfecting models that predict track and train failure. They search for patterns, test and prove theories, and build models that learn constantly, crossing off factors, adding new ones. Their mission is to *describe* the true state of rails and trains, *predict* congestion on the tracks or failures of rolling stock from "imminent" to "possible," and *prescribe* the best next moves that keep things moving and safe. That's the quintessence of a mature precision system in action.

"We tested a variety of data points, things like air temperature, whether the rail ties are wood or concrete, and rail age," recalled one systems engineer. "We were looking to see what combination of information best predicted rail failure mathematically. Air temperature, for example, didn't test well. It wasn't a good predictor, so it is not included in the final version."

The more precise the measure, the more precise is the fix. Three cars per day now trip BNSF's acoustic bearing sensors and can be flagged for maintenance or come to a dead stop out on the tracks. In January 2019 BNSF reported that over its first four months in use, the machine vision system had found cracks in fourteen wheels. Pointing to an image of one, a BNSF boss said, "That's pretty much a derailment."

Even on routine inspections, things go better and faster. Once in the hands of BNSF's mechanical department staff (its maintenance group), in past years technicians would ping a few locomotive wheels and listen—with human ears—for the right bell-like tone. Now, acoustic measurements are far more precise and can be applied to *all* wheels.

When brakes failed, someone had to get under the car, look at the brake pads, look at the air brakes, and find and fix the point of failure. That's dangerous work on a live train. Today, temperature sensors track variation in *every* wheel's heat at it passes trackside sensors. When there's no heat increase after braking (there should be *some*), that flags that car's brake system and axles to maintenance.

Instead of replacing sections of rail—*all* rail—on a fixed timetable, BNSF scans for and replaces just the highest-risk rails and puts the lower-risk rails on a watch list. Tracks that are tagged "red" by models in the system get fast short-term repair and are then taken out of service within thirty days.

All this is made important by the nature of the business of railroading. "What railroads have going for them is the 'coefficient of friction'—how little rolling resistance there is for steel wheel on steel rail, and therefore how cheaply freight can be moved across long distance compared with trucks," Marty Schlenker told us. Schlenker led 110 analysts and 150 contractors for BNSF's unified data and advanced analytics team. "But the other side of that bargain is—you can't change lanes. And there's a lot of miles of single-track mainline left in the country. Bottlenecks cause queues to form, and they take a while to unwind."

The investment in sensors out on the main lines mitigates those breakdowns. "The ideal," one industry observer told us, "is that the rolling stock never fails online. It is all maintained proactively in terminals between runs." That sensor investment is important as well for the safety of crews, passengers, and towns because it should mean far fewer human interactions with dangerous machines and better predictions of failure—provided the systems can model data from the tracks, locomotives, and boxcars, and send alerts to operators for just the right intervention precisely where, when, and how its required.

All of this is enabled by advances in technologies, made possible by everything from new metals and materials in rails to new control systems and sensing devices, to AI and machine learning helping make sense of it all. It is only made possible, however, by well-developed collaborations between IT, operations, and mechanical teams, from design to the fielding of these precision systems, changing workflows to gain the promised wins, relying on increasingly well-trained blue- and white-collar employees to manage it all.

Call it the move from reactive maintenance to preventive maintenance, to predictive maintenance. It all costs about $3.5 billion each year to operate. The value is that BNSF has reduced derailments by 60 percent since 2008 and reduced employee injury rates by 50 percent.

"It might be said that the driver[s] of improved outcomes w[ere]—'better steel, fewer track failures'" Schlenker said. "But the reality is that we started with trained employees, augmented that over time with various degrees of technology, including most recently AI. And as sensors are not prohibitively expensive, we're gradually increasing the density of the coverage on the network with those. The uses of artificial intelligence within the railroads that have been most successful so far is in this arena of improving railroad safety, the defect detection realm." There is, in fact, little doubt in anyone's mind that today's systems would likely have detected the flawed axle that triggered the Castleton catastrophe and perhaps prevented the derailment and disaster.

Not so long ago, injury, death, and environmental calamity were thought to be inherent to railroading—just the cost of doing business. Precision systems changed that equation. With sensors fully deployed, managers were able to ring up new wins for safety—now a core part of the BNSF mission—by driving hard on efficiency in the maintenance shed and in the yards. Imagine the realization: "The fewer inspections we make in the yard, the safer we become on the rails *and* the more profitable."

As with any precision enterprise project, making it all work is far from easy. Schlenker and his colleagues helped us understand some of the key elements that contributed to the success of BNSF's precision journey.

## Getting the Data Right

The variety, velocity, and volume of sensor-sourced data posed real challenges for BNSF. In addition to issues of storage and communication, most worrisome, Schlenker told us, is knowing when you've got a set of "bad readings." Different conditions in the wild—rain, snow, wind, heat, cold—can lead to different signatures of incorrectness. Each of those markers of inaccuracy must be recognized and programmed for, and the system made smarter to deal with them. "The system itself doesn't magically know what's happened," Schlenker said.

Among other challenges, this requires assembling data sets for use in training the AI tools that contain all the defects BNSF expects to find. With precision sensors providing microscopic views of problems like defective wheels—a handful of images mixed in with billions of nondefective ones—finding the needle in the haystack is incredibly complex.

"Car wheels can chip, crack, deform, and degrade in myriad ways," Schlenker says. "You need multiple simultaneous machine vision models to detect each of those conditions. If you put the detector out there, turn it on, and stand up your first model, you'll get that exact failure, but you're not going to get all of the failures." Continuous improvement is a given, led by people but executed by machines that learn. System accuracy demands it.

## Keeping Humans in the Loop

"There [are] lots of critiques of people as imperfect," Schlenker observes. But, he adds, "the automated sensing systems have their own imperfections. They improve on humans in some ways, not in every way. People are better at generalized observations: 'Okay, there's something wrong here. Hmm.' Where the machines are better at, 'I'm looking for exactly this type of defect.'"

A venerable railroad like BNSF has a tremendous wealth of human talent to supplement and enhance its digital precision tools. "Our counterparts in the mechanical department have a field education PhD in undercarriage and wheel defects for cars," Schlenker says. "They are invaluable to the education of the people who run the routines to develop image data sets and train image models."

Another task for which human input is essential is sorting true alerts from false ones. A centralized desk monitor at BNSF's operation center in Fort Worth monitors the problem detection system. As the system ingests data, it heads straight to models that can sound alerts.

The system is set to be generous, meaning it will accept a large number of false positives in order to avoid overlooking true positives. Humans step in to review all the "presumptive positives" that come back from the image systems. The operations team alerts train crews only after several levels of review confirming, "Hey, this is likely a true positive."

## Using Precision Data to Improve Maintenance Regimens

Inspections of track and rolling stock by optical systems, ultrasonic and force measures, and acoustic and ground-penetrating radars have increased, making the entire inspection regime faster, safer, and far more effective. Even when trained maintenance teams have their hands on locomotives in the yards from time to time, transformed inspection systems generate a much more voluminous flow of data—which means that repair people have a much richer report card of everything that's happened to the engine since it was last in maintenance. As a result, it's much easier to define and perform the work needed to keep the locomotive in service for a longer period of time.

## Using Precision Data to Investigate Failures

Recently, a broken wheel on a loaded BNSF coal car caused a derailment. That coal car had passed an image system not long before. BNSF investigators looked at the images and they were inconclusive—it wasn't an obvious miss. In the past, the investigation would have ended there. Today's far more extensive system for continually monitoring equipment permits a much more thorough analysis. A BNSF engineer can say, "Here's the last place where an automated detection system might have been able to find this failure. Let's go back and look at the history of this train and this axle passing this detection system." The result may be further improvement of the system that can make a future failure even less likely.

### Maintaining the Precision System in the Field Continually

Sensors in the field are part of a system that has been hardened to withstand the rigors of weather, vibration, and all manner of disturbance. Keeping them in good working order is a huge job. At BNSF, the field operations team consists of tens of thousands of people, most members of craft unions, many of them out in the field executing maintenance operations each day. Without this constant vigilance, the reliability of the data and the accuracy of the decisions based on that data would soon deteriorate.

## Using Precision Systems to Solve an Age-Old Railyard Problem

We've seen how precision systems are particularly powerful at providing customized and personalized services and solutions to customers and to the organizations that serve them. BNSF has found that these benefits can be used in a unique way to help solve an age-old challenge facing railroads: how best to balance a railcar's active service with essential maintenance downtime. It's a classic last-mile problem that requires systems designers to integrate their new precision offerings into the daily workflows of employees who are accustomed to traditional methods and likely to resist change.

In railyards like those run by BNSF, train crews have, since the dawn of railroading, broken down trains and reassembled cars in new combinations as trains grow closer to their destination, switching some for others until the right combination ultimately rolls toward each customer. Switching has always been costly, reliant on guesswork, subject to delay and failure, and dangerous.

Today, however, BNSF applies AI-enabled optimization techniques to the recombination of cars, reducing the number of switches required. That's good news for customers and for BNSF and, it turns out, for workers. Turning this potential into a reality is one of the challenges that Schlenker sponsored in his service planning role.

As in most last-mile problems, the devil is in the details. Every switch requires what's called a pin pull, in which a yard crew team must get between two cars and physically pull the pin that connects them. Thousands of cars are switched and pin-pulled in the BNSF yards every day—work that is physically demanding and exposes employees to risk and to the elements. Reducing the number of pin pulls would offer many benefits. The switch crews would have more time for other tasks, making the job of sorting the freight in the yard faster and more efficient. It would reduce the risk of injury—one of BNSF's priorities.

But reducing the number of pin pulls is easier said than done. The trick here, as explained by Dasaradh Mallampati from the BNSF operations research

(OR) team, is to concentrate on empty cars that BNSF itself owns rather than cars that a shipper owns. Shipper-owned cars have to go to a specific customer location. Switchers have to couple or decouple them physically, by hand, then rebuild a new train with other shipper-owned cars headed in the same direction. But the BNSF-owned cars can be controlled logically by BNSF yard teams. That eliminates physical switching, including pin pulls, speeding trains through the yard and reducing the risk of injury.

"The economics of avoiding that sorting in the middle of the route are compelling," Schlenker said.

The OR team, led by data scientist Pooja Dewan, devised a plan to use its optimization methods to make enhanced sorting possible. But they knew that convincing the yard managers to go along with it would be a hard sell. For these veteran railroaders, all change is a disruption. Dewan was very deliberate about how she presented the idea to them. First, she framed the win around benefits she knew the yard managers would value: faster, better service to customers; reduced congestion in the yards; less pressure to expand the yards; fewer headaches for yard managers; and a reduction in manual effort as well.

Second, the operations team would customize a pilot program for each railyard based on its unique physical footprint and constellation of cars and customers. Some yards, for example, were configured so that only one switch crew could work at a time. Any system that could take work off switchers' plates, enabling faster work on the cars that remained, would be welcome. The customization plans Dewan and her team developed would make this possible.

Finally, the OR team personalized the appeal to the yard managers. They showed how the new system would make their lives easier and help them hit their key performance indicators (KPIs), both by increasing the number of trains they could push out every day and by reducing the number of accidents and injuries suffered by their yard crew members.

BNSF piloted the enhanced optimization in five yards for several months. The new precision switching saved over 20,000 pin pulls. It was axiomatic that, when scaled, this would translate to reduced workforce injuries, a corporate priority. The gains to overall efficiency were also obvious—freight sped to customers faster and reduced system congestion as promised.

How did managers make this win happen at BNSF? Leadership that said do both—boost efficiency *and* boost safety—gave cover to an opportunistic team manager to go a little out of lane. The project energized a highly skilled OR team, giving them a chance to do something beneficial to all while remaining focused on the most important goals of the corporation. And when it came to implementing the project down to the last mile, customization of the pilot programs and personalization of the pitches to the yard managers both paid dividends. The precision system project at BNSF worked: for the customer, the project, and the enterprise.

## Precision at the *New York Times*: A Unique Recipe for Personalization

For our third example of what it takes to organize an entire enterprise around precision, let's return once again to the story of the *New York Times*. As you'll see, personalization would play an important role in this story, in a significantly bigger and deeper way than in the story about BNSF. Along the way, personalization would also generate a level of conflict that the leaders of BNSF never experienced.

As we recounted earlier, the forward-looking report, *Journalism that Stands Apart: The Report of the 2020 Group*, crafted by seven reporters and published in 2017, encouraged the blossoming of the *Times*'s digital future. By 2018, a move was well under way toward a new way of building content for the *Times*, away from constructing same-for-everyone pages like those found in the print edition, and instead moving toward collecting stories that would be drawn from the *Times*'s many "Desks," from sports to news, each commanded by an editorial team and staffed by reporters, with the interests of individual readers in mind. The initial vision was of a personalized home page for each reader, just like the print edition front page, but assembled uniquely for every individual.

As the process began, everything seemed to be going smoothly. But in a 2019 speech taking stock of the moment, the *Times*'s vice president of engineering, Brian Hamman, described that moment as, in fact, a calm before a raging storm. "We succeeded, right?" he asked an audience, somewhat grimly. "We have blocks, we have programs, we have lists, we are . . . done."

But all was not well. Hamman recalled:

> The thing that we did next, which is very, very dangerous, is we put the names of things on our pages for the first time—"U.S. Politics," "Opinion," "Diversions," "Editors' Picks." And as a result, we really pissed off the newsroom. We got an angry letter from the arts desk, which had an entire meeting to make sure that they had collected their thoughts in the way that was the angriest, that by putting arts and other content into something called "Diversions," we were becoming barbarians. Keep in mind, we had yet to personalize anything. We were still just in a germination phase. And yet we created a war within the building.

"We kept hitting a dead end," recalls Nick Rockwell, Hamman's boss. "And we kept finding resistance—legitimate resistance to monkeying with the layout of this page because it has curatorial value." The personalization team seemed to be taking editorial control away from the groups that had traditionally held it—the various desks at the *Times* and the editors who ran them—and to be giving it to algorithms. It had worked for Netflix and for Amazon. Surely it could work for the *Times*.

What Netflix and Amazon did not face as digital natives was 160 years of cultural history embedded in the lore, identity, architecture, and business model of the *Times* and its Desks, reporters, and editors. No one at Netflix or Amazon had ever created content more than a column label. They'd never made a call about what *should* be important to movie lovers or music fans—just who was winning the popularity contest that minute, and even that they turned over to a machine, as well they should. But for a proud news organization like the *Times*, there was 650 volts of cultural current running through the desks and their decisions. Touch that third rail at your peril.

Complicating matters further (or clarifying them, depending where you stood), a survey of the *Times*'s readers suggested that the move to home-page personalization was off the mark. "The news that you 'like' is actually not the value of the *Times* in readers' minds," said Laura Evans, head of data and insights for the *Times*. " 'Tell me what's the most important news that I should know about.' That's the *Times*'s value proposition. To drive toward a fully personalized home page would actually *fail* the reader."

That was true across all demographics. "It wasn't like, 'Oh, the old newspaper reader wants one thing and the young generation want something else,' " Evans said. " 'Tell me what the most important news is, right now and today.' This was universally wanted by people."

Even so, precision insight to paywall behaviors were revealing major differences among reader segments. Two big groups stood out—*nonsubscribed* readers just coming to the *Times*, and the *already subscribed* readers scheduled for renewal. Since the introduction of its paywall, the *Times* had treated all readers alike. But now new, precise data would shift the paper's thinking. Evans explains it this way:

> At the beginning of the ten-article paywall, it was just, "Get them all to read another article." But if it's your first time with us, for example, even if you read all ten articles right away, you probably aren't going to subscribe. The best call to action was *not* for a subscription, not to go *deep* at that one moment. It was really about getting them to come back again, to *broaden* engagement.

The implications were clear. For that fist-time reader, the *Times* would use "product" to encourage returns—a newsletter sign-up, for example, or a suggestion that they follow a particular story or author. "We wanted to get you to ten articles in a way that would make you more likely to subscribe," Evans says.

For the *already subscribed* reader, the strategy was different. In this case, *breadth* meant exposing readers to content they never realized was cool to have—like the cooking section. "You don't have many unsubscribed readers seeking out the *New York Times* for its Cooking section," Evans says. "But the people that do get to see it start really engaging." Other sections, like the

Games section, where the daily crossword puzzle appeared, and Wirecutter, the paper's home for consumer product information and evaluations, brought unique value to the *Times*'s readers far beyond the insights provided through breaking news stories.

These data-driven insights let the *Times* fashion very different offers based on where readers were on their individual journeys. Evans explains the approach:

> It's very *personal*, but not *personalized*—not full-on machine learning, algorithmically driven at the individual level. And those insights helped us develop new ways of thinking about cohorts of readers, the inflection points in their journeys, how we get someone there, the KPIs that count, and what moves KPIs at each point on that cohort's journey.
>
> You'd look at return rate for the next day or week, for example. Frequency, depth, and breadth—how many sections people were consuming, what percent of nonsubscribers were consuming that many. Those were metrics we'd look at.
>
> These insights didn't rule out algorithmic personalization. But rather than being targeted to the home page, it would be based on the article that brought you here. Most people enter the product at the article. *That* was the most important screen of the product. Your article page is your home screen. So how can you rethink *that*?

The article page could be a very appropriate spot for recommending additional content—but that content would be mediated by a human curator. At the time, machine learning-driven recommendations were especially prone to so-called evergreen blunders—pulling up articles that were "oldies but goodies" when what the reader wanted was all about *right now*. Evergreen content might be good for a media company like Disney, where 50-year-old Mickey Mouse and Snow White were still favorites, but not for a news organization, whose value lies largely in the ability to change continually in response to the events of the day.

Evans explains that machine learning (ML) could be ill-suited to dealing with trends and events that were brand-new. "All of this ML and algorithms had been developed *not* for something that you can't expect," she says. "It's great for a recipe, absolutely. But if we used ML, would the Harvey Weinstein story have been as big? Because we didn't have a rich history of people being extremely interested in things like that. You just have to be careful because ML uses the past so much. And a lot of people aren't careful about that. They're looking for the silver bullet. This is why there always has to be a human interlocutor to ensure that the ML call is not the end game—that should certainly happen before you expose it to the public."

Armed with this important new insight, the *Times* rolled out its own answer to the personalization challenge—one that also served to deescalate the culture war between the desk editors and the personalization team. Rather than giving every reader a unique home page, there would be a separate tab on the *Times* app called *For You*. That was an articles page where all the personalization would live, with stories curated by a combination of algorithms and humans. Top stories would stay separate under its own tab, curated by the newsroom in a largely traditional manner.

By 2021, soaring subscription revenues suggested that the *Times*'s precision strategy was working, perhaps validating the notion of helping the *Times*'s readers to appreciate the full richness of the paper's contents. In a midyear presentation to industry analysts, Meredith Kopit Levien, CEO of the *Times*, speaks about how readers were engaging across "a broader range of storylines" than in the past. "We're leaning into that breadth," she said, "both within our core news experience and across our adjacent products like Cooking, Games, and Wirecutter, by experimenting more aggressively with programming to expose more of our audience to the full value of the *Times*."

All this digital "leaning in" still made the stalwarts nervous, concerned the *Times* could yet lose its way. In 2022, retiring executive editor Dean Baquet told his reporters to turn the Twitter volume down—"Tweet less," he insisted in a note to staff, "tweet more thoughtfully, and devote more time to reporting." (Baquet himself joined Twitter in 2011, posted two tweets in 2014, but no tweets @deanbaquet since.)

"They spend all that money on editors and then people just write stuff willy-nilly online?" chided *New York Magazine*. "Whatever for?!"

Let's take a step back and consider the decisions that made the *Times*'s successful precision transition possible. Everyone knew what the goal was: CEO Mark Thompson had established the "north star" metric of new and renewed subscriptions – what Sean Ellis, a pioneer of "growth hacking," described as "the single metric that best captures the core value that your product delivers to customers." For Airbnb it was "nights booked," for example. Facebook's was "daily active users." For the *New York Times*, it would be "new and renewing subscribers."

The confusion at the *Times* centered on how to achieve those metrics. The technologists believed that readers would become subscribers and renew as they became engaged by stories targeted by algorithmic recommendation based on prior clicks. The editors believed that readers would become subscribers and renew as they became engaged by stories curated by editors based on their professional judgments.

This disagreement reflected a clash of long-standing cultural values with new technology values. From the perspective of a veteran news editor, the technologists lacked legitimacy to challenge the newsroom's hegemony over the home page and its content. In this view, the proper role of technologists was to provide back-office systems that kept the paper and payroll moving. From the perspective of the technologists, the lessons of Netflix and Amazon held out fast, proven solutions to personalization.

The conflict at the *Times* that resulted drew from different impressions of what customers wanted. But the emotional current of technologists, reporters, and editors illustrates some major factors that any executive or manager seeking to build a precision enterprise must consider. One is culture shock. As suggested by corporate culture guru Edgar Schein, technologists inherently tend to favor solutions that can be scaled by the millions and repeated with as little human intervention as possible. "A key theme in the culture of engineers is the preoccupation with designing humans *out of the systems* rather than into them," Schein writes. That was certainly the promise of the *Times*'s personalization by algorithmic recommendation as the technologists understood it.

By contrast, a core element in the culture of "operators," as Schein calls them—in this case, the *Times*'s editors—is a belief in the essential involvement of humans in the crucial work of the enterprise. "Because the action of any organization is ultimately the action of people," Schein says, "the success of the enterprise depends on people's knowledge, skill, learning ability, and commitment." Certainly, the reporters and editors of the *Times* had those qualities in spades, and they were justly proud of the *Times*'s reputation for journalistic quality built and hard won over generations.

The move to bring personalization to the *Times* based on precision brought these two cultures into collision. As Schein goes on to explain, when an organization seeks to "reinvent [itself] because the technologies and environmental conditions have changed drastically," as the *Times's Report of the 2020 Group* signaled they had, a clash between technologists and operators is highly likely. To their credit, the leaders of the *Times* took the disagreement seriously and worked hard to find a creative, constructive solution.

Another crucial factor is status quo bias on both sides. In this book, we've seen a number of examples of status quo bias—the familiar reluctance of organizational veterans to embrace a change that appears wildly less valuable than their present way of doing things. The resistance to losing their control over the content of the newspaper coming from the news desks is a classic example.

But the technologists also suffered from a form of status quo bias—what Harvard Business School's John Gourville calls "the developers' curse." Technologists were wedded to the idea of algorithmic recommendations—that is the kind of tool that digital-first organizations typically use. But the *Times*'s technologists likely overestimated the utility of algorithmic recommendations, perhaps just as much as the editors underestimated it. The technologists may not have appreciated that the kind of algorithm Netflix or Amazon uses to recommend superhero movies or romance novels might not work for a news organization trying to convey what matters in a world where stories, events, and trends are changing every day, hour, and minute.

Fortunately for the *Times* and its readers, the dustup between technologists and editors was resolved by the *Times*'s customers who, research showed, were pretty much as the newsroom suggested—happy with some algorithmic recommendation but eager for the recommendations of editors, too. This discovery reinforced the essential elements of the customers' engagement with the *Times*'s brand. In the discovery, it resulted in a mature, sophisticated withdrawal of the technologists from the fray of personalization—in effect, a strategic retreat—as the enterprise charted a path forward toward a unique precision future combining both personalization and expert curating. Laura Evans reflects on how the *Times* achieved this successful bridging:

> We had a really great management setup, with Nick [Rockwell] being head of tech, myself as head of data, and Meredith Kopit Levien as head of product and revenue. It was a three-spoke stool, a cross-functional approach at the highest level, this group working cohesively together. And the newsroom is that person sitting at the top of our stool—Joe Kahn, with an invaluable systems perspective for how news works and how we could be helpful in that. All the spokes of that stool there to support what news was trying to do. The newsroom *is* the mission of the organization.

There's little doubt that the "war" over the home page has left a lasting imprint on the *Times*. Today, the *Times* has become a news organization ready for a hybrid future, agile in developing and testing new products and services, having made the shift to an emergent culture of precision based on experimentation, continuous testing and learning, collaboration across disciplines, and personalized engagement. At the same time, the *Times* continues to value and elevate the skills and insights of the newsroom and its editors, no matter what version of the *Times* readers turn to—whether print, desktop, or mobile, perused at the suburban breakfast table or on the Choptle lunch line—the *Times*'s engineers' take on the typical new, Gen X, Y, or Z *Times* reader.

The business leadership of the CEOs and the collaboration among top lieutenants mattered. "All were of a mindset of, 'We should be evidence-driven and data-driven.' Without that," Rockwell said, "nothing would have been received with the enthusiasm it was."

## Summing Up: Lessons for Leaders of Enterprise Precision

Precision systems that scale up across the enterprise exist above radars rather than emerging as rogue offshoots or as skunkworks. They are visible, and they operate across multiple corporate domains and activities. Thus, they necessarily involve the CEO and senior executive staff members in crucial decisions about strategy for enterprise precision: its purposes, stewardship, staffing, readiness, viability, and effectiveness, and what investments to make next.

The essential work of enterprise precision systems comprises description, prediction, prescription, and continuous improvement, which involve the operational activities of scooping up data, analyzing them, making sense of them, formulating hypotheses about what might happen next, testing and proving, injecting the components of change, monitoring for impact, and continuously improving results. Big data, AI, and ML are central to these operations; in the hands of managers, they make precision change possible at an unfamiliar speed and scale. It is possible to select, from among the entire universe of voters, for example, those whom to target for change; from among all newspaper readers, whom to offer certain articles and subscription rates; from among millions of seedlings on the farm, which to select for elimination as weeds and which to nurture.

Precision systems can therefore be understood to offer among the highest levels of corporate performance possible, offer new gain with far less pain, and create more winners and fewer losers than status quo systems and strategy. Precision strategy sets forth the means by which the organization will achieve these goals and measures its success by whether it achieves them precisely as intended, hitting precisely the right targets with precisely the right treatment, achieving just the outcomes intended, nothing more, nothing less.

Although the advent of big data, AI, and machine learning make such radical change possible, the reform of enterprise relationships and capabilities is essential for these potentials to be recognized. Out of necessity, enterprise precision strategy cuts across and unifies diverse corporate domains and disciplines of engineering, marketing, and data sciences on behalf of a single, narrow, and bounded business purpose. Such transformation therefore requires developing new cross-boundary, cross-discipline teams and management skills, the reform of workflows and job design, the institution of tools and technologies for continuous learning, and laddering all via new team objectives and key results (OKRs) and KPIs on new performance metrics and outcome measures.

Across an enterprise, snippets of these activities and capabilities may already exist in pockets and stovepipes. However, leaders or managers of stovepiped precision systems may not recognize their systems as precision systems *qua* precision system or as pieces of a bigger corporate puzzle. Thus, they may be hampered in making the appropriate investments in that strategy or in understanding the critical linkages across all for desired outcomes, and they may in fact operate at cross-purposes.

Thus, the role of leadership of precision becomes paramount. The central task of the leader of enterprise precision is, first and foremost, to hold a mirror to the enterprise precision strategy so that all can understand its essential corporate purposes, the possibilities for precision and its value, their roles in it, their contributions to it, and overall progress toward it. The senior executive's responsibility is to light a beacon that guides all unequivocally to the goal and resolves conflict and confusion along the way. It is then to ensure the readiness of units central to the core activities; manage the collaborations of units and managers toward that end; measure all by their contribution to the single, shared, and bounded business purpose; and refresh leadership at all levels with awareness of the precision strategy and, if need be, with new talent ready, willing, and able to compete on precision.

# CHAPTER 10

# RED ZONES OF PRECISION

For a manager or executive embarking on a precision project today, good intentions are not enough; a myopic view of the project that fails to consider broader organizational or even societal impacts is an incomplete view. Fortunately, discussion of the many risks and challenges that face those who venture into precision is robust today. Thought leadership on so-called red zones ranging from bias to privacy abounds, and public understanding of these issues is rapidly increasing. In this chapter, we touch upon a series of risks facing those who are launching a precision initiative: what they are, how to identify them, and how to prevent them.

The impacts of being ill-prepared for or ignorant of the red zones of precision are normally mild in the scheme of things: a missed quarterly target, some disappointed customers, or the failure to launch a new product. But in other cases, the impact can be very significant, in ways that those who design the system may fail to imagine. Often the impact can be seen only when looking beyond the data and into the implementation with the real world.

A cautionary tale comes from the Children's Hospital of Pittsburgh, where a shift was underway from doctors giving traditional pen-and pad prescriptions to a computerized physician order entry (CPOE) system that promised huge gains in efficiency and significant reductions in adverse drug events from scribbled instructions and misunderstood dosing. By all accounts, the transition achieved these goals.

But something else also happened. The new CPOE disrupted the old workflow, and children arriving via medical helicopter started to die at higher rates.

Why? New delays cascaded from the start. Over the years hospital staffs had figured out the shortest, fastest path for the child facing crisis, readying the orders midflight, and having medicines at the ready upon landing. But the new process, they waited until the child reached the emergency room and only then started the CPOE process. Doctors went from taking seconds to minutes and multiple clicks of the mouse to enter the order. Bandwidth jams over the hospital wireless system froze screens and clicks. Medications sat in a new centralized dispensary rather than on a cart in the intensive care unit, where nurses could grab them as needed. The result was an astounding spike in infant fatalities.

Similar problems have occurred at other medical centers where CPOE was introduced. "We found that a widely used CPOE system facilitated twenty-two types of medication error risks," a University of Pennsylvania research team wrote of one such case. "Examples include fragmented CPOE displays that prevent a coherent view of patients' medications, pharmacy inventory displays mistaken for dosage guidelines, ignored antibiotic renewal notices placed on paper charts rather than in the CPOE system, separation of functions that facilitate double dosing and incompatible orders, and inflexible ordering formats generating wrong orders. Three quarters of the house staff reported observing each of these error risks."

Implementation improvements ultimately fixed the University of Pittsburgh's workflow problems, and child fatality numbers decreased. But some irreversible damage had been done, and learning occurred the very hard way. The biggest lesson was that simply inserting a new digitized workflow in place of a carefully developed—albeit imperfect—manual system without adequately analyzing the changes made at each step doesn't always work.

Technology reporter Hayden Field recently asked several leaders in artificial intelligence (AI) a straightforward question: "What's the single biggest challenge that those building and working on algorithms will need to grapple with in the next five years?" The question itself betrayed some truths. First, there are many challenges, enough to merit a debate about which one tops the list. Second, the time period of five years highlights how quickly things move in AI. Beyond five years, predictions are nothing more than blind guesses.

The nine interviewees' answers shared distinct common themes. Six of them highlighted the powerful and insidious ways human bias can compromise AI's potential for good. The other three focused on the difficulties of scaling and deploying AI solutions in the real world. All illustrated how precision projects that are thoughtfully conceived, intelligently planned, and carefully executed can still go wrong because of unforeseen complications arising from the irreducible complexities of organizational leadership and of life itself.

For managers of precision systems, each of the common red-zone failure points that we will discuss merits attention and planning. Sometimes the mistakes that happen when implementing a precision system bring harm to your organization, your customers and stakeholders, or society at large; sometimes they simply prevent a project from ever getting launched, leading to wasted time, money, talent, and other resources. Beginning a new project by scrutinizing the failures of others and learning from them in order to avoid repeating them is a good use of time.

## Red Zone 1: Bias and Ethical Problems

In late 2020, a trending topic of conversation about Twitter exploded on the app itself. Users had discovered some behavior by the app that was unexpected, reliably reproducible, and apparently racist. Whenever a user uploaded a photo of multiple people and post it in a tweet, Twitter would automatically crop the image for presentation in the home screen Twitter feed thumbnail in a way that selected and centered white faces and cropped out any nonwhite faces—no matter how the faces were arranged. The app worked that way even on light- and dark-furred dogs and on black-and-white cartoon characters.

The photo-cropping tool was an attempt by Twitter to deploy entity recognition—specifically, facial recognition—to make the humble administrative task of creating thumbnails for images slightly easier. The intention was no doubt to detect a human being automatically in a photo and center them, on the assumption that the human was the main subject of the photo. But the intention did not matter because the execution failed.

When the criticism exploded on Twitter, the company responded by acknowledging the issue while refusing to explain its cause. Otherwise, the tech giant did not enter the conversation, promising only to fix the problem—and assuring users that the feature had gone through extensive bias testing before its release.

Less than six months later, as the COVID-19 vaccination rollout dominated headlines in the United States and around the world, Stanford University Medical Center became headline news for the wrong reasons. Leadership had developed and deployed an algorithm intended to prioritize which staff members should receive the vaccine first, considering variables like age, workplace level of exposure to COVID-19, and guidance from the California Public Health Department. Somehow, the algorithm selected for early vaccination only seven of the center's 1,300-plus resident physicians who worked on the front lines of COVID-19 patient testing and treatment, instead prioritizing administrators and doctors working from home. The outrage was swift and justified. And like Twitter, Stanford University Medical Center apologized and promised to remedy the injustice, but it did not enter any discussion about how the blatantly inequitable algorithm had been released to begin with.

In her book *Weapons of Math Destruction*, Cathy O'Neil refers to this as the problem of "having hard conversations." Algorithms, she asserts, came into use to help us avoid difficult social choices—"Who deserves a loan?" she asks. "Who deserves homeless shelters? When is a teacher a 'good' teacher?" Today those questions are answered by algorithms and, she asserts, "not for any particularly good reason" and not by any consensus of values we debate and share. She forecasts, hopefully, that as algorithms proliferate, the hard conversations needed to address these issues will become more frequent and cause constructive friction in debate.

O'Neil guides us to ask: when is an algorithm simply being used to avoid hard discussions? If banks, health-care providers, and other gatekeepers of important goods and services can calculate individuals' worthiness for those goods and services via an algorithm (usually under the guise of process automation), then hide behind the algorithm's decision when challenged to explain it, individuals and society collectively suffer. This is not only an issue of bias but also of explainability.

Researchers at Harrisburg University withdrew a study claiming that their facial recognition software could predict criminality after it was quickly criticized in academic circles for several serious problems. As Harvard fellow and University of Pennsylvania lecturer Momin Malik noted in his summary of the failings: criminality is not an inherent trait that can be predicted; it is a judgment imposed on behavior. The study had also confused data about arrests and convictions with data about actual crimes perpetrated—which, as Malik notes, does not reliably exist. Finally, claims made in the study such as "80 percent accurate and with no racial bias" are simply implausible or categorically impossible.

Commercial applications of AI often don't extensively dedicate thought or time to these questions; they're pursuing other goals. But why don't precision projects do what they possibly can to prevent bias every time? The reasons are often mundane. Transformative projects tend to be developed under time pressure. Lean project management frameworks are designed to strip out unnecessary steps in order to ship the product as fast as possible. Sometimes this emphasis on speed is justified. Whether the goal is to be first to market with a compelling product, first to society with a lifesaving vaccine, or first to a crime scene before the perpetrator flees, there's often good reason to focus on getting a precision system built quickly so it can produce value for people rather than sitting on a shelf, gathering dust. But when the desire for speed leads teams to skip the essential steps of testing for bias and designing equitable, transparent, accountable systems, the positive impact may end up overwhelmed by the negative.

Data and systems that produce and perpetuate inequitable treatment can cause significant harm, often for many years. As we discussed in chapter 4, this is what Dr. Simukayi Mutasa learned when he investigated the system powering pediatric bone-age assessments and discovered that only white children from mid-century Cleveland had been included in the original data set. Correcting the bias in that data set was impossible, so Mutasa took the route of creating a new, more representative data set that produced arguably less biased and more accurate patient evaluations.

In other cases, the best solution may be to abandon the attempt to build a precision system altogether. Many cities and states are rejecting public facial recognition and machine-based recidivism predictions. And in 2021, Twitter released its answer to the problem of racially biased, automatic thumbnail creation—a revised photo-cropping system in which the user who uploads the image can design the desired thumbnail themselves, with no help from AI. Sometimes, simpler is better.

## Avoiding the Red Zone

Of course, even better than taking action to fix data or systems that produce unfair or harmful results is to prevent them from being implemented in the first place. What are the steps that precision systems managers can take to avoid this red zone?

A good way to start is by challenging your own assumptions about which kinds of AI work require investigation through the bias and ethics lens. People working in or around precision often assume—perhaps unconsciously—that no such investigation is needed regarding a particular AI project they are working on. Sometimes, this is because the purpose of the precision system seems like a fundamentally harmless one. That assumption can lead to serious problems later on.

The first data scientists and engineers working on Facebook's news feed product may have assumed that choosing which posts from a user's family, friends, and other sources to show first couldn't possibly pose dilemmas of ethics or bias. Because the choice didn't involve subjecting users to financial or other direct harm, the system designers felt free to focus on their core metric of success—engagement measured through clicks, shares, and likes. Yet somewhere along the line, Facebook's news feed became a vehicle for the rapid spread of misinformation, a market shaper for legitimate news publishers, and a driving force behind the bitter political polarization of the last decade. Whether that happened deliberately or through mere negligence, the societal consequences were real.

The people behind Stanford's botched vaccine rollout surely did not intend to displace frontline workers from their rightful place at the front of the vaccine distribution line. Most likely, the team was actually seeking to achieve *greater* equity by calculating eligibility according to a consistent framework rather than having humans—who were themselves also waiting for the vaccine—make the decisions based on their personal and professional judgments. But in their rush to implement the plan, they cut corners by minimizing prerelease testing and review—a mistake that negated the team's positive original intentions. In retrospect, prioritizing vaccine recipients was arguably a scenario in which using an algorithm was inappropriate. Rigorous ethical discussions incorporating the voices of various affected stakeholder groups could have achieved a more just outcome and avoided the uproar.

All this may perhaps be subsumed under John Gourville's concept of the "developer's curse," discussed earlier—a form of status quo bias in which systems developers who have sweated the hard work of innovation overvalue their invention, and underweigh its costs.

The lesson is a simple one: Regardless of whether you believe your precision project has ethical implications, it should undergo ethical investigation early and often. Twitter's racist thumbnail problem might have been avoided with what Desmond Patton calls "social work thinking" for AI. "Social work thinking underscores the importance of anticipating how technological solutions operate and activate in diverse communities. Social workers have long been the trained 'blind spotters' that user experience and AI design needs." When assembling a team to work on a precision project, data scientists and engineers are not the only ones needed—considering the range of users and stakeholders who will be affected is a skill of its own.

The academic arena of data science is also tackling the problem of bias and how to measure it. Bo Cowgill of Columbia University and Catherine Tucker of the Massachusetts Institute of Technology (MIT) have introduced "counterfactual evaluation" as a new approach to measuring bias in algorithms, without needing access to the algorithm itself. "If a company uses a new algorithm, will the change cause outcomes to be more biased or unbiased? . . . The question of algorithmic bias is also inherently marginal and counterfactual. In many settings, the use of a new algorithm will leave lots of decisions unchanged. [For example, a] . . . new algorithm for approving loans may select and reject many of the same candidates as a human process. For these non-marginal candidates, the practical effect of a new algorithm on bias is zero. To measure an algorithm's impact on bias, researchers must isolate marginal cases where choices would counterfactually change with the choice of selection criteria."

Although not all precision is AI, and not all AI is machine learning, machine learning does occupy an outsized role in discussions of risk because outputs created by machine-learning algorithms are far removed from human decisions. An emerging discipline of explainable AI seeks to rein in the distance by creating processes and methods to explain machine learning (ML) results. Explainable AI goes hand in hand with its partner accountability, and together they enable people to understand and trust machine-learning outputs.

An example of doing both poorly can be found in Apple's launch of their credit card product—Apple Card. To assign applicants with a unique credit limit, Apple used a machine-learning algorithm to predict credit risk. Applicants weren't told that this was how their limit would be calculated, and they weren't told how the algorithm worked; which factors it considered; or, most important, why they received the limit that they did. Curious customers who questioned the company were given stock answers by customer service agents that they were not responsible for the algorithm's output, couldn't explain it, and there was no appeals process to change it. Customers felt locked out of a decision that had a great impact on their personal finances, and they felt the company eschewed responsibility by hiding behind an algorithm.

To do better than Apple did, there are two main imperatives—even if you're not using machine learning. First, assess whether the use case you're solving is a low-stakes use case for explainability (for example, explaining why a certain show was recommended in a list of recommended shows) or a high-stakes use case (anything affecting people's health or livelihoods). Use that to develop appropriate explainability controls or even to change the approach if you're in the red zone. In the case of the Apple Card, the company could have generated a lightweight calculator on their application page, allowing applicants to enter different inputs and see what their credit limit would be based on those inputs. At the very least, they could have explained the different factors contributing to the algorithm and how they were weighted.

Second, create accountability structures. Just because a prediction, decision, or other output is made by a system doesn't mean that humans can or should be excluded from owning the impact of those outputs. In Apple's case, it could have provided an option for applicants to call and discuss or even appeal their assigned credit limits, giving customer service agents responsibility to adjust limits within certain ranges.

## Red Zone 2: Issues of Privacy and Governance

The world of digital technology has recently been dominated by discussions of privacy. Companies gather, analyze, and deploy data about us in order to achieve business results—more precise targeting of advertisements, product recommendations you're more likely to click on, service delivery methods that are more efficient, and so on. Throughout this book, we've showcased these kinds of results to illustrate what can be achieved through the skillful use of data. But now more than ever, people are becoming aware of how the data about them is collected and used. They're demanding more robust rights and controls regarding these data. They've become sensitized to the various forms of data abuse, ranging from the merely annoying (a custom-selected advertisement following you around the internet) to the dangerous (a government agency deploying mass-scale facial recognition to monitor and control the movements of citizens).

As a result, issues surrounding how privacy protections should be designed into a precision project have been raised to the highest priority. Leaders must understand and be held accountable for balancing the objectives of their organization and the technological projects they pursue to achieve them with the individual's right to privacy. In some cases, privacy concerns may rank very low on the list of priorities. When the subjects and objects of analysis aren't human—for example, in the case of Blue River applying image recognition to differentiate weeds from crops—privacy rights aren't at issue.

But at the other end of the scale, some applications involve data about people that almost anyone would consider highly sensitive. Personally identifiable information (PII) such as names, phone numbers, email addresses, Social Security numbers, and more, as well as sensitive personal information regarding topics like health, religion, and sexuality require high standards of security. Even when PII is removed, it can still be possible using statistical techniques to reliably reidentify the person the data originally referred to. In the case of Patternizr at the New York Police Department (NYPD), many steps had to be taken in the data preparation phase to remove not only identifying information about the victims, perpetrators, and officers involved in each crime but also any secondary information that could be triangulated to deduce identities too.

In chapter 9, we told the story of Unilever's AI-based hiring tools. In early 2021, the HireVue—whose system it was—announced they would no longer use facial recognition in its predictive hiring practices because of increasing concerns about the ethical implications and, in fact, a decreasing need to incorporate it anyway. The system had become accurate enough with natural language processing (text analysis) alone.

Between the two extremes of data about objects like plants and personally identifiable data about humans lie data about people or their behavior that are anonymous and contain no sensitive information. These types of data are subject to lower privacy demands. For example, in the case of article recommendations at the *New York Times*, no PII was kept in the data stores that scientists used to optimize the product experience. All that was left behind was pseudonymized clickstream data, unreadable to the human eye.

Privacy concerns extend not only to the data collected about people but also to the secondary data products created from them. Machine learning can be especially adept at creating privacy problems through its superhuman ability to draw deductions. For example, a software tool with machine-learning capabilities could observe an individual's internet-browsing behavior regarding causes and treatments for HIV/AIDS and deduce that they likely suffer from the syndrome. Such derived insights, whether confirmed or mere probabilistic estimates, are also considered sensitive personal data.

To complicate matters of privacy further, any organization that operates internationally needs to be familiar with its privacy obligations under multiple different regional frameworks. These laws, like most, are written in language that can be vague or interpreted in a variety of ways when applying them to the reality of any specific situation. Enforcement agencies such as Data Protection Authorities (DPAs) in the European Union or the Federal Trade Commission (FTC) in the United States are expected to provide clarity on how exactly these laws will be enforced, partly through working examples of actual enforcement. As this book goes to press, such examples remain sparse; regulators have made less progress in bringing enforcement action than the law's authors would have liked. Many companies are taking a conservative approach, choosing to apply the most stringent requirements to their businesses globally, to avoid fracturing operations.

The rollout of consumer privacy laws worldwide may be spotty, but consumer demand for such laws remains strong. People are increasingly aware, self-educated, and concerned with how data about them is collected and used, and they increasingly expect organizations to prove that they're following the rules. The large and growing public appetite for protection in the digital world applies to any precision project that concerns itself with people in any capacity—so precision leaders should give this privacy-by-design way of thinking center stage from day one.

## Avoiding the Red Zone

Organizations that fail to follow strict regulatory guidelines regarding the use of personal data may find themselves subject to legal sanctions. Of course, they also run the risk of serious reputational injury. The relevant laws can be complex, but there is a path to avoiding the red zone that can be more intuitive than it might seem. Here are four key steps on that path:

- *Be aware of relevant laws and comply with them.* Different jurisdictions have varying laws that govern what you can and can't do with people's data. The most widely known is Europe's General Data Protection Regulation (GDPR), which mandates an opt-in model under which people must explicitly consent to the collection and use of their data. The California Consumer Privacy Act (CCPA) is an American cousin that allows data processing on a slightly less strict opt-out basis. Laws specific to certain industries, such as the Health Insurance Portability and Accountability Act (HIPAA) in the United States for health-care data, may apply additionally, as may laws about data relating to children, like the Children's Online Privacy Protection Act (COPPA) or the European GDPR-K. When it comes to legal compliance, spending money on external counsel is usually a good investment.

- *Develop clear data usage policies and communicate them widely.* Privacy violations can happen unintentionally when employees use data without understanding their obligations. Developing policies about the use of data and spending time to train team members on them, in addition to making these policies publicly available, is an excellent way to calibrate expectations and prevent costly and dangerous errors. Try hard to articulate your rules and guidelines in a user-friendly way that's easy for everyone to understand.
- *Don't use people's data in ways they wouldn't reasonably expect.* Even in a field that involves myriad legal and policy obligations, there's still room for common sense. Most consumers understand the concept of trading some data in return for a service they deem valuable, but data-handling organizations often forget to hold up their end of the tacit bargain. Using my click data to predict a book I might want to purchase? That's a common paradigm of personalization that most people find acceptable. Selling that data to another company I have no relationship with so they can target me with advertisements? That's potentially an uncomfortable surprise that could make me want to end our relationship. It's also worth remembering that the law and consumers' expectations may differ, with the latter being sometimes more demanding than the former requires. It's best to stay in touch with how people actually feel in addition to the legal requirements and consider meeting whichever bar is higher.
- *Upskill on data privacy techniques for analysis.* In the relatively new field of *differential privacy*, analytical techniques are used to generate insights while preserving privacy. For example, Google's advertising technology framework called the Privacy Sandbox establishes a minimum size for any interest-based cohort large enough to prevent any single user from being identified personally, even across multiple cohorts. You or a team member may want to learn basic concepts of cryptography like *one-way hash functions*, which are frequently used to encrypt data with the goal of improving privacy. Whether or not these data privacy techniques are appropriate or warranted is a case-by-case decision (and this certainly does not constitute legal advice), but it is worth knowing that simply having data or not having it is not the only way to control for privacy.

The more you and your organization know about how to implement privacy pragmatically, the more confident you can feel about using precision systems to work with large amounts of data while still avoiding the red zone in which basic principles of privacy are violated. Legal, policy, and user research experts can support technical teams by bringing knowledge about best-practice privacy preservation to the table, which usually leads to decreased risk of making bad mistakes and to increased confidence in the quality of the precision system.

## Red Zone 3: Failing to Keep Humans in the Loop

Precision systems offer an advantage that comes with a significant risk: They allow humans to outsource the hard work of thinking to a technological tool. The problem is that, once the system is operating, the humans who made it often stop thinking about it themselves, with consequences that can be benign or disastrous.

This risk in AI, called *automation bias*, occurs when humans rely too heavily on systems to help them make decisions and consequently ask fewer questions, pose fewer challenges, and generally stop thinking independently. David Lyell and Enrico Coiera describe this phenomenon: "when users become over reliant on decision support, which reduces vigilance in information seeking and processing." They studied it in health-care settings, where automation bias was mainly observed in simpler tasks, for example, diagnoses, rather than more complex tasks, like ongoing monitoring, and made recommendations for health-care providers to avoid this risk by decreasing the cognitive load of the health-care workers operating the decision support.

Laura Evans of the *New York Times* had to beat the drum about this problem at the news giant. The product, data, and engineering teams collaborated over years to develop sophisticated algorithms to personalize each reader's experience and thus show each reader articles they're most likely to want to read. But that doesn't mean editors can take their hands off the wheel completely.

Algorithms can only perform based on what they've been taught, either through supervised training or through autonomous, unsupervised learning. The training data sets are crucial. If they're too heavily weighted toward one user type, other types won't be well catered to. In the case of news, the problem is the short-lived nature of the key artifacts—namely, stories. Other content types, such as Netflix original series or the novels sold on bookselling sites like Amazon, can be consumed months or years after production, but news goes stale quickly. That makes it hard for a model to draw reliable insights from users' reactions to it.

To cope with this problem, Evans's team made sure that humans stayed in the loop the whole time. "Since recommending older news content is actually not appropriate," Evans says, "and machine learning is not that great at handling time, we knew we had to use ML to push recommendations to the human curator, who could then make sure it's correct and push it forward."

In the world of real estate, Zillow recently faced the repercussions of overly relying on AI for what is still a mostly human process: home buying. After its iBuying program went on an automated spree of bidding aggressively on thousands of homes and buying them on behalf of the company in order to flip them, the market cooled off more quickly than Zillow's iBuying AI had predicted. As a result, the company now owned overpriced properties and had to begin selling them at a loss to institutional buyers. Zillow's CEO publicly attributed the error to inaccurate predictions about the market's continued growth, but in reality the failure was in the management of AI. Why were bots able to make these offers autonomously? Who was reviewing them, checking for accurate appraisals? Why did Zillow leadership go all-in on this AI in a way that put the whole company at risk? When the Zillow stock price dropped 30 percent after the publication of these troubles and 25 percent of the workforce was subsequently let go, should there have been leadership accountability for that?

Sometimes the stakes are even higher. At BNSF headquarters, leading engineers sit side by side with the data folks who are preparing the training data sets and creating the models, guiding them about what specific mechanical faults look like. In this way, the railroad's precision system combines the best of both worlds—the power of data and the wisdom of real-world expertise. After train machinery is flagged by the company's expansive sensor systems as needing proactive repair, a human engineer must still be involved to approve the machine before it goes back into circulation because the safety of BNSF workers and the communities surrounding the railroad lines may be affected by any error in judgment that the precision system might commit.

## Avoiding the Red Zone

Incorporating human judgment into the preparation, development, testing, and optimizing of a precision system—whether it relies on AI or not—is essential to integrating a precision system successfully into the real world. It helps precision systems gain the acceptance of the practitioner so that the decision-making power is ultimately hers. It assures that existential decisions, from the next play the coach calls on a football field to whether an armed drone will shoot to kill a target, stay in the unique domain and responsibility of humans. Evans explains the philosophy behind this:

> In the creation of [a precision system], there absolutely has to be a human. You've got your training data set. You have your outcome that you're trying to optimize. There should be an intercept there first, certainly before you expose it to the public. There should be that moment of asking, "Was this what we would have wanted, what we would have expected, what should have been?" You need a person looking at the outcomes and telling you, "Success, failure, success, failure, success, failure." You continue to apply that to your training set until you get all successes and no failures. Only then do you release your system. But people skip that step too often.

Furthermore, the performance and accuracy of those final decisions made by humans need to be evaluated continuously. Periodic testing coupled with user testing and qualitative surveys can help to identify and mitigate automation bias. In other words, for decisions that matter, humans must be responsible. That doesn't just mean they're clicking the button; it means they are engaged, understand what's going on, and take accountability for it.

## Red Zone 4: The Last Mile Problem

Andrew Ng, cofounder of Google Brain and deeplearning.ai and a well-known thought leader in the field of AI, focuses on another red zone problem—namely, "bridging the gap between proof of concept and deployment." Eric Topol calls it the "AI Chasm" problem—"the gulf between developing a scientifically sound algorithm and its use in any meaningful real-world application."

The challenges of bridging this gulf are many: generalizing from an algorithm that works well on a small data set to other populations "in the wild," transforming experimental code produced as proof of concept into code for production systems that pass regulatory muster, and the financial and time commitments necessary to mature systems to this next level. Never mind the difficulty of moving from pristine research labs to rough-and-tumble hospital settings where a little bit of jiggle in a machine setting can throw the imaging off. Bad data, old systems, wrong people, poor timing, insufficient support, invalid conclusions: there are a thousand ways a precision system can fail in bridging the last mile between a worthy project and successful deployment. Anyone who's worked in data can tell you at least a handful of horror stories.

Landing the last mile means correctly understanding what type of problem you've set out to solve and committing the effort needed to solve it all the way through, from end to end. "Despite many 'demos' and proof-of-concept implementations," Ng sums up, "getting an AI system deployed in a factory, hospital, or farm today is still a labor of love." But just as hope is never a strategy, labors of love call for heroes, who can be in short supply.

As we saw in chapter 6, Joel Rose, founder and CEO of Newclassrooms .org, has witnessed a similar problem in education technology. School systems around the country are filled with smart, determined people working hard to fix various problems in children's education, but they are failing repeatedly for one reason: No amount of improvement in the early stages of instruction design can save a system that fails at the last step—the delivery of precisely the right content, at the right level, and the right moment to each individual student. He is pursuing personalized learning as a solution to that biggest and most fundamental problem, having recognized that, until the last mile of personalized curriculum is fixed, education debt will continue to mount, and with it failure for children.

The skills required to build data into value—through AI or otherwise—are demanding, hard to develop, and rare. The supportive tools required are not yet fully commoditized, although companies like Amazon, Google, and others are striving to build them, meaning that data scientists still need to be part engineer in order to get anything done. When a precision system does make it live, it still requires ongoing maintenance and optimization in ways that traditional software engineering companies haven't been able to support fully.

As Ng puts it, the tech industry "needs to build out the tools, processes, and mindsets to turn AI creation and deployment of systems into a systematic and repeatable process." As this happens, presumably over the next few years, AI will become increasingly commoditized, and the gap between organizations who can bridge the last mile and those who can't will continue to shrink. In the future, when AI has been done away with coding to the extent that traditional software engineering has done already, those organizations that still haven't made the transition to precision will have no excuses left.

## Avoiding the Red Zone

If there were a simple way to avoid the perils of the last mile, people like Ng would already have found it. The obstacles are varied, unavoidable, and big.

In chapter 2, we talked about creating a good-enough precision system, and we used the analogy of the skateboard versus the Ferrari to explain what that means. The same analogy helps to explain why the last mile tends to be bumpy: a skateboard is not really very similar to a Ferrari. It's not even *part* of a Ferrari, ready to be incorporated without changes into the design of a high-end sports car. It is a totally different solution to the problem of getting from point A to point B, whose only purpose is to demonstrate that the trip is, in fact, possible. It will be discarded once it has served that purpose and you start working on building the full car solution. The path of a precision system from early proof of concept to real-world-ready system is similar: very often, the first iteration is not fit for release and must be completely reimagined.

The differences between the proof-of-concept prototype and the real system may be quite varied. Perhaps the data you used to power the analysis was initially pulled from databases as a favor by an information technology (IT) staffer, and now you need to build the real database export. Perhaps you tested it on a handful of willing test participants, and now you need to collect a large number of real-world users who will consent to be part of your next experiment. Perhaps you hacked together the results into a one-off PowerPoint, and now you need to create dozens of dashboards reflecting real-time data. Whatever the reasons, what you built for the proof of concept won't get you through the last mile. There's no quick and easy way to sidestep the push through the last mile, but there are some predictable requirements that can be foreseen and planned for:

- *Develop a realistic schedule.* The move from the lab to the real world is often harder and more time consuming than the initial win on the lab workbench. Leave just as much time for last-mile iteration as for the initial research and development (R&D), if not more.
- *Expect to require an array of new skills.* Perhaps the initial R&D work was heavy on pure science, while the journey to production emphasizes engineering. New models will need to be trained and monitored, experimental results will need to be evaluated, and fresh data will need to be pulled in for updates. This helps to explain why an emerging discipline called machine-learning operations (ML ops) aims to equip data scientists with traditional engineering skills and tools.
- *Focus on metrics-based milestones rather than timelines.* Avoid guaranteeing that you'll get the new system up and running in four weeks, three months, or any specific time frame. The deployment process is fundamentally nonlinear and doesn't respond well to traditional linear project management frameworks. Instead, you can communicate (for example) that the model can go live once it's reached 80 percent recall accuracy, or a 20 percent click-through rate, or whatever measure that matters to the goal.
- *Train the users.* The people who will be using your system need to be brought along on the journey, whether they are nurses who will use your system to improve their triage methods, safety inspectors who will use it to prioritize buildings in need of review, or talent scouts who need to find new artists to sign. Don't plan on tossing a finished solution over the fence and hoping that users will pick it up and use it successfully. Instead, plan for thorough training, continual communication, and much advocacy to ensure that the system is understood, accepted, and implemented effectively.

In a discussion for deeplearning.ai's YouTube channel, Ng looks into the future of machine learning and delves further into ML operations, proposing the heuristic that getting a model to approximately 70 percent accuracy is easy enough, while getting it above that threshold is significantly harder. Suggesting that too many practitioners home in on optimizing the model as a way to increase results, he proposes that instead they should turn their attention to the data itself. Cleaning, augmenting, and generally improving the data, Ng says, requires a new way of thinking—a systematic approach centered on the concept of the life cycle of a machine-learning project. Over the next few years, knowledge about how best to manage machine-learning products will continue to expand along with best practices and architectures, resulting in the eventual conquering of this domain and significantly easier deployment, optimization, and management of precision systems.

Until then, the last mile remains fraught. Plan accordingly.

## Red Zone 5: The Dangers of Mismanaged Debt

When innovating, whether as a lean start-up or an R&D unit within an established company, the push to get a good-enough product out the door is usually the top priority. Teams have to prove that an idea works before investing heavily in it because the cost of failure rises along with the cost of sunk effort. In order to prove the concept quickly, teams can and should take on certain kinds of debt. However, we aren't referring to financial debt, like a loan; we're referring to the deferred cost of decisions made during technology development in the interests of speed and progress.

Examples include *technical debt*. When shipping a minimum viable product, teams often take technical shortcuts. They hack together a quick workaround to a problem instead of spending the weeks required to build a solution in a scalable and enduring way—as they should. The lightweight solution appropriate for the proof-of-concept stage is not suitable for wide adoption or use. Before the final rollout, the debt must be paid. Leaders who are unaware of this, or willfully blind to it, will pay the price with crippling interest costs later—for example, the heightened cost of development for every new feature due to poorly designed foundational systems or the even greater costs resulting from outright system failure.

Another example is *decision debt*, derived from the choice to defer decisions to later dates. "We can figure that out after milestone X"—does that language sound familiar? Even worse, sometimes decisions are deferred so long that the people who have the knowledge required to finalize the outcome have moved on, leaving those who join the team with an inherited decision due and no tools to make it. Wisely accruing decision debt is an important tool for leaders as long as they don't lose track of all the decisions awaiting their eventual attention and as long as they maintain the information and context required to make the decision accessible to those who will eventually need it. When ignored long enough, deferred decisions turn to denial, and projects suffer as new decisions rely on old ones that still haven't been made.

Finally, there's *organizational debt*, which is especially important for executives. In the interests of an urgent test-and-prove effort, leaders will often pull skilled people from existing assignments to form a temporary team. Once the project has proven its value and is up and running, however, the temporary team is too often simply left in place, the preexisting teams aren't made whole, and the new team responsible for maintaining and improving the precision system isn't thoughtfully or deliberately rounded out. This results in all the organizational teams being weaker and less productive, with cascading negative effects.

## Avoiding the Red Zone

Examining technical, decision, and organizational types of debt through the lens of traditional financial debt yields commonsense insights about how to manage each one. It starts with recognizing that you're using debt to move quickly and understanding exactly what the debt is. Then it requires a plan for tackling the debt while keeping the original plan in mind throughout your future resource planning debates. There's no such thing as a credit score for technical, decision, or organizational debt, but many companies would be in for a rude shock if there were. Don't be one of them.

Jorge Bernal Mendoza reflects on the smart strategies needed to deal with debt for precision development based on his experiences transforming the digital organization of Pemex, Mexico's, nationalized oil company (discussed in chapter 4):

> So often, companies want all the whiz-bang benefits of AI or digital or whatever instantly, but they're not willing to pay the price. You must have a payment plan to pay down the technical debt you accrue. You can decide to prioritize small, consistent payments through dedicated foundational engineering time, or you can pay it all in a big chunk by doing an all-in rewrite. It doesn't matter, so long as you don't ignore the debt.
>
> And don't forget about the cash cow—the major legacy business that needs upkeep. For many companies, the new precision system will not pay the big organizational bills for years to come. So you need to invest in new technology without neglecting what powers your business today. Allocate your resources accordingly.

Again, machine learning raises unique concerns when it comes to technical debt. Researchers at Google examined technical debt in machine-learning systems, arguing that "ML systems have a special capacity for incurring technical debt, because they have all of the maintenance problems of traditional code plus an additional set of ML-specific issues. This debt may be difficult to detect because it exists at the system level rather than the code level." Some of the issues they noted were highly technical and mathematical risks; others were more accessible to the layperson—like "changes in the external world."

The concepts of technical debt apply to any technical system. The level of debt incurred and its complexity depend on a multitude of factors like the speed of progress and the number of shortcuts, the choice of technology and use of machine learning, and dependencies on external frameworks and tools. Going into a precision project with eyes wide open about how debt is accrued and having a strategy for servicing that debt are imperative.

## Red Zone 6: Problems of Measuring Impact

In November 2020, the one battle that mattered for the Democratic National Committee (DNC) was won: Democrats triumphed in the presidential election, and Nellwyn Thomas, Nate Rifkin, and their teams saw their tireless, data-driven transformation efforts pay off. They had managed databases of millions of Americans' contact details, honoring individual opt-out preferences for each. They had ensured that persistent campaign outreach did not turn into unwanted harassment while also keeping all those contact details securely stored to prevent significant privacy breaches. They had dealt with the last-mile challenges posed by COVID-19, which made traditional outreach via knocking on real doors impossible. And they had invested significant resources in dealing with debt, especially technical debt. Thomas's decision to devote years of effort to transforming the underlying infrastructure powering the DNC had been a strategic judgment based on the estimated cost of continuing to ignore that debt—which would be losses in 2020 and beyond.

Yet after all that, and despite the happy outcome on Election Day, Thomas and Rifkin's teams faced one more type of red zone that merits acknowledgment. There was no way to prove what worked to change hearts and minds or to hold fast the voters who might have wavered and voted Republican. National elections are incredibly complex phenomena. They are swift. There are many variables that help to explain why one candidate wins and another loses, but it is difficult if not impossible to test them methodically one by one to find out which ones were decisive. There is correlation, yes—we can see that people texted by the Tech for Campaigns (TFC) group, for example, turned out to vote 0.7 percentage points more than their counterparts in the same districts who did not receive texts. But can they prove causation? In the end, no one can be certain how any voter voted or why. That rules out randomized control trials, the gold standard of measured effect, and makes it "challenging to know if your texts are having an impact," TFC wrote. But TFC found that "response rates and sentiment are a strong proxy for later voter turnout." The same might be true of other metrics—the volume of volunteers signing up, the number of qualified citizens persuaded to register, and the percentage of phone calls resulting in donations are all, potentially, *proxy* metrics for what works.

In any political race, those proxies may be as good as it gets. And they may be good enough. In one highly competitive race, for example, Liz Snyder, a TFC client candidate in the Alaska House, won her 2020 race by only eleven votes. Texting could have made all the difference, perhaps driving even unexpected, unlikely voters or swing voters or younger voters to cast a ballot. All showed higher turnout rates in 2020 if texted compared to similar voters who were not. With texting so cheap (some platforms cost only $.01 per text)—who could afford *not* to text?

Underneath it all, research points to the risks of confirmation bias in action. In essence, campaigns have used techniques from direct mail to text for so long there is little doubt that they must "work." COVID-19 ultimately left little choice but to "believe" in online and phone messaging. In the final three months of the 2020 election, Tech for Campaigns estimated that Americans received more than 11.6 *billion* political messages by various digital and electronic means.

The problem of measuring the impact of precision systems is not simply knowing what works in the classic and limited sense. For example, the failure of computer-aided detection systems for mammography was a large-scale, multidecade disaster, as Eric Topol recounts: it was approved by the Food and Drug Administration (FDA) in 1998, used broadly, but then discovered twenty years later to fail utterly at improving diagnostic accuracy. It was costly to insurers but without "established benefit to women," as one major study found.

The problem of measuring the impact of precision systems is also knowing that precision works to reduce collateral impacts from the old ways of bringing about change while achieving precision where, when, and how managers want. That is the promise of precision, for example, for COVID-19 tracing in Singapore to be swift to develop, easy to deploy and use, comprehensive across the universe of possible patients, and efficient in producing scalable outcomes that manual tracing would miss. Did Jason Bay's TraceTogether do that—deliver accuracy, privacy, and speed as it scaled to millions and tied off the pandemic? Those measures count in assessing whether precision delivered on its full promise.

The same, then, is true for precision policing systems—if they promise early detection of patterns, as they must and can, did they deliver patterns accurately, quickly, and to scale? In doing so, did they reap the harvest of improved perceptions of police and policing from less indiscriminate policing—and if not, why not? If See and Spray promised reductions in pesticide use and improved crop yields by applying pesticides more accurately—did the See and Spray system achieve this optimization? If it did not, See and Spray managers are leaving value on the table. Did Zipline deliveries of vaccines mean that immunosuppressed patients would no longer have to traipse through infected populations to get vaccinated but could vaccinate in place with far reduced risk? That's a promise of precision drone delivery—did Zipline's system deliver that? If it didn't, that's worth knowing and fixing.

Those new possibilities differentiate precision systems from all others—the good that comes from achieving better outcomes and less harm—more of what you *do* want and less of what you *don't*—compared to the systems they replace. New precision systems must be given new metrics and measured to account for the harm avoided. In tallying the wins for precision, do the harms reduced count just as much as the positive gain? Together they should add up to astounding value. Did they? And note that with fewer negatives and broad-based positives managers can bring many more supporters to their side. Did they? Prudent managers can capitalize on that.

Another lens on value measurement and its purpose—continuous improvement—relates specifically to machine-learning products used in regulated settings like health care. Regulatory bodies around the world are just beginning to understand how to regulate AI. As we said in chapter 2, AI refers to computers programmed to make humanlike decisions. Machine learning, on the other hand, is enabling computers to learn independently and make decisions in ways that humans never programmed them to do originally.

These are the situations where regulation falls short today, and so, as Pearse Keane and Eric Topol describe, regulations often inhibit machine-learning systems from actually learning in practice. "Although these systems typically learn by being trained on large amounts of labeled images, at some point this process is stopped and diagnostic thresholds are set . . . after this point, the software behaves in a similar fashion to non-AI diagnostic systems. That is to say the auto-didactic aspect of the algorithm is no longer doing 'on the job' learning. It may be some years before clinical trial methodologies and regulatory frameworks have evolved to deal with algorithms capable of learning on a case-by-case basis in a real-world setting."

This reality forces delays and halting progress on those developing precision systems for health care: they can only improve the system until it must be shipped to customers. And even then, the frequency with which customers update their version may be out of their control too. This means that improvements made in the lab may remain invisible and unmeasurable for a long period of time, stretching out the design-test-prove loop.

## Avoiding the Red Zone

The reality is that measuring the impact of precision systems can be technically difficult and operationally challenging. Proxy metrics help where the direct benefits are in fact unobservable. But expect the breadth of precision wins to be breathtaking and challenging to grasp, even as they scale fast and well, but of unparallel value to count, and count well.

The difficulties of measuring the true impact of a precision system should be considered by leaders when evaluating whether to invest or not, and whether to invest further in fixes. It's another reason why those who hope to be members of the precision vanguard need both strong leadership skills and a greater-than-average level of personal commitment, even courage. Precision systems work—but don't count on finding any single number that proves the fact beyond a reasonable doubt.

* * *

Stumbling into a red zone doesn't necessarily mean failure. Every successful case study offered in this book attests to that fact because every precision system experiences red-zone challenges from time to time. Red zones are to be recognized, identified, mitigated, and avoided where possible. Failure to do so can result in failure for the project, but the reverse is also true—managing red zones well can result in greater success.

Teams embarking on precision projects for the first time will realistically experience failure risks throughout the course of development. Most data science efforts fail for want of budget, purpose, data, personnel, or platform before ever operating as part of a system. A clear business purpose, talented people, and strong political support are key to making it out of the lab and onto the field of play. Perseverance, thoughtfulness, and openness to feedback and change are crucial qualities for anyone seeking success with a precision system.

Recognize that the very definition of success evolves over time. As we get better at identifying the systemic shortcomings of today's precision systems, we incur a new responsibility to reexamine our old work to see how it can be improved. This should be a routine part of the ongoing maintenance of precision systems. As organizations scale and use of their products increases, no precision project can ever really be considered finished. With the potential for positive impact growing continually, we should never be completely satisfied with what we believe we have achieved. We should instead test and prove that our precision system is hitting its mark, that is, producing change exactly where, when, and how we want it, nothing more and nothing less, and ensure such great value again and again, at greater scale.

# CHAPTER 11

# SUMMING UP

## Some Final Lessons for Leaders

In this book, we've recounted stories of a wide range of precision systems. They were drawn from various industries and served different purposes; faced many kinds of challenges; and were driven by the talents of people with diverse backgrounds, personalities, and leadership styles. We chose these stories because of the lessons they offer for other leaders who hope to apply the powers of precision systems to the problems and opportunities they recognize in their own fields of endeavor. We hope you've enjoyed learning about the members of the precision vanguard whose struggles, successes, and occasional failures we've described—and we hope you've gathered some valuable nuggets of practical wisdom along the way.

Now we'd like to leave you with some final lessons. They take the form of four sets of imperatives that we think all aspiring leaders of precision systems should take to heart:

- Innovating a technology of precision
- Building to precision: on-ramps
- Managing to precision: the enterprise
- Leading for precision: transformation

We'll discuss each set of imperatives separately in this chapter.

## Innovating a Technology of Precision

If you are or want to be a leader of a precision system, your first imperative is to innovate what Clayton Christensen has referred to as a new *technology* of people, data and platforms, processes, and performance. That technology should work as a system with a purpose: to achieve precisely the change you require, where, when, and how you want it.

Data are the fuel of precision. It flows across your platforms. Respect it, handle it with care—but don't assume its quality. As must any fuel, data must be refined, purified, rinsed of bias, blended, and ensured that the data are the right grade for your systems. Poor data quality can cause precision's downfall.

If data are the fuel of precision, platforms are its beating heartbeat. Whether the platform is a BNSF command center and its sensor network, a tractor and its imaging setup, a ground control station for drones halfway around the world, or a K–12 classroom accelerating learning, each one is mission built. Platforms should start out lean but will inevitably grow more complex with sophistication. It's important to remember that developing new platforms with data isn't linear—progress can stall or even halt before a breakthrough.

When it comes to technology, more isn't always better—all that matters is whether the platform is fit for its purpose. But that fit can be hard to find. New technologies are emerging rapidly that offer new ways to solve old problems, so striking the balance between future-proofing your technology investments against starting somewhere and diving into real applications is delicate and crucial.

Leadership for precision ensures that the data and platforms align, creating a system that will work and work and work—technically, as a system; operationally, producing results; and for the organization, engaging teams and achieving value as promised. The data is instrumented, accurate, and compatible. Data technology powers precision best when it's usable by those who should use it, streamlined for those who manage it, and explainable for those who must pay for it.

It's a big job. In our ever-changing world. it never gets done once and for all. It must be continually revisited, reevaluated, and reworked to meet a continually evolving set of challenges, demands, and needs.

## Building to Precision: On-Ramps

As a leader of precision, your second imperative is to understand and master the complex array of processes involved in conceiving, building, implementing, and maintaining a precision system. Most often, the journey to precision begins when you spot an opportunity—a persistent problem, a looming crisis, an urgent goal, a charge from the boss. You look to your bureaucratic compass—your bosses, subordinates, colleagues, external stakeholders. Can you go it alone or, as is usually the case, do you need to involve others? Your first move is the first call you make around your network touching each of the four points of your compass as needed, spreading out across your network to call a coalition into existence. It includes supporters and sponsors, contributors and givers, customers and patrons. Above all, there is a problem that needs fixing or an opportunity that needs seizing. You frame it as a goal, as simple as you can make it, around which all can rally (or at least refrain from opposing).

Then you set to work with a core group, a few trusted colleagues from the precision vanguard—risk takers like you who will make common cause, contribute their ingenuity, excitement, authority, and skill sets. Together, you define the problem to be tackled and the hypothesis driving the solution, and begin to set up the framework of design, test, and prove, which you'll carry throughout the life of the project. You define success and the specific measures that must be met to claim victory.

Once the problem and goals are understood, think about the kind of system you need: Will it focus on pattern matching? Image recognition? Geotracking? All three and/or others? Is the work doable by spreadsheet and simple statistics, or do you need artificial intelligence (AI) and machine learning? Do you have the data? Is the right kind of platform available? Cobble together a vision that works, a plan that is practical, to be built using whatever capacity you have on hand or can easily acquire. And connect with superiors and allies to make sure you have enough authority to move around the network comfortably, even stepping out of lane or across boundaries as needed without ruffling too many feathers.

Then the detailed work begins. Assemble technical teams to model early solutions, then do some early proof-of-concept work. Identify the gaps between what you have and what you need, define more precisely the opportunity within reach, and begin to surface your package in anticipation of scaling. As needed, you reach out for bigger contributions of time, data, people, and budget.

Take stock of how far and how fast you can move without losing support and bumping into opposition. Frame the initiative to hit the CEO's sweet spot of strategy—and leave everyone whole, creating new well-being or benefits without diminishing what's already in hand. *Skip the endowed*—those strongly wedded to the status quo—and make it easy for the others to see the big benefits your plan will offer. *Shrink the ask* to reduce technical complexity and operational change so that winning support is easier and faster. Strive to *get value in the hands of users fast*, whetting their appetites and building support for more.

And at every step, keep asking, "*Will it work, and work, and work*"—technically as a system, operationally to address the problem, and for the organization, its people, and its purposes?

## Managing to Precision: The Enterprise

When scaling a precision enterprise, your third big imperative is to manage your precision system so that it integrates seamlessly and powerfully with the entire organization. The following seven principles will make this integration easier—although still far from easy:

- *Focus obsessively on outcomes.* Guided by the CEO's north star of strategy, engineer the change—test it, prove it, and get your metrics and measures right. Manage precision end-to-end, from data to outcomes. Always ask, Does the precision system in the field and in the hands of humans work to produce the promised change? Be prepared to rework the system as needed to produce the promised outcomes.
- *Never stop testing and proving.* Your operational methods can always be more streamlined, more efficient, more understandable; your data can always be more accurate, more thorough, more usable; your platforms can always be made more user-friendly, more accessible, more versatile; your assumptions can always be further challenged. The day after you launch your precision system is the first day to begin work on identifying its shortcomings and building the improvements that will evolve it.
- *Choose your technologies but keep in mind their limits and their power.* The virtue of AI and machine learning is not that they do a better job than humans. At best, they do no worse. But they scale to handle volume, go fast, and standardize treatments—treating like cases alike, no matter what, for better and for worse. And together, humans and machines will always do better than either alone. Above all, *always* design humans into the decision chain for crucial decisions no machine can, or ever should, make as well, and know that no decision system is free of bias because systems are built by humans.
- *Keep the user experience of precision systems as simple as you can.* Precision systems change the technology *and* the routines of work. That's the point—they *transform* if you let them. But if you change *what* people do and *how they* do it all at once you may be in for the long haul. If you don't have the time (and sometimes you won't) *shrink the ask.* Use systems that are backward compatible, keep users and operators winning with systems they are comfortable with and find easy to use.
- *Build better teams.* Precision systems can't happen without cross-disciplinary teams working together collaboratively and productively—which means that team composition, team solidarity, and team management should be topics every precision manager thinks about day and night. Reset objectives and key results (OKRs) and key performance indicators (KPIs) to roll up to precision outcomes and hit targets with their cumulative effect so that every team

member is pulling the right oar and all the oars are pulling in the same direction. And then, as soon as you think you've assembled the best possible team, start planning how you will reconfigure it and shift the team leadership to restress the next steps in the process.

- *Change your organizational routines and workflows to take advantage of precision.* You can't simply bolt precision onto your existing workflows. In most organizations, the processes used for delivering value—whether through products, services, information, or activities—can require change—new workflows, job designs, performance metrics, even new rules and regulations. That may mean stepping out of your lane to win support from organizational veterans who are comfortable with the traditional ways of operating. Work around them and with them as best you can—and make sure that the added value you can produce through precision is so big and undeniable that even the die-hard resisters will ultimately have to yield.
- *Never forget that the world is dynamic.* What makes your precision system incredibly powerful today may vanish tomorrow—and with it, whatever advantage today brought. Don't become a victim of sunk cost bias by falling in love with the system you just spent months and countless hours creating. Instead, give your heart to the even better system you and your team will begin building tomorrow.

## Leadership for Precision: Transformation

Precision changes the calculus on the battlefield, in the classroom, in the ecommerce marketplace, on the farm, on the football gridiron, and in the newsroom. But transformation is not automatic. It must be foreseen, aspired to, engineered for, tested, proved, monitored, and supported continuously with improvements and protections over time. Here's a list of five principles to help you become a more effective leader of precision transformation:

- *Boldly redefine your goals.* Assume today that, without precision, you are leaving value on table—ad revenues not picked up, for example. Or you're causing needless damage with blunt force change—spraying entire fields instead of one weed at a time. Know that precision systems can fix that. Open your eyes to a new world of possibilities, a world of full-spectrum effects in which you can achieve multiple, seemingly contradictory goals at once; satisfy multiple potential and actual customers; deliver higher-than-ever direct value; and carry lower-than-ever overhead. Then go for this new universe of possibilities. The virtue of precision systems is that they handle a new totally addressable universe of possibilities—all 50 million potential voters in seventeen battleground states for Biden, 200 million ad placements possible for the news app *upday*, 100 million track geometry measurements every one kilometer of all 22,000 miles of BNSF-owned track—every weed, every child, every news reader, one at a time and by the millions. Plan for it, test it, and prove it. Then do it.

- *Reconfigure your performance metrics and delivery systems.* Precision lets managers reach huge new target populations with greater impact and accuracy than ever. That shifts the denominator and cries out for new performance metrics that reflect loftier goals than you ever dared to set. Then re-form your value delivery systems to match the new goals, managing all for end-to-end outcomes and impacts in which there are new definitions of what counts as a win.
- *Hold fast to the new future as it emerges.* Deliberately look for ways to define the organizational culture as a culture of precision. Consider what you measure and control; how you react to critical incidents and crises; how you allocate budget, rewards, and status; how and what you teach and coach; how you define and communicate organizational values and beliefs; and the stories of success (and failure) that you choose to tell and retell. Use the new culture of precision as a power lever to keep change progressing.
- *Be comfortable in the several roles of leadership for precision.* As an admiral, you may lead a fleet of initiatives; your principal task will be allocating scarce resources across all. As an orchestra leader, you'll lead a single complex project involving many specialists; your central task will be to ensure that everyone is in tune and playing from the same score. And as a pit boss, you'll lead your own team through its day-to-day challenges; your core responsibility will be to keep the group focused and on task, executing against the plan as required. Recognize how these roles differ from one another, and learn to move adroitly from one role to the next as the demands of the work evolve.
- *Remember that change means cultures in conflict.* Engineers, operators, and managers will all react differently to the pressures generated by the move to precision. Know the third rails of status quo bias; recognize the realities of who will win and who will lose; and be clear about your theory of change: how do things work for the change you want, and who has to do what to achieve it?

Above all, refine your strategy to be a *precision strategy*, aligning people, platforms, and processes on new performance possibilities. Prove, test, and scale. Then relight your beacon for precision; keep that north star brightly lit and in your sights. It will guide all, and they will need it.

# ACKNOWLEDGMENTS

Early on, before there was ever a book, we asked colleagues Lewis Shepherd, Christopher Tucker, James Waldo, James Ledbetter, James Mintz, and Thomas H. Davenport whether a book about precision systems made sense. Having clarified its theme and message for all, each said, "Yes, write it." We are grateful for this invaluable support, as their vote was the green light that set us on our path.

Getting to a finished book was a challenge because in March 2020, all travel and in-person contacts ceased as the COVID-19 pandemic blanketed the world. That changed the entire approach to collaboration for us and to the research we would conduct. Everything and everyone became pixilated; but it also meant that a new norm of research emerged that was entirely digital—no in-person interviews, at-conference workshops, or other face-to-face work. And there was power in that because the technologies of Zoom, the web, email, Teams, and the internet gave us quick and global entrée to an astounding array of practitioners and their data, all of which we engaged digitally.

We are therefore grateful to our researchers Rebeca Juarez and Lorenzo Leon, who worked from Mexico City and then Strasbourg to provide a backbone of data before and after interviews. Cody Want of London helped get us started and on course as our first editor; Karl Weber followed as our principal editor, closer to us in New York—but still we never once met in person. Karl helped us construct the book, chapter by chapter, and unified the voices throughout. We could not have put one foot in front of the other and end here, with this finished book, without him. Nick Glynn gave us insights to the publishing world. Ian Botts, Mike McNeeley, and Jason Cicchetti reminded us of the power of stories and our great opportunity to tell them here. We are grateful to Alice Martell, our authors' representative; Brian C. Smith, our editor at Columbia Business School Publishing; and Myles Thompson, our publisher who, with reassuring confidence, said, "Yes, definitely," to our proposal for *Precisely*.

We interviewed many for this book and transcribed speeches, panels, webcasts, and podcasts of many more whom we never met but who we quote. We are grateful to all our interview subjects and, from afar, all presenters, panelists, and participants whom we may know only from YouTube. But among the interview subjects, some stand out for the time and patience they extended, making simple their complex tales. They include Nathan Rifkin, Nellwyn Thomas, Jose Nunez, and Nina Wornhoff for the DNC accounts; Dean Wise, Allan Zarembski, and Marty Schlenker for BNSF; Daniel Stern and Eric Eager for the sports analytics account (Eric warned us the NFL would be hard to crack, and it was); Jason Bay for Singapore and tracebacks; and General John Jumper for Predator's oft-told tale, now updated and reset for the precision world. Laura Evans, Josh Arak, Nick Rockwell, and Tiff Fehr of the *New York Times* helped us see their world as we never would have otherwise. Dr. Joshua Weintraub and Dr. Alexander Baxter anchored our radiology accounts. A special thank-you goes to Brett Danaher, David Boyle, Sameer Modha, and Nolan Gasser, whose time and insight to music and the arts benefited us profoundly; to Lori Pollock and James Byrne of the New York Police Department (NYPD), and on reflection and with great respect to the NYPD's William J. Bratton, Dermot F. Shea, and Philip Banks III. Ben Singleton, then of JetBlue; Alex Dichter of McKinsey; and Charles Cunningham of Southwest Airlines shared with us their great insight into precision systems in airline operations.

We owe a great debt and gratitude to teachers and colleagues at universities along the way, notably Mark H. Moore, Stephen Goldsmith, Steven Kelman, and Jerry Mechling of Harvard Kennedy School, Mitchell Epstein and John Gourville of Harvard Business School, and William Eimicke of Columbia University's School of International and Public Affairs. We hope that, in writing *Precisely*, we have paid forward some of this debt so that others can benefit from their great wisdom, as we have.

# CHAPTER NOTES

## 1. Introduction: From Digitization to Precision

For the Ravens's analytical edge, we drew principally from our background interview with Eric Eager, of PFF Sports, then stories reported by Kapadia (2019), Brock (2019), Oyefusi (2019), Hartley (2020), Hensley (2019, 2020, 2021), Dubow (2021a, b) and McLaughlin (2019) (See *National Football League [NFL]*, below, for additional sources.). The New York Police Department (NYPD) account draws principally from our interviews with Alexander Cholas-Woods, Lori Pollock and Michael Lipetri, and others, and from news accounts reported by Dorn (2019), Moore (2019), Kilgannon (2019), Lampden (2020), Fondren (2022), and Miller (2020). Chung (2020) references "the first day is the worst" for machine-learning models—they improve over time and with use. Korea's preparation for COVID relied on Watson (2020), McCurry (2020), Sohn (2020), Sternlicht (2020), *Japan Times* (no author; 2020), Gottlieb (2021), the editorial board of the *Washington Post* (2022), Worldometers (2022), and Fisher (2020). SailGP was drawn principally from Schmidt's (2021), Smith's (2019). and Springer's (2022) accounts.

We draw the reference to "technologies of people, processes and platforms" from Clayton Christensen's use of "technology" as he defined it in *The Innovators Dilemma* (2011)—"the processes by which an organization transforms labor capital, materials, and information into products and services of greater value. All firms have [such] technologies." Tom Davenport's "The Need for Analytical Service Lines" (2011) is essential reading. Michael Tushman's *The Ambidextrous CEO* (2011) gave us further insight into the embrace of tensions between new and old corporate job design, workflows, and teams as impactful for innovation. Ackoff (2001) and Meadows (2008) on systems, and Kotter (2012) and Kanter (1985) on change are foundational. The "big data/small data" discussion is based on Ng (2021), Chahal (2021), Ravishankar (2018), Ray (2022), Macaulay (2022), Kaye (2022), and Waddell (2018). "Nudge" and its issues can be found in Thaler and Sunstein (2008), Thaler (2017), and Chater and Lowenstein (2022).

*Zipline*

We relied on Zipline's own accounts, including statements by its cofounders Rinaudo and Wyrobek (Zipline, 2021a-c). Khanna and Gonzalez's Harvard Business School case (2021) and article (2018) helped set the stage further. Reporting by Farley (2020), Peters (2020), Dijkstra (2021), and Toor (2016) helped us fill out the picture. We drew data and insights from reports by Greve et al. (n.d.), Guerin et al. (n.d.), Stokenberga and Ochoa (2021), Droneblogger (2021), Reuter and Liu (2021), Zhu's interview of Nick Hu (2021) and Flyzipline.com (n.d.)

*Zest*

Zest caught reporters' eyes early. We benefited from Crosman's (2012) work and from reporting by Fuscaldo (2019), Rao (2010, 2012), and Zax (2012). We learned about Zest-like FinTech startups like Mahindra Finance and FIA Global from Mishra (2018) and Sharma (2022) respectively. Zest's subsequent legal troubles were well documented by Aquino (2020) and Brueggemann (2019), among others.

*National Football League (NFL)*

In addition to the sources cited above, we drew from Siefert (2021) on NFL fourth-down offenses; NFL clock measures in Sault (2019) and Silverman (2018); Hayhurst (2020) on RFID chips in uniforms; EdjSports on Staley (2021); Cosentino (2021) on "think like a coach"; Fortier (2020), O'Connell (2022) on the NFL move to analytics; Joyce (2019) on Gettleman; McLaughlin (2019) on Daniel Stern; Tomlinson's quote is from Hensley (2021); Schatz (2022) on the aggressiveness index; Lundberg (2021) on Jackson's offensive style; Fox (2021), and Walder (2021) comparing NFL analytics units. The Cooper Kupp account is from Bishop (2021). Pagnetti's account is from Fox (2021). We drew additionally from reporting by Hayhurst (2020), Levin (2021), McCaskill (2021), and Togerson (2022). NFL valuations are on Kurt Badenhausen's beat (2022).

*Why We Wrote This Book*

Kara Swisher's reporting on Silicon Valley has long proved itself invaluable. Her coinage (2021) pronouncing the digitized future inevitable is timeless. Marc

Andreessen's notion of liquifying markets is offered in Morris (2013). Tom Davenport's "Artificial Intelligence for the Real World" (2018) inspired our hunt for precision system types. From our interviews, Alex Dichter's insights (2021) to "bolting old systems onto new" at airlines, and Dr. Joshua Weintraub's at hospitals (2021), were eye-opening, further buttressed by Coiera on clinical disengagement (2018), and Han (2005), Sittig (2006), and Koppel (2005) on medical error from disrupted routines. Davenport's exhortation to try fixing broken processes first, squarely put in his 2018 *Harvard Business Review* (HBR) article, gave us courage to suggest the same when designing precision systems. The concept of brand equity, mentioned here, recurs throughout the book. It is Kevin Lane Keller's (2001). The fourth Industrial Revolution is sourced to the World Economic Forum's Klaus Schwab (2015). The "magic bullet" theory of information technology (IT) transformation is Markus and Benjamin's (1997). The "precision chasm" borrows from Keane and Topol's "AI chasm" (2018). Tom Goodwin's brilliant capture of our times is from his Techcrunch blog (2015).

## 2. Six Kinds of Precision Systems

Sources of insight and inspiration include Sara Brown's machine-learning writings for MIT Sloan's *Ideas Made to Matter* blog (2021), Paul Ford's thoughts on what "real" programming is (or isn't) for *Wired* (2020), and Khodabandeh, Ransbotham and Shervin's study of Gina Chung's account of artificial intelligence (AI) development at DHL (Chung, 2020). Malone, Rus, and Laubacher (2020) offer a very accessible definition of what AI is and the other disciplines it builds from that are used to illustrate what precision systems broadly can do. Davenport on AI journeys (2021) was invaluable, as were calls to action for a move to precision ecosystems in medicine by Gruen and Messinger (2021), and Bloomfield (2021).

Dima Shulga's list of most important data science must-knows for Towards Data Science (2018) further illustrates how multidisciplinary the field of data science really is, and goes some way to explaining why good data scientists are in short supply and in high demand. Kate Crawford and Vladan Joler's "Anatomy of an AI System" for Virtual Creativity (2019) and Vincent Granville's (2016) extensive list of quantitative techniques for Data Science Central informed the mathematical components of the six kinds of precision systems framework and demonstrate why a framework for grouping these techniques into useful buckets was necessary.

More generally, Donald Schon's writings on teaching and education (1990), Russell Ackoff's approach to solving societal problems through systems (1974), and Churchman's editorial "Guest Editorial: Wicked Problems" for *Management Science* (1967) offered sage connections back to the world before and beyond data—the technology may be new, but the problems are not.

*Patternizr at the NYPD*

The reference to Jack Maple's Compstat is from Bratton and Tumin's "Collaborate or Perish" (2012). Contemporary content came from invaluable interviews with Chief Lori Pollock, Chief Michael Lipetri, Evan Levine, Alex Cholas-Wood, and Zach Tumin, all currently or formerly of the New York Police Department (NYPD). In sharing their experiences with data science in the police force, they shed light on challenges both unique and shared across industries.

Other material was drawn upon, for example, reporting from *New York Magazine*, NBC Connecticut, and Spectrum News on Compstat, pattern recognition, and the operationalization of Patternizr. These were primarily used to fact check the Patternizr story and to include third-party perspectives on its rollout and impact.

Critique of predictive policing, both as a concept and regarding specific applications and their respective failures, was highly valuable in moderating the breathless enthusiasm of its proponents. We referenced writings by Will Douglas Heaven (2020), and Levine and Cholas-Wood's published paper about Patternizr (2019), which offers substantial perspectives on the limitations of Patternizr both built in deliberately and incidental.

*Systems for Finding Patterns, Personalization*

Allen Yu's study of Netflix's expertise in AI and machine learning for the blog Becoming Human (2019), as well as Lucia Moses's reporting for Digiday about the *Washington Post*'s robot reporter (2017), gave us crisp and tangible examples of how content gets shaped and surfaced by precision systems today in ways one may not be aware of.

General background on recommendation systems that informed the inclusion of this archetype into our list of six included Schrage's "Recommendation Engines" (2020) and Google's publicly available introduction to recommendation systems (2022). Schrage's book sheds light on not only how recommendation systems work and what they are used for but also touches on the risks that they can create when mismanaged—including creating echo chambers of information that can have broad societal implications.

The reference to "genomic approach" as an alternative to collaborative filtering in the instance of the cold start problem was derived from Deldjoo, Dacrema, Constantin, et al.'s writings for the journal *User Modeling and User-Adapted Interactions* (2019).

*Systems for Predicting Values*

Ben Singleton (then of JetBlue Airways) was generous with his time and taught us about the myriad potential applications of AI within air travel, especially for purposes of dynamic pricing and customer experience. Further insight into airline dynamic pricing practices was found in Altexsoft's blog post about the topic, which clarified in useful terms the distinction between rule-based and dynamic pricing.

Dr. Peter Chang of the Center for Artificial Intelligence in Diagnostic Medicine (CAIDM) at the University of California, Irvine, also shared his time and insight about the early days of the COVID-19 pandemic and how he was able to triage patients quickly based on likely mortality using a straightforward algorithm. That work is also publicly referenced by the Radiological Society of North America online (2020), which was used for verification.

The power and breadth of application for prediction algorithms was introduced to us thoroughly through Agrawal's "Prediction Machines" (2018), which posits that all machine-learning endeavors can be essentially boiled down to continuous, large-scale efforts at prediction—no matter the specifics or the application. HBS Online's "What Is Predictive Analytics?" (2021) added helpful introductory definitions for the topic.

*Systems for Entity Recognition*

Marty Schenkler of BNSF Railway taught us much about predictive railroad maintenance, informing our account of the railway's transformation from "old industry" to cutting-edge deployment of machine learning and effective use of big data. Jessica Holdman's account of BNSF's camera flaw detection for the *Jamestown Sun* (2019) provided extra details.

We had several teachers on the topic of AI in medicine, including Dr. Simukayi Mutasa, who is cited in a later chapter, and the writings of Dr. Peter Chang. But on the topic of image recognition in cardiography specifically, we found riches in Dr. Eric Topol's conversation with Daphne Keller for the Stanford AIMI Symposium (2020).

Mozur and Clark's illuminating account about facial recognition deployment by the Chinese government against citizens for the *New York Times* (2020) offered a sober warning about the risks of AI. Finally, *Health Imaging* (2020) summarized how an AI-powered pathology system personalizes cancer patient care today.

*Systems for Geotracking*

Gillette Stadium's advanced networking system, and the New England Patriots' strategy for optimizing home-viewing experience through technology, illustrate best-in-class geotracking precision systems. This account was sourced from publicly available information on the topic, including publications from Matthias (2012, 2013), Huang (2013), and NECN (2014).

The account of how cellphone tracking was used by the U.S. Justice Department to track down and prosecute the January 6 insurrectionists came from publicly available information including Klippenstein and Lichtblau (2021); Warzel and Thompson's (2021) unmissable reporting for the *Intercept*; Timberg, Harwell, and Hsu's reporting for the *Washington Post* (2021); and Madison Hall's work for *Business Insider* (2021).

During the writing of this book, many of the leads that originated from social media content that the insurrectionists uploaded themselves turned into actual justice impacts, including imprisonment. Trials are still ongoing as this book goes to press, but the power of precise geolocation data for geotracking and justice enforcement has already been proven.

*Systems for Modeling Scenarios*

Sohrabi et al.'s "An AI Planning Solution to Scenario Generation for Enterprise Risk Management" (2018) discusses private sector applications of scenario modeling that were referenced for budgeting and planning and was also recapped by the same authors on IBM's own blog, referenced for summary (2018).

The description was informed by *Digit*'s description of increasingly sophisticated combat simulation within the Defense Advanced Research Projects Agency (DARPA). DARPA's own blog describes the combat simulation trials, and Digit's reporting (2020) was used as a reference.

The concept of computer generated forces and the way they can be used to expedite various combat scenario simulations came from the Netherlands Aerospace Centers information about their artificial intelligence for military simulation (AIMS) product offering, which sells this as a service to military purchasers.

References to DeepMind, Google AI's AlphaZero chess simulator, were sourced through DeepMind's publicly available blog, which contains riches for those interested in creative and ethical pursuits of artificial intelligence.

*Shared Principles Across Precision Systems*

The structure of this section is based on insights derived from Madeleine Want's own experience. Steven Miller's framework of three types of precision systems for Singapore Management University (2018) is referenced as an alternative; it groups precision system types purely by the type of benefit offered to humans rather than by the nature of the technological application.

Chris Liu's perspective on engineering and data science for Coursera Engineering via Medium (2019) substantiated the part focused on how data science and traditional engineering differ—as did Peter Skomoroch's must-knows for AI product management (2019, 2020). Skomoroch focuses on the nonlinearity of data science development in contrast to traditional software engineering.

Roman Yampolskiy's account of incident 43 for the Intelligence Incidence Database (1998), and Joy Buolamwini and Timnit Gebru's unmissable research on gender bias in classification (2018) provided much-needed perspective when we looked at the risks of bias in AI. The twenty-year gap between those two publications only serves to highlight the similarity of the concerns they both raise, which offers a lesson on how bias persists throughout technological advancements because it originates with humans.

## 3. Design, Test, and Prove: Key Steps in Building a Precision System

### *Blue River, John Deere, and Precision Agriculture*

Jorge Heraud of John Deere (formerly of Blue River) gave a firsthand account of his and Lee Redden's journey to bring machine learning to the ancient field of agriculture. Everything from the initial idea through the design, testing, proving, and scaling of the See and Spray system was covered in these interviews.

Many publicly available sources bolstered and offered further perspective about this remarkable achievement: Data Collective's account of the acquisition (2017), Louisa Burwood-Taylor's podcast AFN (2017), Scott Martin's account for the NVIDIA blog (2019), Redden's own interview recordings on Soundcloud (2021), Paul Schrimpf's description of See and Spray for PrecisionAg (2021), Caroline Osterman's reporting for the Berkeley Masters of Engineering Blog (2019), Mike Murphy's Future of Farming for Protocol (2020), Chris Padwick's study of how PyTorch that was used by BlueRiver for the Robot Report (2020) and in general how the implementation of computer vision for See and Spray took place (primarily for a technical audience), as well as *Forbes* articles by Bernard Marr (2019) and Jennifer Kite-Powell on Daniel Koppel; (2019).

Terry Picket for John Deere Journal (2015); Peter Fretty for Industry Week (2020); and Leslie, Siegelman, and Kiesseg for Stanford School of Business (2013) rounded out an in-depth understanding of the topic. General perspective on precision farming came from the National Museum of American History archives (2018).

### *Design, Test and Prove: The Core Steps*

John Doerr's classic *Measure What Matters* (2018), which introduced the concept of objectives and key results (OKRs); Jim Brikman's account of what a minimum viable product is (2021); and Henrik Kniberg's attempt to capture the same (2020): all these sources fundamentally offer proposals for how precision project leaders can get focused and stay focused on the most important objectives.

Ron Kohavi, Diane Tang, and Ya Xu wrote *Trustworthy Online Controlled Experiments: A Practical Guide to A/B Testing* (2020), which offers the best and most thorough capture of corporate digital experimentation. Wilhem Kornhauser's adaptation of success measurement to machine learning for Towards

Data Science (2021) helped bridge the gap between building for predefined outcomes and building for undefined or variable outcomes, as artificial intelligence (AI) often necessitates. This was supported by Amy Gallo's summary of A/B testing, where it came from and how it's used today, for the *Harvard Business Review* (2017), as well as a more statistical summary of the five steps involved in hypothesis definition and testing that came from the National EMSC Data Analysis Resource Center (NEDARC) (2011).

Agraal, Gans, and Goldfarb's "A Simple Tool to Start Making Decisions with the Help of AI" (2018) as well as Thomke and Manzi's summary of business experimentation for the *Harvard Business Review* (2015) helped lay the foundation for the discussion of experimentation that follows.

This section also references the cases of New York Police Department's (NYPD) Patternizr and Douglas Merrill's Zest Cash, which are cited in chapter 2 and chapter 1, respectively. The example of usability testing for AI products focused on a hypothetical voice assistant was informed by Debjani Goswami's writing for Qualitest via Medium on the topic (2021).

Sylvia Engdahl's introduction to dynamic A/B testing for machine-learning models for Amazon Web Services' (AWS) blog (2008) is an invaluable and early description of what the practice is and why it's useful, and dives deeper into the techniques and measurement frameworks that power it at AWS.

## 4. Getting Started with Data and Platforms

Eisenmann's framework is discussed in "Platform Envelopment" (2011). Sources for comparing various data science platforms and commercially available technology included "Gartner's Magic Quadrant for Data Science and Machine Learning Platforms" (2021), and subjective musings like Altexsoft's comparison of Microsoft, Google, and IBM's cloud machine-learning offerings. For a broad yet accessible introduction to all concerns related to beginning machine-learning projects, Veronika Megler's paper for Amazon Web Services (2019) is invaluable. For the importance of habit for digital newsrooms, see Jacob (2019). Brand equity, as before, is Keller's (2001). Hamman (2019) discusses a photo of senior *Times* editors debating headline graphics. Schaufeli (2012) is foundational for research into work engagement. "Walk Through Walls" is John Linder's characterization of cops in Bratton's thrall in 1994–1995 (Smith, 1996).

### *Replatforming the* New York Times

Nick Rockwell's introduction to the *Times* is sourced to his interviews with us. Thompson's recollections are found in Taqqu (2020) and Pulley (2011), Bloomberg's data in Patterson (2017), and Iliakopoulou's assessment in her authored article (2020). Josh Arak, formerly of the *New York Times*, generously provided a firsthand account of his time developing and scaling an experimentation framework for the digital arm of the business over several years. He also published some of his work, through the NYT Open blog, about ABRA (Araka and Kaji 2017)—the name of the experimentation framework, which we used to verify the account. Hassan (2018) gave additional insight to the *Times*'s effort to match its readers' interests to stories, and Heideman (2017) on tweaking the *Times*'s page performance. Mill (2018) offered detail on measuring the *Times*'s readers' subscription behavior; Marin and Yakhina (2019) on testing while "designing for growth" at the *Times*; Feinberg (2019) on design for better product at the *Times*; Gayed et al. (2019) on reengineering the *Times*'s paywall; Locke and More (2021) on reengineering mobile A/B testing; and Podigil, Arak, and Murray (2017) on technology moves to streamline the *Times*'s workflows.

### *The First Step in Getting Started: Assess the Data*

Kirk Semple's (2017) reporting on fuel thefts jump-started our PEMEX inquiry. Jorge Bernal Mendoza recounted the challenges and successes he/she/they experienced as senior information executives of PEMEX, including the development of a sophisticated oil pipeline sensory network to detect and prevent oil theft (2020). Huawei (n.d.) and BNAmericas (2020b) give background on PEMEX digitization initiatives. PEMEX describes the risks facing its business due to theft in its publicly available Securities and Exchange Commission (SEC) filings (2018); Wadlow and Pepper (2018) offer complimentary insight. For perspective on energy insecurity and politics and their implications for PEMEX, see Agren (2019), Avendano (2016), and Payan and Correa-Cabrera (2014). PEMEX's war on fuel theft and its toll on Mexico is profiled by Cunningham (2019), Jones and Sullivan (2019), McDonnell (2017), Navarro (2017, 2015), Ralby (2017), and BNamericas (2020a). Avelino, Paiva, and Silva (2009) help us count the ways and costs of fuel theft. Rucker (2010) reported the "inferno" of San Martin Texmelucan.

For outstanding technical and operational overviews of leak detection systems, see Baroudi, Al-Roubaiey, and Devendiran (2019) and Duhalt (2017). Emerson's paper on best practices in leak and theft detection (2016) cannot be surpassed. Similar overviews with different perspectives can be had from Geiger, Werner, and Matko (2015); Grey Owl Engineering (2018); and Fiedler (2016). Technical papers of interest included Sandberg, McCoy, and Koppitsch (1989), and Schneider Electric (2018).

### *Diagnosing Problems with Data Availability and Quality*

Information about best-practice data quality management abounds. For issues in pipeline industries, including outstanding technical and operational overviews of data management in leak detection systems, see Baroudi, Al-Roubaiey, and Devendiran (2019); Duhalt (2017); and Emerson's paper (2016). Among the best of our many non-PEMEX sources were Evan Kaplan's "The Age of Instrumentation Enables Better Business Outcomes" (2018), and Laura Sebastian-Coleman's "Data Quality Assessment Framework" for the Fourth MIT Information Quality Industry Symposium (2013). Both aided us in conveying the importance and nuances of data quality to precision, as were Krasadakis's thoughts on data quality assessments for the Innovation Machine via Medium (2017) and Brennan's summary of "The Ten Fallacies of Data Science" (2021). "Division" is @mrogati's "favorite data science algorithm" (2014). See Cook (2019), among others, for an account of the *New York Times*'s use of Excel for pandemic tracking.

*Righting the Wrongs in Bone Age Assessment*

Dr. Simukayi Mutasa, Dr. Peter Chang, and Dr. Joshua Weintraub spoke with us about these matters. Mutasa and Chang's unique perspectives as both physician and self-taught "physician software engineer" and data scientist were invaluable. Mutasa's published work on deep learning for bone-age assessment (2018), Pan et al.'s (2020) challenge of the old Greulich and Pyle method, and Larson et al. (2018) on the equivalence or better of machine reads of pediatric bone-age X-rays contributed to our understanding. Davenport and Dreyer's "AI Will Change Radiology, but It Won't Replace Radiologists" (2018) was additionally helpful. Topol's assessments are sourced to Keane and Topol (2018), Topol (2019 a,b), and Park (2019). The assessment of the superiority of machine learning for some diagnostics is Topol's (2020); see also Ridley (2020). Di Ieva is quoted by Topol (2020) and sourced to di Ieva's correspondence (2019). Chang and Chow's research was reported by Bassett (2020). See also peer-reviewed studies by McGuinness (2021); Ha, Chin, and Karcich (2109); and Ha, Mutasa, et al. (2019). Data and views of the Human Genome Project are sourced to Molteni (2022). See also Johnson (2018) for Chang's work at the Center for Artificial Intelligence in Diagnostic Medicine (CAIDM) at the University of California, Irvine; ColumbiaRadiology.com (2020) for questions and answers with Richard Ha; and Cannavo (2021) and Ridley (2021c) for regulatory and insurance reimbursement issues. The assessment that ancestry and clinical histories are poor predictors of future breast cancers is Ha's (2021). For models of organization decision making in the era of artificial intelligence generally, see Shrestha, Ben-Menahem, and von Krogh (2019). Yee (2021) reporting on advances in AI and radiology was on point for us. Gruelich and Pyle's standing as Stanford University Press's most profitable title in its 120-year history was mentioned to us by Mutasa and sourced to Shashkevich (2017).

*Tools for Data Management*

Nell Thomas, CTO of the Democratic National Committee (DNC), spoke to us directly over multiple interviews and educated us about the mammoth task of strengthening the DNC's data platform and strategy in the run-up to the 2020 election. We learned as well from her Medium.com articles (2020, 2021, 2022). Many other publicly available sources bolstered the account and generalized the insights to be drawn for modern campaigning, including Catalin Cimpanu's study of campaign dark patterns for *ZDnet*, Gabriel Debenedetti's recount of the election night for *Intelligencer*, and Reid Epstein's reporting for the *New York Times*. Also referenced were Green, Gerber, and Nickerson's 2003 writings on door-to-door canvassing for the *Journal of Politics*; Tate Ryan-Mosley's account of the 2020 election for *MIT Technology Review* (2020); and Schipper, Burkhard, and Hee Yeul Yoo's writings about political microtargeting for the *Quarterly Journal of Political Science* (2019). The PhD dissertation that Thomas relied on to de-duplicate the DNC voter file is Wortman's (2019).

## 5. The Precision Vanguard

Steven Kelman's *Unleashing Change: A Study of Organizational Renewal in Government* (2005) is an indispensable guide to research on change and on conceiving the "change vanguard" as a cadre of ready-to-go activists. Together with the foundational work of authors from Rogers (2003) to Berwick (2003) (and see Gourville [2005]), it sets the stage for our own thinking on the precision vanguard. The status quo bias is amply discussed in Gourville (2006). We borrow the notion of "creators and distributors" from Bernard Rosenberg and Ernest Goldstein's book of that same name (1982), and the notion of "wicked problems" from Rittell and Webber (1973) and Churchman (1967). We misapply Cathy O'Neil's title *Weapons of Math Destruction* (2017) to the general problem of insurgency triggered by data analytics, inspired also by Marc Andreessen's notion of liquefying markets (Morris, 2013) as essential to gaining power and insight from data loosened from its origins and moorings. Shirky (2009) and Anderson (2008) write wonderfully of the sea change in technology and its sociotechnical implications that has given rise to the power of billions. Hubert (2015) provides eye-opening explication of Scratch/Viacom's Millennial Disruption Index. Duggan's *Napoleon's Glance: The Secret of Strategy* (2004) gives further insight to Napoleon's oft-repeated adage, "Engage, then see," which, although never (apparently) fully authenticated, is useful all the same.

*The Precision Vanguard Today*

We were helped in describing the makeup of the precision vanguard by speeches, articles, and presentations of many, including Davenport (2018), Rockwell and Wiggins (2019), Siddhartha and Rockwell (2021), Schmidt (2015), Goldsmith (2017) and Hansen's interview of IDEO's Tim Brown (2011). The notions of Python and Scala as the precision vanguard's lingua francas, and lean and agile as its project method were suggested, further, by Blue River Technology's Chris Padwick (2020); Eric Reis (2011); and AltexSoft's best-in-class guide (2020), "How to Structure a Data Science Team: Key Models and Roles to Consider." The abiding faith in data to fix all, referred to now as technosolutionism, is amply critiqued by Morozov (2013) and Cegłowski (2016). We studied and learned from cases and articles on firms recruiting *from* and *for* the precision vanguard, including LinkedIn (Shaw and Schifrin, 2015), Netflix (McCord 2014), Google (Garvin and Wagonfeld, 2013), and many others published by *Harvard Business Review*, *MIT Sloan Management Review*, Harvard Business School, and Stanford Graduate School of Business. We base our explication of culture here on Schein's foundational work (2016). We borrow the construct "unruly messes" from Ackoff (2001). For our concepts of "beacons" and "north stars" of strategy, we draw from Rumelt (2011), Man (2020), Sull and Eisenhardt (2016), and Beck (2001).

*Theresa Kushner at Cisco, Maddy Want at Axel Springer*

We drew from Theresa Kushner's INFORMS conference presentation (2102) and our several interviews with her over the years, as well as many published reports of Cisco's trials and tribulations under CEO John Chambers, including Blodget (2011), Bent (2014), Chambers (2016), and Ebersweiler (2018). The concept of "humble yet optimistic" leadership is Spear's (2010). Madeleine Want's account of her moves at Axel Springer for the upday product development were further buttressed by Nowobilska (2018) and Caspar (2018, 2019). Davis (2017), Kedet (2021), Quinn (2021), Tse (2017), and Chokshi (2019) all provided further background on programmatic advertising. Kushner and Hayirli (2022) discuss *purpose-built* coalitions.

## 6. The Political Management of Precision

Gourville (2006) discusses "backwards compatibility" as a workaround to status quo headwinds. Bos (2003) gives a fuller description of "backwards compatibility." See Whitaker (2012) for how Google tests its software. Henderson and Clark (1990) discuss "radical innovations" and its challenges. Mark Moore and Archon Fong's concept of calling a public into existence (2012), itself borrowed from John Dewey (1954), buttresses our accounts of coalition building that begins chapter 6. Throughout *Precisely* we are guided by the work of Kahneman and Tversky (1979); Kahneman, Knetch, and Thaler (1990); Knetch (1989); Thaler (1980); Thaler and Sunstein (2009); Samuelson and Zeckhauser (1988); and others, each describing different traps in decision making, together explaining why innovation is fraught, and all best summarized by Gourville (2003, 2006). Grove's (2003) explanation of his 10X formulation was later amplified by Andy McAfee's wonderful account of what he called "the 9X email problem" (McAfee, 2006). We borrow the concepts of different kinds of intelligence from Goleman (2004), who describes many, among them *emotional*; from Joseph Nye (2010), who focuses on *contextual*; and from Joel R. DeLuca (1999), on *political* savvy. Kanter (1986) always has important formulations—"corporate entrepreneurs" is a good example. Weiss (2019) offers his perspective on the entrepreneurial manager, acknowledging his debt to Stevenson (2006).

*Joel Rose of Teach to One: Navigating a Complex Value Network*

The maps and the compass of managers are best described by Richard Haass (1999) and do important work for us here. The importance of networks is based on Burt's (2007) work on social capital, and Goldsmith and Eggers's (2004), incorporated here. The notion of three conversations comprising the political game being played in a bureaucracy as "like 3-dimensional chess" is Lynn's (1982). Christensen's value network concept (2016) helps explain the great power of the status quo to confound innovators.

Our interviews with Teach to One's Joel Rose and Chris Rush (all via Zoom, 2021) were essential to our account, as were their numerous presentations and publications, and Teach to One's monograph series on Medium.com (2020a–2020f)

Many reporters, writers, and scholars aided our understanding of the politics of education and innovation. Evaluations by Ready (2013), Rockoff (2015), and TNTP/Zearn (2021) showed us the power of an idea—"don't remediate, accelerate"—when backed by data and analysis. Accounts by Haimson (2019), Fleisher (2014), Johnson (2017), Reich (2020), and Woo (2017) helped show why such initiatives can be controversial. Smith's (2005) account of Anthony Alvarado's time as a reformer in New York, Ravitch's (2011) account of Alvarado in New York and San Diego, and Litow's Brookings report on the politics of big city education (1999) were instructive on the forces at play for superintendents who are intent on reform. The notion of "education debt," specifically attributed to COVID-19 by Dorn (2021), Hawkins (2021), Rose (2021), and Larrier (2021) helped us understand the argument for acceleration. Heymann (1987) kept our eyes open to the issues of legitimacy that Joel Rose and his team battled throughout. We teased apart other cases on education reform to divine their truths, and learned from them, especially Honan, Childress, and King's (2004) account of superintendents' reforms in San Diego.

*Getting to Yes/Imperatives for Managers*

The subtitle "Getting to Yes" is borrowed from Fisher, Ury and Patton's book of the same name (2011). "Holding the mirror" to an organization is an essential diagnostic tool for any change measure. Among the best we know is Cameron and Quinn's competing values assessment framework (2011). "Shrink the ask" is a "Gourvillian" concept—although it finds expression in many shapes and forms throughout the literature on organization change. "Customize the offering" is Christensen's admonition (2016)—to customize the system to the environment, then personalize the offering to the individual. "Know what counts as a win" is basic marketing advice—the currencies of payment (Bratton and Tumin, 2012) are many and varied depending on your target segments. Our discussion of the "endowed" versus "unendowed" is again drawn from Gourville (2006) and by implication and reference, Knetsch (1989). The superintendent's "sweet spot of strategy" is borrowed in part from Sull and Eisenhardt's various writings on "simple rules" (2012, 2016). Shoring up legitimacy is derived from Heymann (1987), and Moore and Fong (2012), as voiced repeatedly in their writings. "Be part of the conversation" is drawn from Hill and Lynn (2015). Kaplan and Mazique's observations come from *Trade Routes: The Manager's Network of Relationships* (1987), which has shaped our thinking since its publication nearly three decades ago. "Diversify metrics" is John Doerr and Larry Page (2018), through and through, among others, and evident every day in news accounts like Esquivel's (2021) of teachers sifting through the wreckage of past practice to find meaningful new measures of their students' achievements besides letter grades.

## 7. Bootstrapping Precision

The notion of a manager on a change mission having a "hunting license" that is in limited scope and expires is Neustadt's (2000), mentioned also in Heymann (1987). Michael Krieger introduced us to the importance of taking first steps; "going against a problem" rather than solving it; and formulating a good problem statement, "simple in the extreme," as among a collaborations greatest and most difficult efforts. The strategic triangle is central to Mark H. Moore's conception of the task of managers, public and private, over and above the merely administrative (1995).

*Jason Bay and Contact Tracing in Singapore*

We were introduced to Jason Bay from Mitchel Weiss's Harvard Business School case (2010), and Weiss's book (2021) on similar topics. We interviewed Jason Bay, reviewed his many presentations, read his copious Medium.com blog posts, all to get the fullest picture we could of the person and the task. We built the six characterizations of his requirements of a foundation provided by Calvino in his extraordinary Norton Lecture series, *Six Memos for the Next Millennium* (1988) and, in a more technical vein, Bert Bos's design principles for the World Wide Web Consortium (W3C; 2003).

Throughout, we drew from numerous published accounts, blog posts, and GovTech Singapore government documents (2020, a-f). Among them, Alkhatib (2020) presents the best technical explanation we have seen for generalists of how Bluetooth contact tracing works. Dehaye offers an even deeper dive (2020). Ball (2020) explains what could—and, for the United Kingdom, did—go wrong with contact-tracing apps. Bourdeaux (2020) shares cautions on limits and possibilities for contact-tracing apps. Brown (2020) worries about digital divides in the pandemic for half the world's inhabitants who have no internet access. Landau (2020) cautions against the gilded promises of techno-solutionism. Newton (2020) worries about the risks of government overreliance on Apple-Google for such critical infrastructure. O'Neill (2020) cautions against the "flood" of coronavirus apps tracking people. Rotman (2020) probes more deeply to the fundamental causes of technology failures in containing COVID-19. Nan, Ang, and Miller's Singapore Management University case study (2022) provided further detail.

*Bootstrapping Predator, the Precision Drone*

We began our investigation of drone warfare in 2008 with a three-day session of twenty-five senior military commanders and industry leaders at Harvard on the future of "unmanned" and robotic warfare after Iraq, reported by Tumin and Oelstrom (2008). Our foundational sources comprise a series of interviews we subsequently ran with General John Jumper (2012–2021), coupled with accounts by Walter J. Boyne (2009), Richard Whittle (2015), P. W. Singer (2009), Jane Mayer (2009), Mitt Regan (2022), Dario Leone (2020), Arthur Holland Michel (2015), and others. We relied further on reporting by Detsch (2022), Fahim (2020), and Witt (2022) for their broad sweep of analysis. We relied further on reporting by Detsch (2022), Fahim (2020), and Witt (2022) for their broad sweep of analysis.

Telenko (2022) introduced us to the importance of software mapping for drone guidance. Cramer (2021) and Hoppe (2021) alerted us to the use of autonomous drones. Owen (2022) described the drone-through-a-sunroof event.

Altman (2022), Jimmy (2022), Kilner (2022), Parker (2022), and Pleasance (2022) provided details of the several other attacks described. Peterson (2022) offered lessons on the war. Kramer (2022) reported on Ukrainian innovations in fast and light drone warfare. Miller (2022) concluded the brief history of drone warfare with the account of Ayman al-Zawahri's assassination.

## 8. Creating and Managing Teams of Precision

### *Distributed Organizing for the Biden-Harris Campaign: Crafting Victory with Teams of Precision*

From Hackman and Walton (1986) to Wageman (1997) and Weich (1999), Katzenbach (2015) to McChrystal (2015), the literature on teams is vast. For *Precisely*, our sources comprised interviews, presentations, articles, and books—essentially the body of current knowledge on making change happen—by seasoned technologists, business leaders, product managers, and data scientists in this age of artificial intelligence (AI), machine learning, and precision.

*Foundational.* Our multiple interviews of Rifkin, Thomas, Nunez, and Wornoff (all by Zoom, in 2021) were foundational, and all amplified by articles they wrote, presentations they gave, and panels they joined. Books giving us rich insight included Bond and Exley's *Rules for Revolutionaries: How Big Organizing Can Change Everything* (2016); Green and Gerber's *Get out the Vote: How to Increase Voter Turnout* (2019); Teachout and Streeter's *Mousepads, Shoe Leather, and Hope: Lessons from the Howard Dean Campaign for the Future of Internet Politics* (2007); and Issenberg's essential *The Victory Lab: The Secret Science of Winning Campaigns* (2013), among others.

*Technical Matters.* We learned from numerous articles, including by Alaniz (2018) on polling place locator apps, James (2020) on testing new data products at the Democratic National Committee (DNC), McDonald (2020a, b) on building the DNC data science clearinghouse, Kravitz (2021) on tech for relational teams, Leonard (2022) on Ossoff's relational campaign, Woods (2021) on AI used in the Biden campaign, Wortman and Reiter (2019 on deduping the DNC data file, and others—including Matt Hodges's tweets.

*People Issues of Recruitment, Training, and Development.* We were helped by Marx (2017) on "Calling All Data and Tech Innovators: We Have Elections to Win," Mates (2020) on her decision to join the DNC tech team, Dillon (2020) on managing the DNC's remote tech teams, and Cardona (2020) on developing the cultures of the DNC tech teams.

*Strategic Decisions, Including Tools and Tactics.* Leighninger (2020) provided research on relational organizing in the 2020 Buttigieg campaign, Krasno and Green (2008) on television and voter turnout, Khanna (2020) on building tools for rapid response to voter suppression incidents, Broockman and Kalla (2014) on campaign ground games, O'Keefe (2018) on P2P texting in campaigns, Sankin (2016) on the Hustle text app, and Kalla and Broockman (2018) on tools and technologies of political persuasion.

*Operations Issues.* DeMartini (2020) offered a first-person account of a "Dem Dialer." Goh (2020) defined Zoom best practices for campaigns. Nunez (2020) discussed building a grassroots organization in the middle of the pandemic. Somaiah (2018) shed light on keys to the game for "deep canvassing." Tarsney (2021) gave insight into engaging voters remotely during pandemic. Issenberg (2016) wrote powerfully on Sanders's team bringing the lessons learned about distributed organizing to his campaign, among others. Wornhoff evaluations of the tech tools is in Shomikdutta (2021). See also Augustine-Obi (2022).

*Mobilizing Teams of Precision at the* New York Times. Our interviews of Rockwell, Evans, Hamman, Arak, Bauer, Fehr, and Young (all 2021, by Zoom) were foundational. We supplemented these with their numerous presentations and panel discussions and with the articles they wrote. Chris Wiggins's panels and presentations were equally foundational.

*Reforming the* Times*'s Core Strategy, Digital Platform, Product Offerings, and Technology Organization.* Sources included Doctor (2017, 2019) on Mark Thompson reforming the *Times*'s strategy; Lee (2012) on the outlook for print at the *Times*; Moss (2021) on the *Times*'s move to the cloud; Pownall (2018) on the redesign of the *Times*'s website chasing its mobile offering; Prabhat (2020) customizing the *Times*'s programmatic advertising; Sverdlik (2017) and Podojil (2021) on the *Times*'s move to the cloud; Benton (2014) on the leaked *Times* innovation report; Benton (2017) on the *Times*'s digital strategy; Motley Fool (2020a) for Meredith Kopit Levien's Q3 Earnings Call comments; Johnson, Howard, and Sandlin (2021) on building the *Times*'s subscription-first strategy; Edmund (2012) on the loss of advertising revenues; Fischer (2021) and Lee (2020) on the results of the *Times*'s subscription-first strategy; Owen (2021) on plans to move the *Times*'s *Wirecutter* behind a paywall; Edmonds (2019 a, b) on the *Times*'s readying newsroom expansion plans; Mullen (2021) on the *Times*'s "Project Feels"; Alpert (2017) on the future of the *Times*'s newsroom; LaForme (2018) on reversing decline of the *Times*'s home page; Doeland (2019) on the *Times*'s digital transformation generally; Spayd (2017) on the move to personalization at the *Times;* Sullivan (2014) on personalization at the *Times*; Fehr (2019), Fehr and Williams (2021), and Cook (2019) on collaboration for COVID reporting and a

Pulitzer; Smith (2017), Pulley (2011), and Peters (2011) on new subscription plans, making the paywall tighter, and lifting revenues; Scire (2020b) on Thompson's industry forecast; Scire (2020c) on the *Times*'s 2020 revenue performance; and Lupesko (2021) on the challenges of personalization at scale, generally.

*The* Times*'s Teams in Action Optimizing Ads, Subscriptions, Paywalls, and Product for Revenue Performance.* Sources included Coenen (2019) on the *Times*'s experiments with recommendation algorithms; Iliakopoulos (2019) on engineering for news recommendations and Iliakopoulos (2020) for developing the *Times*'s messaging platforms; Yew (2018) on pitching new projects; Donohoe (2020) on managing the *Times*'s huge store of publishing assets; Douglas et al. (2018) on improving performance of the *Times*'s home page; Hassan (2018) on improving matching the *Times*'s readers' interests to stories; Heideman (2017) on tweaking the *Times*'s page performance; Locke and More (2021) on the *Times*'s mobile testing; Mill (2018) on measuring the *Times*'s readers' subscription behavior; Tameez (2021) on the *Times*'s exploration of Slack to personalize news; Bambha (2017) on using "T-shirt size" project estimations; Ciocca (2018) and Ciocca and Sisson (2019) on engineering collaborative editing, Marin and Yakhina (2019) on "designing for growth" at the *Times*; Feinberg (2019) on design for better product at the *Times*; Nayak (2018) on the *Times*'s teams management; Burke (2021) on engineering for site reliability at the *Times*; Gayed et al. (2019) on reengineering the *Times*'s paywall; Locke and More (2021) on reengineering mobile A/B testing; Podigil, Arak, and Murray (2017) on tech moves to streamline the *Times*'s workflows; Ferrazzi (2014) on getting teams right; Ellis (2020) on the north star metric; Goldsmith (2017) on "the perfect team;" Mandel (2020) on building data science teams; Skomoroch (2019) on product management for AI; Siddhartha and Rockwell on team design, performance, and laddering (2021); Olavson (2012) on Google data science teams; AltexSoft (2020) on building great data science teams; and Liu (2019) on data science and engineering collaborations, and numerous NYT Open Teams Medium postings (2021a–f).

*Election Day Preparation.* We turned to the following articles and authors for insight into preparing for Election Day at the *Times*: Scire (2020a); Khanna (2020); Mamet (2018); Merten (2019); Patel (2019); Shaheen, Araula, and Bower (2021a, b); and Dolias and Wan (2019).

## 9. Building the Precision Enterprise

*Bill Bratton Brings Rough Precision to the NYPD* For the foundation of chapter 9, we built upon numerous interviews, conversations, and personal communications of Bratton and Tumin (2008–2021); Bratton and Tumin's *Collaborate or Perish* (2012); Bratton's *Turnaround* (1998); and Buntin's Harvard Kennedy School case (2000).

Gladwell's (1996) concept of the "tipping point" typifies magical thinking in speaking of crime. Guzman (2018) accurately describes Bratton's regrets. Additional historical accounts of interest are Kilgannon (2010) on squeegees, Lardner and Reppetto (2001) on the New York Police Department (NYPD) in "the bad old days," Levitt (2010) on NYPD history and culture, and Maple (1999) on being a cop. Pooley (1996) typifies reporting of the day explaining crime drops, Sparrow (2015) critiques Compstat for what it is and is not, while Wesburd (2004) tracks Compstat's "me too" embrace by other police departments. Berger and Neuhaus (1989) developed the idea of "mediating institutions" of civil society as essential to order-making, and hence vital to protect.

*Bratton in Los Angeles (LA).* The Associated Press (2016) sums up Bratton's tenure at the Los Angeles Police Department (LAPD), the *Los Angeles Times* (2007) interviews a cross section of Angelinos as Bratton announces his resignation. The *Los Angeles Times* (2007) and Sousa and Kelling (2010) review (favorably) the impact of the LAPD under Bratton on a troubled public park, and Wood (2007) further assesses Bratton's tenure.

*Bratton at NYPD (Second Round).* O'Brien (2019) sorts through the mythology and facts of broken windows theory and practice. Rayman (2017) reports on NYPD rank-and-file opposition to DeBlasio, Rose (2020b) reports on Michael Bloomberg's embrace of stop and frisk and its political consequences, Sisak (2021) describes the federal monitor's oversight of stop and frisk, Smith (2018) describes the nuts and bolts of a Compstat session, Weichselbaum (2016) sums up the "Bratton era" at the NYPD. Susan Crawford's Harvard Law School report on Bratton and technology (2016) was enormously helpful. The New York Civil Liberties Union counted the number of NYPD stop-and-frisk incidents (2012). Lillis (2022) reported on the State Department polling for sentiment.

*NYPD Post-Bratton.* DeStefano (2021) wonders about the simultaneous drop in crime and the number of stop-and-frisk incidents since 2013, Kennedy (2017) offers a popular treatment of his sophisticated crime reduction methods, Moore and Braga (2003) review and reimagine an improved Compstat, CS360 (n.d.) reports on a new Compstat method, NYPD Project Management Office (2017a, 2017b) shows how the NYPD learned to track citizens' responses to individual police officers' work incident by incident.

*Melissa Gee Kee at Unilever.* Tom Davenport's body of work in human resources (HR) analytics is foundational, starting with his 2010 *Harvard Business Review* article (with Harris and Shapiro) "Competing on Talent Analytics." Nesterak (2014) provides a useful overview of behavioral science and "people analytics." Bock (2014) previews the changing nature of work and people management at Google. Human Resources Online (2014) shares Kee's emerging view of "employee branding" as essential to recruitment. Recruiter.com (2017) reports on shifting tides of recruiting millennials. Useem's panelists (2015) comment on "the biggest mistakes I've made in people analytics." Shaw and Schifrin (2015) and Yemen and Davidson (2009) discuss LinkedIn and JP Morgan Chase efforts, respectively, to shift to new modalities of recruitment. Shellenbarger (2019) reports on the dangers of hiring for cultural fit.

*The Tools of Precision.* Skillroads.com (2021) discusses the impact of artificial intelligence (AI) and facial recognition as "game changers" for recruitment. Feloni (2017) and Gee (2017) report on Unilever's embrace of "brain games" and AI in recruitment. Bestpractice.ai (n.d.) reports favorably on the introduction of a new process at Unilever; Lemouche (2021) describes how to "game" the new process. More considerations of predicting employee performance can be found in Dullaghan (2015).

*Making Her Moves/Shifting to Precision.* Melissa Gee Kee discusses the Unilever initiative in Bailie (2018). HireVue (n.d., 2020) offer fact sheets and Melissa Gee Kee comments. Vlastelica further assesses all in light of emerging diversity requirements (2021). Schellmann (2021) discusses potential for bias in hiring algorithms. Hirevue (n.d.) and Maurer (2021) report on Hirevue discontinuing facial analysis screening.

*Sensored Precision—BNSF Section.* Authors' interviews (2021) with Dean Wise, Martin Schlenker, and Dasaradh Mallampati are foundational, as are BNSF's Annual Safety Review (BNSF 2019a) and maintenance strategy (2019b), Norfolk Southern's (2020) use of predictive data for customer service, Garcia (2018) on BNSF's use of the Internet of Things (IoT), Ghanci (2020) on BNSF sensor use, Hickins (2012) on Union Pacific using software to predict and prevent train derailments, Reilly (2018) on how IoT is making railways safer, Zarembksi (2019) on big data trends in railroading, Baillargen (2018) on the FRA's predictive analytics program, Stern (2017) on the rail sector's move to digital in maintenance operations, and Alawad (2019) on machine learning from accidents at railway stations. The orange juice account is found in Lariviere (2013).

*Technical Challenges of Streaming Sensor Data.* Sources included Holdman (2019) on imaging to detect rail flaws, Yin (2016) on deep learning for fault diagnosis on high-speed trains, Zhou (2020) on using multisensory arrays for "flat wheel" recognition, Bladon (2015) on the challenge of integrating multiple monitoring technologies, Lederman (2017) on data fusion in track monitoring, Larionov (2019) on "multiphysical systems" that monitor track condition, Wang (2018) on big data fusion for track geometry inspections, Zarembksi (2014) on using data to address track geometry impacts on rail defects, Al-Jarrah (2019) on the challenge of sensors sending data to a fusion center, Jardine (2006) on implementing condition-based maintenance, Lee (2017) on anomaly detection in air leakage from train braking pipes, Li (2014) on using machine learning to improve rail network velocity, and Xu (2011) on predicting track quality.

*Operating Challenges.* Sources included Williams and Betak (2018) on using text and data analytics in railroad operations, McKay (2018) on integrating human intelligence and expertise, McKinsey (n.d.) on the benefits of digital maintenance activities, Popke (2019) on rail systems using data analytics for predictive maintenance, Thompson (2018) on using AI on ERP data, and Dewan (2019) on automated train identification requirements.

*Optimization for Safety and Efficiency.* See Sashihara (2011) and Poona Dewan on Houlgin-Vera (2018) panel.

### *Precision at the* New York Times

Authors' interviews of Nick Rockwell, Laura Evans, and Brian Hamman; and Rockwell's and Hamman's numerous podcasts, presentations, writing, and panel discussions were essential and foundational. Additional insight came from Schein (1996) on three cultures of management, Gourville (2006) on the status quo bias, and Motley Fool (2020) for the *Times*'s CEO's commentary. On Baquet pushback on Twitter at the *Times*, see the articles by Benton (2022) and Sicha (2022), and Bacquet's Twitter account.

## 10. Red Zones of Precision

The works of O'Neil (2017), Thomas (2022), and Malik (2020) are foundational for us. Risks and issues in computerized physician order entry (CPOE) insertions are detailed in Del Beccaro (2006), Han (2006), Leape (1995), Koppel (2005), and Bates (1995). Field (2021) interviewed artificial intelligence (AI) leaders.

### *Red Zone 1: Bias and Ethical Problems*

Twitter's disastrous foray into autocropping thumbnails was reported publicly by the BBC, and Twitter's own engineering leader Rumman Chowdhury shared their perspective on the company's engineering blog (2021). Desmond

Patton's "social work thinking" (2020) is introduced as a countermeasure. Cathy O'Neil on algorithms as substituted for "difficult conversations" is quoted in Field (2021). Stanford Health Care's botched vaccine rollout prioritization algorithm was first reported by Massachusetts Institute of Technology's *MIT Technology Review* (2021), which we referenced along with Drew Harwell's reporting for the *Washington Post* (2020). Harrisburg University's withdrawn report is discussed in Malik (2020) Dr. Simukayi Mutasa is referenced again here, from the same interviews that was mentioned for chapter 4 (2021). General musings on the topic of bias in AI that we found helpful include Manyika et al. (2019) for *Harvard Business Review* and Richard Waters for the *Financial Times*, as well as Jacob Grotta's keynote for Moody's Analytics (2018) for a deeper focus on how algorithm performance, degradation, and bias contributed to inaccurate house prices, problematic loans, and later the 2008 financial crisis.

### *Red Zone 2: Issues of Privacy and Governance*

The European Union's (EU) General Data Protection Regulation (GDPR) offers a description of "privacy by design" on its website (accessed 2021), and Deloitte has developed material on the same topic for commercial application (accessed 2021)—both were referenced for this section. Cameron Kerry's writing about the implications of AI on privacy for the Brookings Institution (2020) and Kashmir Hill's reporting for the *New York Times* (2021) on mass facial recognition tracking provided real-world examples of problems often still thought of as theoretical. Finally, Christian Rudder's delightful book *Dataclysm: Who We Are (When We Think No One's Looking)* (2015) was a useful source on the inference of private information, and Google's Privacy Sandbox blog provided information about the company's efforts to reshape internet advertising for more privacy (2021). Zeynep Tufekci's "Engineering the Public: Big Data, Surveillance, and Computational Politics" and her general thought leadership on this topic were formational influences and highly recommended reading (2014).

### *Red Zone 3: Failing to Keep Humans in the Loop*

Laura Evans, formerly of the *New York Times*, shared with us her experience leading data for the institution and the importance of curation and human input in digital news consumption (2021). The Index hack is reported by Beam (2022). Marty Schlenker of BNSF was cited in chapter 2, and his insights were used again here, along with a piece by BNSF about smart rail safety (2019a). Lyell and Coiera's research on automation bias (2017) helped introduce us to that crucial topic and balance the risk of bad model decisions with bad human decisions.

*Red Zone 4: The Last Mile Problem*

Topol on the "AI Chasm" is quoted from Keane and Topol (2018). Andrew Ng's response to Field's question forms the basis of this section (2021) and is bolstered by Joe Williams's description of Ng's strategic guidance for AI published by Protocol (2021). Grubhub's troubled promotion was reported by Weekman (2022), Carman (2022), and Chan (2022).

*Red Zone 5: The Dangers of Mismanaged Debt*

Debt in machine learning was treated by Holt, Golonic, Davydov, Phillips, Ebner, Chaudhary, Young, Crespo, and Dennison for the International Conference on Neural Information Processing Systems; their work offers a comprehensive dive into the topic (2015). Also used were Alexandra Shaheen's reporting (with Araula and Bower) about the *New York Times* data science organization for Nieman Lab (2021a, b) and Martin Zinkevich's writing about the rules of machine learning for Google (2021).

*Red Zone 6: Problems of Measuring Impact*

For politics see Tech for Campaigns's texting report (2020), Miller (2021) for studies on the impact of texting, Datres (2020) for GOTV messages "proven to work," the Institution for Social and Policy Studies (2021) on the survey of GOTV research, Green and Gerber for the fundamental GOTV research (2004), and Nornhoff's research in Shomikdutta (2021). See Bill Franks's summary of the risks of measuring analytics programs for LinkedIn (2020), Simon Schreiber's advice on measuring AI products for The Startup on Medium (2020), and Agrawal et al.'s "How to Win with Machine Learning" for the *Harvard Business Review* (2020); all contributed concepts to this section. For Topol on AI and the Food and Drug Administration (FDA), see Keane and Topol (2018).

# BIBLIOGRAPHY


Abbeel, Pieter. April 2021. "Keenan Wyrobek Discusses How Zipline's AI-Drones Are Reshaping Africa." The Robot Brains Podcast. Accessed January 25, 2022. https://shows.acast.com/the-robot-brains/episodes/keenan-wyrobek-discusses-how-ai-drones-are-reshaping-africa.

Ackerman, Spencer. 2020. "It's Probably Too Late to Use South Korea's Trick for Tracking Coronavirus." *Daily Beast*, March 26, 2020. https://www.thedailybeast.com/too-late-to-use-south-koreas-coronavirus-tracking-technique.

Ackoff, Russell L. 1974. *Redesigning the Future: A Systems Approach to Societal Problems*. New York: John Wiley.

Ackoff, Russell L. 2001. "Brief Guide to Interactive Planning and Idealized Design." https://pdfs.semanticscholar.org/f252/3e87a10a7385248573e174f766d45f67ee61.pdf.

Agrawal, Ajay, Joshua Gans, and Avi Goldfarb. 2018. *Prediction Machines: The Simple Economics of Artificial Intelligence*. Boston, Massachusetts: Harvard Business Review Press.

Agrawal, Ajay, Joshua Gans, and Avi Goldfarb. 2018. "A Simple Tool to Start Making Decisions with the Help of AI." *Harvard Business Review*, April 17, 2018. https://hbr.org/2018/04/a-simple-tool-to-start-making-decisions-with-the-help-of-ai.

Agrawal, Ajay, Joshua Gans, and Avi Goldfarb. 2020. "How to Win with Machine Learning." *Harvard Business Review*, September 2020. https://hbr.org/2020/09/how-to-win-with-machine-learning.

Agren, David. 2019. "Mexico Explosion: Scores Dead After Burst Pipeline Ignites." *The Observer*, January 19, 2019, sec. World News. https://www.theguardian.com/world/2019/jan/19/mexico-explosion-deaths-burst-pipeline-gasoline.

Alaniz, Mica. 2018. "Data-Driving over 2.3 Million Voters to the Polls." *DNC Tech Team* (blog), *Medium*, November 27, 2018. https://medium.com/democratictech/data-driving-over-2-3-million-voters-to-the-polls-ccod1523b18c.

Alawad, Hamad, Sakdirat Kaewunruen, and Min An. 2019. "Learning From Accidents: Machine Learning for Safety at Railway Stations." *IEEE Access* PP (December 2019): 1–1. https://doi.org/10.1109/ACCESS.2019.2962072.

Al-Jarrah, Mohammad A., Arafat Al-Dweik, Mohamad Kalil, and Salama S. Ikki. 2019. "Decision Fusion in Distributed Cooperative Wireless Sensor Networks." *IEEE Transactions on Vehicular Technology* 68, no. 1 (2019): 797–811. https://doi.org/10.1109/TVT.2018.2879413.

Alkhatib, Ali. 2020. "We Need to Talk About Digital Contact Tracing." *Ali Alkhatib* (blog), May 1, 2020. https://ali-alkhatib.com/blog/digital-contact-tracing.

Allegretti, Aubrey. 2020. "Coronavirus: 'World-Beating' Track and Trace System Ready for Schools to Reopen, PM Promises." *Sky News*, May 21, 2020. https://news.sky.com/story/coronavirus-world-beating-track-and-trace-system-ready-for-schools-to-reopen-pm-promises-11991606.

Allison Marsh. 2018. "John Deere and the Birth of Precision Agriculture—IEEE Spectrum." February 28, 2018. https://spectrum.ieee.org/tech-history/silicon-revolution/john-deere-and-the-birth-of-precision-agriculture.

Alpert, Lukas I. 2017. "New York Times Lays Out Plans to Restructure Newsroom." *Wall Street Journal*, January 17, 2017. https://www.wsj.com/articles/new-york-times-lays-out-plans-to-restructure-newsroom-1484676531.

AltexSoft. 2020. "How to Structure a Data Science Team: Key Models and Roles to Consider." *AltexSoft* (blog), June 30, 2020. https://www.altexsoft.com/blog/datascience/how-to-structure-data-science-team-key-models-and-roles.

AltexSoft. 2021. "Comparing Machine Learning as a Service: Amazon, Microsoft Azure, Google Cloud AI, IBM Watson." *AltexSoft* (blog), April 25, 2021. https://www.altexsoft.com/blog/datascience/comparing-machine-learning-as-a-service-amazon-microsoft-azure-google-cloud-ai-ibm-watson.

Altman, Howard. 2022. "Debacle on the Donets: How Russian Forces Got Obliterated Trying to Cross a River." *The Drive*, May 12, 2022. https://www.thedrive.com/the-war-zone/debacle-on-the-donets-russian-forces-got-obliterated-trying-to-cross-a-river.

Alvarado, Anthony. 2005. *District Wide Reform: San Diego, CA*. "Making Schools Work with Hedrick Smith." PBS. http://www.pbs.org/makingschoolswork/dwr/ca/alvarado.html.

Alves, Bruna. n.d. "Number of Fossil Fuel Thefts in Mexico 2019." *Statista*. Accessed June 19, 2020. https://www.statista.com/statistics/709121/number-fuel-theft-detect-mexico/.

Anderson, Chris. 2008. *The Long Tail: Why the Future of Business Is Selling Less of More*, rev. ed. New York: Hachette.

Andreessen, Marc. 2013. Marc Andreessen on the Age of Context (Er, the Future of the Tech Industry). Interview by Robert Scoble. https://www.youtube.com/watch?v=YiQyDhXiU4s.

AndroidGuys. 2020. "Teach to One Builds Math Proficiency with Tailored Approach Offering Digital Personalized Roadmaps." *AndroidGuys* (blog), November 17, 2020. https://www.androidguys.com/promoted-news/teach-to-one-builds-math-proficiency-with-tailored-approach-offering-digital-personalized-roadmaps.

Apple Developer. n.d. "Exposure Notification: Implement a COVID-19 Exposure Notification System That Protects User Privacy." Accessed July 17, 2020. https://developer.apple.com/documentation/exposurenotification.

Aquino, Alyssa. 2020. "Ex-Google Exec's Co. to Pay $18.5M to End Payday Loan Suit." *Law 360*, February 11, 2020. https://www.law360.com/articles/1242920/ex-google-exec-s-co-to-pay-18-5m-to-end-payday-loan-suit.

Arak, Josh. 2021. Interview by Zachary Tumin and Madeleine Want.

Arak, Josh, and Kentaro Kaji. 2017. "ABRA: An Enterprise Framework for Experimentation at *The Times*." *NYT Open* (blog), *Medium*, August 30, 2017. https://open.nytimes.com/abra-an-enterprise-framework-for-experimentation-at-the-times-57f8931449cd.

Arnold, Amanda, Claire Lampen, and Brooke LaMantia. 2020. "Everything We Know About the Tessa Majors Murder Case." *Cut*, February 19, 2020. https://www.thecut.com/2020/02/tessa-majors-barnard-student-death.html.

Arumilli, Gautham. 2020. "What Is the Voter File?" *DNC Tech Team* (blog), August 31, 2020. https://medium.com/democratictech/what-is-the-voter-file-a8f99dd07895.

Associated Press. 2016. "Former LAPD Chief William Bratton to Resign as New York Police Commissioner." *Los Angeles Times*, August 2, 2016. https://www.latimes.com/nation/nationnow/la-na-nypd-bill-bratton-resigns-20160802-snap-story.html.

Augustine-Obi, Sean and Russell Mindich. 2022. "Political Tech Landscape Report 2021." Higher Ground Labs. https://docsend.com/view/bbcj4j3g98bcqnb2.

Avelino, Álvaro, Jose Paiva, Rodrigo Silva, et al. 2009. "Real Time Leak Detection System Applied to Oil Pipelines Using Sonic Technology and Neural Networks." In *35th Annual Conference of IEEE Industrial Electronics*, 2009, 2109–14. https://doi.org/10.1109/IECON.2009.5415324.

Avendaño, L. L. 2016. "The Political, Economic, and Social Roots of Energy Insecurity in México." *Journal of Economics, Business and Management* 4, no. 7 (2016). doi:10.18178/joebm.2016.4.7.435.

Awbrey, Jake. 2020. "The Future of NFL Data Analytics." Samford.Edu. June 9, 2020. https://www.samford.edu/sports-analytics/fans/2020/The-Future-of-NFL-Data-Analytics.

Badenhausen, Kurt. 2022. "NFL Team Valuations 2022: Cowboys Rule at $7.6B as Average Tops $4B – Sportico.Com." Sportico.Com. August 1, 2022. https://www.sportico.com/valuations/teams/2022/nfl-team-valuations-2022-cowboys-1234684184/.

Bailie, Ian, Bernard Marr, Laszlo Bock, and Mel Gee Kee. 2018. "CogX 2018—The Impact of AI on HR and Recruitment." Panel discussion at the CogX Festival in June 2018, Tobacco Dock, London. Published June 26, 2018. *YouTube*. https://youtu.be/o6-Fc2K3YFY.

Baillargeon, Jay P. 2018. "Update on FRA's Predictive Analytics Program." Presentation at the Big Data in Railway Maintenance Planning Conference, University of Delaware, December 2018.

Ball, James. 2020. "The UK's Contact Tracing App Fiasco Is a Master Class in Mismanagement." *MIT Technology Review*, June 19, 2020. https://www.technologyreview.com/2020/06/19/1004190/uk-covid-contact-tracing-app-fiasco.

Bambha, Abhishek. 2017. "Project Estimation Through T-Shirt Size." *Radius Tech Blog* (blog), *Medium*, September 22, 2017. https://medium.com/radius-engineering/project-estimation-through-t-shirt-size-ea496c631428.

Baquet, Dean (@deanbaquet). n.d. *Twitter*. Accessed May 5, 2022. https://twitter.com/deanbaquet.

Baroudi, Uthman, Anas A. Al-Roubaiey, and Abdullah Devendiran. 2019. "Pipeline Leak Detection Systems and Data Fusion: A Survey." *IEEE Access* 7 (2019): 97426–97439. https://doi.org/10.1109/ACCESS.2019.2928487.

Barshay, Jill. 2019. "New Study Casts Doubts on Effectiveness of Personalized Learning Program." *Hechinger Report*, February 25, 2019. http://hechingerreport.org/new-study-casts-doubts-on-effectiveness-of-personalized-learning-program-in-new-jersey.

Bartells, Lynn. 2014. "In Colorado, the GOP Shifted the Ground Game." *Denver Post*, November 8, 2014. https://www.denverpost.com/2014/11/08/in-colorado-the-gop-shifted-the-ground-game.

Bassett, Mike. 2020. "Radiologists Develop AI Tool to Identify Risk of COVID Complications." Radiological Society of North America. Last modified August 27, 2020. https://www.rsna.org/news/2020/august/ai-predictive-tool.

Bates, David W., et al. 1995. "Incidence of Adverse Drug Events and Potential Adverse d\ Drug Events: Implications for Prevention. ADE Prevention Study Group." *JAMA* 274, no. 1 (1995): 29–34.

Bauer, Anne, and Anna Coenen. 2020. "AI in Action E142: Anne Bauer, Director of Data Science and Anna Coenen, Lead Data Scientist at New York Times." In *Alldus Podcast*. Published November 3, 2020. Anchor podcast. https://alldus.com/blog/podcasts/aiinaction-anne-bauer-anna-coenen-new-york-times.

Baxter, Alexander. Interviews by Zachary Tumin and Madeleine Want.

Bay, Jason. 2019. "Leading Through Crises of Reputation." *Medium* (blog). April 29, 2019. https://medium.com/gsbgen317s19/leading-through-crises-of-reputation-a2e83e54f5a9.

Bay, Jason. 2020a. "Automated Contact Tracing Is Not a Coronavirus Panacea." *Medium* (blog). April 10, 2020. https://blog.gds-gov.tech/automated-contact-tracing-is-not-a-coronavirus-panacea-57fb3ce61d98.

Bay, Jason. 2020b. "One Month On." *Medium* (blog). April 20, 2020. https://blog.gds-gov.tech/one-month-on-6c882875b9ba.

Bay, Jason. 2020c. "TraceTogether at One." *Medium* (blog). March 20, 2020. https://blog.gds-gov.tech/tracetogether-at-one-7ed20de20dfo.

Bay, Jason, Joel Kek, Alvin Tan, Chai Sheng Hau, Lai Yongquan, Janice Tan, and Tang Anh Quy. 2020. "BlueTrace: A Privacy-Preserving Protocol for Community-Driven Contact Tracing Across Borders." White Paper, Government Technology Agency, Singapore, April 9, 2020. https://bluetrace.io/static/bluetrace_whitepaper-938063656596c104632def383eb33b3c.pdf.

*BBC News*. 2021. "Twitter Finds Racial Bias in Image-Cropping AI." May 20, 2021. https://www.bbc.com/news/technology-57192898.

Beam, Christopher. 2022. "The Math Prodigy Whose Hack Upended DeFi Won't Give Back His Millions." *Bloomberg.com*, May 19, 2022. https://www.bloomberg.com/news/features/2022-05-19/crypto-platform-hack-rocks-blockchain-community.

Beck, Martha. 2001. *Finding Your Own North Star: Claiming the Life You Were Meant to Live*. New York: Three Rivers.

Behr, Michael. 2020. "AI Pilot Defeats Human in DARPA Fighter Combat Simulation." *Digit*, August 24, 2020. https://digit.fyi/ai-pilot-defeats-human-in-darpa-fighter-combat-simulation.

Bent, Kristin. 2014. "Cisco's Big Challenge: Will Chambers Stay to See It Through?" *CRN*, August 7, 2014. https://www.crn.com/news/networking/300073623/ciscos-big-challenge-will-chambers-stay-to-see-it-through.htm.

Benton, Joshua. 2014. "The Leaked New York Times Innovation Report Is One of the Key Documents of This Media Age." *Nieman Lab* (blog). May 15, 2014. https://www.niemanlab.org/2014/05/the-leaked-new-york-times-innovation-report-is-one-of-the-key-documents-of-this-media-age.

Benton, Joshua. 2017. "This Is the *New York Times*' Digital Path Forward." *Nieman Lab* (blog). January 17, 2017. https://www.niemanlab.org/2017/01/this-is-the-new-york-times-digital-path-forward.

Benton, Joshua. 2022. "The *New York Times* Would Really Like Its Reporters to Stop Scrolling and Get Off Twitter (at Least Once in a While)." April 7, 2022. *Nieman Lab* (blog). https://www.niemanlab.org/2022/04/the-new-york-times-would-really-like-its-reporters-to-stop-scrolling-and-get-off-twitter-at-least-once-in-a-while/.

Berger, Peter L. and Richard John Neuhaus. 1989. *To Empower People: The Role of Mediating Structures in Public Policy*. Ex-Seminary Library edition. Washington: AEI Press.

Bergman, Artur. 2018. "Open Questions: A Conversation with Fastly CEO Artur Bergman." Interview by Nick Rockwell. *NYT Open* (blog), *Medium*. September 7, 2018. https://open.nytimes.com/open-questions-a-conversation-with-fastly-ceo-artur-bergman-9c5e023baf57.

Bernal Mendoza, Jorge. 2020. Transcript: Interview by Zachary Tumin and Madeleine Want.

Berwick, Donald M. 2003. "Disseminating Innovations in Health Care." *Journal of the American Medical Association* 289, no. 15 (2003): 1969–1975. https://doi.org/10.1001/jama.289.15.1969.

BestPractice.ai. n.d. "Unilever Saved 50,000 Hours in Candidate Time over 18 Months." Accessed November 16, 2021. https://www.bestpractice.ai/studies/unilever_saved_over_50_000

_hours_in_candidate_interview_time_and_delivered_over_1m_annual_savings_and _improved_candidate_diversity_with_machine_analysis_of_video_based_interviewing.

Best Practice Institute Staff. 2015. "The Unique Challenges of HR in Asia." *BPI Modern Talent Magazine*, May 12, 2015. https://www.bestpracticeinstitute.org/blog/the-unique-challenges-of-hr-in-asia.

Bhatia, Katy, A. M. Gaythorpe, et al. 2020. *Report 25: Response to COVID-19 in South Korea and Implications for Lifting Stringent Interventions*. May 29, 2020. London: Imperial College London. https://doi.org/10.25561/79388.

Bishop, Greg. 2022. "Inside Cooper Kupp's Unique Origin Story and Training." *Sports Illustrated*. January 21, 2022. https://www.si.com/nfl/2022/01/21/cooper-kupp-unique-origin-story-of-unlikely-superstar-daily-cover.

Bladon, Paul, Kim Bowling, Hark Braren, Jean Deslauriers, and Denis D'Aoust. 2015. "The Challenges of Integrating Novel Wayside Rolling Stock Monitoring Technologies: A Case Study." Paper presented at the International Heavy Haul Association Conference, Perth, Australia, June 2015. https://docplayer.net/100232156-The-challenges-of-integrating-novel-wayside-rolling-stock-monitoring-technologies-a-case-study.html.

Blodget, Henry. 2011. "The Truth About Cisco: John Chambers Has Failed." *Business Insider*, May 13, 2011. https://www.businessinsider.com/cisco-csco-john-chambers-has-failed-2011-5.

Bloomfield, Claire. 2021. "Ecosystem Approach Will Bring AI Benefits to the Clinical Front Line." *Healthcare IT News*, June 18, 2021. https://www.healthcareitnews.com/news/emea/ecosystem-approach-will-bring-ai-benefits-clinical-front-line.

BNamericas. n.d. "Mexico's PEMEX Steps Up War on Fuel Theft." *BNamericas.com*. Accessed June 19, 2020a. https://www.bnamericas.com/en/news/mexicos-pemex-steps-up-war-on-fuel-theft.

BNamericas. n.d. "PEMEX Launches Digitization Drive." *BNamericas.com*. Accessed June 19, 2020b. https://www.bnamericas.com/en/news/pemex-launches-digitization-drive1.

BNSF. 2019a. "Artificial Intelligence: A Look at How BNSF Creates a Safer, Smarter Railroad." Safety. Last modified March 27, 2019. https://www.bnsf.com/news-media/railtalk/safety/artificial-intelligence.html.

BNSF. 2019b. "BNSF 2019 Annual Review." Safety. https://www.bnsf.com/about-bnsf/bnsf-review/2019/safety.html.

BNSF. 2019c. "Making the Grade: How BNSF Maintenance Is Planned and Performed." Last modified June 26, 2019. https://www.bnsf.com/news-media/railtalk/service/maintenance-planning.html.

Bock, Laszlo. 2014. "Laszlo Bock on How Google Is Changing the Nature of Work." Speech at the 2014 Re:Work with Google event. Published November 10, 2014. YouTube video, 20:10. https://youtu.be/QOXpajH89hw.

Bond, Becky, and Zack Exley. 2016. *Rules for Revolutionaries: How Big Organizing Can Change Everything*. White River Junction, VT: Chelsea Green.

Borchers, Callum. 2013. "Patriots Aim to Enhance Stadium Experience as Home Viewing Keeps Getting Better." *Boston.com*. Last modified January 18, 2013. http://www.boston.com/business/2013/01/19/patriots-aim-enhance-stadium-experience-home-viewing-keeps-getting-better/rvA34GcIlWh5PN6spfu5LI/story-1.html.

Bos, Bert. 2003. "What Is a Good Standard: An Essay on W3C's Design Principles—Backwards Compatibility." *W3c.Org*. https://www.w3.org/People/Bos/DesignGuide/compatibility.html.

Bourdeaux, Margaret, Mary L. Gray, and Barbara Grosz. 2020. "How Human-Centered Tech Can Beat COVID-19 Through Contact Tracing." *The Hill*, April 21, 2020. https://thehill.com/opinion/technology/493648-how-human-centered-technology-can-beat-covid-19-through-contact-tracing.

Bourgeois, L. J., and Kathleen M. Eisenhardt. 1988. "Strategic Decision Processes in High Velocity Environments: Four Cases in the Microcomputer Industry." *Management Science* 34, no. 7 (1988): 816–835.

Boyne, Walter J. 2009. "How the Predator Grew Teeth." *Air Force Magazine*, June 29, 2009. https://www.airforcemag.com/article/0709predator.

Braga, Anthony A., and David M. Kennedy. 2021. *A Framework for Addressing Violence and Serious Crime: Focused Deterrence, Legitimacy, and Prevention*. Cambridge: Cambridge University Press.

Brandom, Russell. 2019. "The NYPD Uses Altered Images in Its Facial Recognition System, New Documents Show." *Verge*, May 16, 2019. https://www.theverge.com/2019/5/16/18627548/nypd-facial-recognition-altered-faces-privacy.

Bratton, William. 1998. *Turnaround*. New York: Random House.

Bratton, William, and Zachary Tumin. 2012. *Collaborate or Perish!: Reaching Across Boundaries in a Networked World*. First edition. New York: Crown Business.

*Breaking Defense*. 2021. "Predicting the Unpredictable: How Raytheon Technologies Uses Modeling and Simulation to Prepare for UAS Threats." October 4, 2021. https://breakingdefense.sites.breakingmedia.com/2021/10/predicting-the-unpredictable.

Breer, Albert. 2017. "How All 32 NFL Teams Handle Analytics vs. Old School." *Sports Illustrated*, June 28, 2017. https://www.si.com/nfl/2017/06/28/nfl-analytics-front-office-old-school-approach-draft-game-planning-charting.

Brennan, Shane. 2017. "The Ten Fallacies of Data Science." *Towards Data Science*, September 17, 2017. https://towardsdatascience.com/the-ten-fallacies-of-data-science-9b2af78a1862.

Bright, Julian. 2021. "Dynamic A/B Testing for Machine Learning Models with Amazon SafeMaker MLOps Project." *AWS Machine Learning Blog*, July 9, 2021. https://aws.amazon.com/blogs/machine-learning/dynamic-a-b-testing-for-machine-learning-models-with-amazon-sagemaker-mlops-projects/.

Brikman, Jim. n.d. "A Minimum Viable Product Is Not a Product, It's a Process." *Y Combinator*. Accessed November 26, 2021. https://www.ycombinator.com/library/4Q-a-minimum-viable-product-is-not-a-product-it-s-a-process.

Brock, Todd. 2019. "Garrett on Analytical Decisions: 'We Don't Use Those Stats Within the Game.'" *Cowboys Wire* (blog), November 25, 2019. https://cowboyswire.usatoday.com/2019/11/25/garrett-on-analytical-decisions-we-dont-use-those-stats-within-the-game.

Broockman, David, and Joshua Kalla. 2014. "Experiments Show This Is the Best Way to Win Campaigns. But Is Anyone Actually Doing It?" *Vox*, November 13, 2014. https://www.vox.com/2014/11/13/7214339/campaign-ground-game.

Brown, Deborah. 2020. "Closing the 'Digital Divide' Critical in COVID-19 Response." *Human Rights Watch* (blog), March 25, 2020. https://www.hrw.org/news/2020/03/25/closing-digital-divide-critical-covid-19-response.

Brown, Eliot. 2017. "Why Big Data Hasn't Yet Made a Dent on Farms." *Wall Street Journal*, May 15, 2017, sec. Business. https://www.wsj.com/articles/why-big-data-hasnt-yet-made-a-dent-on-farms-1494813720.

Brown, Sara. 2021. "Machine Learning, Explained." *Ideas Made to Matter* (blog), MIT Sloan School of Management, April 21, 2021. https://mitsloan.mit.edu/ideas-made-to-matter/machine-learning-explained.

Brueggemann, Brian. 2019. "Former Google Exec's Company Gets Caught in 'Rent-A-Tribe' Class Action Lawsuits; Interest Rates Allegedly as High as 490 Percent." *Legal Newsline*, April 2, 2019. https://legalnewsline.com/stories/512366003-former-google-exec-s-company-gets-caught-in-rent-a-tribe-class-action-lawsuits-interest-rates-allegedly-as-high-as-490.

Bruno, Giancarlo, Abel Lee, Matthew Blake, Jesse McWaters, Rob Galaski, and Hwan Kim. 2015. *The Future of Financial Services*. Final Report, June 2015. Geneva: World Economic Forum. https://www3.weforum.org/docs/WEF_The_future__of_financial_services.pdf.

Buffett, Howard W., and William B. Eimicke. 2018. *Social Value Investing: A Management Framework for Effective Partnerships*. Illustrated ed. New York: Columbia University Press.

Buntin, John, and Carol A. Chetkovich. 2000. "The NYPD Takes on Crime in New York City. (A) and (B)." Harvard University, John F. Kennedy School of Government Case HKS1557.1, January 1, 2000.

Buolamwini, Joy, and Timnit Gebru. 2018. "Gender Shades: Intersectional Accuracy Disparities in Commercial Gender Classification." *Proceedings of the First Conference on Fairness, Accountability and Transparency* 81 (2018): 77–91. https://proceedings.mlr.press/v81/buolamwini18a.html.

Burke, Tim. 2021. "We Don't Get Bitter, We Get Better." *Medium*, April 15, 2021. https://open.nytimes.com/we-dont-get-bitter-we-get-better-b5d2783d5cd3.

Burt, Ronald S. 2007. *Brokerage and Closure: An Introduction to Social Capital*. Oxford: Oxford University Press.

Burwood-Taylor, Louisa. 2017. "How an Artificial Intelligence and Robotics Startup Got Acquired by John Deere [Interview of Jorge Heraud.] Agfundernews.com. October 3, 2017. https://agfundernews.com/artificial-intelligence-robotics-startup-got-acquired-john-deere.html.

Calvino, Italo. 1988. *Six Memos for the Next Millennium*. Cambridge, MA: Harvard University Press.

Cameron, Kim S., and Robert E. Quinn. 2011. *Diagnosing and Changing Organizational Culture: Based on the Competing Values Framework*. 3rd ed. San Francisco: Jossey-Bass.

Cannavo, Michael J. 2021. "Can Radiology AI Win with a Bad Hand?" *AuntMinnie.com*, June 21, 2021. https://www.auntminnie.com/index.aspx?sec=ser&sub=def&pag=dis&ItemID=132698.

Cantwell, Houston R. 2012. "RADM Thomas J. Cassidy's MQ-1 Predator: The USAF's First UAV Success Story." Master's diss., Air Command and Staff College, Maxwell, AL. https://www.amazon.com/RADM-Thomas-Cassidys-MQ-1-Predator/dp/1288230230.

Cao, Longbing. 2017. "Data Science: Challenges and Directions." *Communications of the ACM* 60, no. 8 (2017): 59–68.

Cardona, Pam. 2020. "Developing Cultural Infrastructure." *DNC Tech Team* (blog), *Medium*, August 4, 2020. https://medium.com/democratictech/developing-cultural-infrastructure-569e2fc72d3f.

Carman, Tim. 2022. "Grubhub Apologizes for 'Free Lunch' Promo That Slammed NYC Restaurants." *Washington Post*, May 20, 2022. https://www.washingtonpost.com/food/2022/05/18/grubhub-nyc-promo/.

Carroll, Diane. 2021. "List of Platform Integrations." Civis Analytics. Last modified May 5, 2021. https://civis.zendesk.com/hc/en-us/articles/115004269763-List-of-Platform-Integrations.

"Case Study: Unilever." Human Resources Online. August 27, 2014. https://www.humanresourcesonline.net/case-study-unilever.

Caspar, Stephanie. 2018. "We Want to Have Influence." Interview by *Inside.mag*. https://www.axelspringer.com/data/uploads/2018/11/Interview-EN.pdf.

Caspar, Stephanie. 2019. "Axel Springer Head of Data and Tech: Media Culture Overhaul Needed to Save Journalism." Presentation at FIPP, March 26, 2019. //www.fipp.com/news/media-culture-overhaul-needed-save-journalism/.

Cavaretta, Michael. 2021. "It's Deceptively Easy to Use AI for Business Insights. It Can Be Done with Descriptive, Predictive and Prescriptive Analytics, but Let's Start with a Simple Descriptive Example." *LinkedIn*, February 19, 2021. https://www.linkedin.com/posts/michael-cavaretta-ph-d-795a965_ai-artificialintelligence-datascience-activity-6768521841276440576-wd2b.

Cegłowski, Maciej. 2016. "Remarks at the SASE Panel on the Moral Economy of Tech." *Idlewords.com*, June 26. https://idlewords.com/talks/sase_panel.htm.

Chahal, Husanjot, and Helen Toner. 2021. " 'Small Data' Are Also Crucial for Machine Learning." *Scientific American*, October 19, 2021. https://www.scientificamerican.com/article/small-data-are-also-crucial-for-machine-learning/.

Chambers, John. 2016. "Cisco's John Chambers on the Digital Era." Interview by Rik Kirkland, March 18, 2016. https://www.mckinsey.com/industries/technology-media-and-telecommunications/our-insights/ciscos-john-chambers-on-the-digital-era.

Chan, Wilfred. 2022. " 'This Can't Be Real': Grubhub Promotion Turns New York City Restaurants into a 'War Zone.' " TheGuardian.Com. May 19, 2022. https://www.theguardian.com/us-news/2022/may/18/this-cant-be-real-grubhub-promotion-turns-new-york-city-restaurants-into-a-war-zone.

Chang, Peter. 2021, various dates. Interviews by Zachary Tumin and Madeleine Want.

Chater, Nick, and George F., Loewenstein. (2022). "The i-Frame and the s-Frame: How Focusing on Individual-Level Solutions Has Led Behavioral Public Policy Astray." March 1, 2022. https://ssrn.com/abstract=4046264 or http://dx.doi.org/10.2139/ssrn.4046264.

Chee, Kenny. 2021. "COVID-19 Tech Heroes Recognised for Racing Against the Clock to Help Fight Pandemic in S'pore." *Straits Times*, March 18, 2021. https://www.straitstimes.com/tech/tech-news/covid-19-tech-heroes-recognised-for-racing-against-the-clock-to-help-fight-the.

Chen, Andrew. n.d. "How to Build a Growth Team—Lessons from Uber, Hubspot, and Others." Accessed May 23, 2021. https://andrewchen.com/how-to-build-a-growth-team.

Chokshi, Sonal, Ryan Caldbeck, and Jeff Jordan. 2019. "A16z Podcast: Who's Down with CPG, DTC? (And Micro-Brands Too?)." February 15, 2019. https://a16z.com/2019/02/15/cpg-dtc-microbrands-grocery-online-offline-commerce/.

Cholas-Wood, Alex. 2021. Interview by Zachary Tumin and Madeleine Want.

Cholas-Wood, Alex, and E. S. Levine. 2019. "A Recommendation Engine to Aid in Identifying Crime Patterns." *Journal on Applied Analytics* 49, no. 2 (March 2019): 154–166. https://doi.org/10.1287/inte.2019.0985.

Chowdhury, Rumman. 2021. "Sharing Learnings About Our Image Cropping Algorithm." *Insights* (blog), *Twitter Engineering*, May 19, 2021. https://blog.twitter.com/engineering/en_us/topics/insights/2021/sharing-learnings-about-our-image-cropping-algorithm.

Christensen, Clayton M. 2011. *The Innovator's Dilemma: The Revolutionary Book That Will Change the Way You Do Business*. New York: Harper Business.

Christensen, Clayton M. M., Jerome H. Grossman, and Jason Hwang. 2016. *The Innovator's Prescription: A Disruptive Solution for Health Care*. New York: McGraw-Hill Education.

Chung, Gina. 2020. " 'The First Day Is the Worst Day': DHL's Gina Chung on How AI Improves over Time." Interview with Sam Ransbotham and Shervin Khodabandeh. Transcript included. *MIT Sloan Management Review*, October 27, 2020. https://sloanreview.mit.edu/audio/the-first-day-is-the-worst-day-dhls-gina-chung-on-how-ai-improves-over-time.

Churchman, C. West. 1967. "Guest Editorial: Wicked Problems." *Management Science* 14, no. 4 (December 1967): B141–B142. http://www.jstor.org/stable/2628678.

Cimpanu, Catalin. 2020. "Political Campaign Emails Contain Dark Patterns to Manipulate Donors, Voters." *ZDNet*, October 17, 2020. https://www.zdnet.com/article/political-campaign-emails-contain-dark-patterns-to-manipulate-donors-voters.

Ciocca, Sophia. 2018. "Building a Text Editor for a Digital-First Newsroom." *NYT Open* (blog), *Medium*, April 12, 2018. https://open.nytimes.com/building-a-text-editor-for-a-digital-first-newsroom-f1cb8367fc21.

Ciocca, Sophia, and Jeff Sisson. 2019. "We Built Collaborative Editing for Our Newsroom's CMS. Here's How." *NYT Open* (blog), *Medium*, August 1, 2019. https://open.nytimes.com/we-built-collaborative-editing-for-our-newsrooms-cms-here-s-how-415618a3ec49.

Clark, Kevin. 2018. "The NFL's Analytics Revolution Has Arrived." *The Ringer*, December 19, 2018. https://www.theringer.com/nfl/2018/12/19/18148153/nfl-analytics-revolution.

Coenen, Anna. 2019. "How The *New York Times* Is Experimenting with Recommendation Algorithms." *NYT Open* (blog), *Medium*, October 17, 2019. https://open.nytimes.com/how-the-new-york-times-is-experimenting-with-recommendation-algorithms-562f78624d26.

Cohen, Jackie. 2020. "Good for What? Humans vs. Computers." *DNC Tech Team* (blog), *Medium*, May 28, 2020. https://medium.com/democratictech/good-for-what-humans-vs-computers-eb3a644fed81.

Cohen, Julie E., Woodrow Hartzog, and Laura Moy. 2020. "The Dangers of Tech-Driven Solutions to COVID-19." *Brookings TechStream* (blog), June 17, 2020. https://www.brookings.edu/techstream/the-dangers-of-tech-driven-solutions-to-covid-19.

Coiera, Enrico. 2018. "The Fate of Medicine in the Time of AI." *The Lancet* 392 no. 10162 (2018). https://www-clinicalkey-com.ezproxy.cul.columbia.edu/#!/content/journal/1-s2.0-S0140673618319251.

Coiera, Enrico. 2019. "On Algorithms, Machines, and Medicine." *Lancet Oncology* 20, no. 2 (2019): 166–167.

ColumbiaRadiology.com. 2020. "Q & A with Dr. Richard Ha: Deep Learning and Breast Cancer Risk Assessment." Columbia Department of Radiology, October 15, 2020. https://www.columbiaradiology.org/news/q-dr-richard-ha-deep-learning-and-breast-cancer-risk-assessment.

CompStat360. n.d. "Implementing CompStat360." CS360. Accessed November 16, 2021. https://www.compstat360.org/implementing-compstat360-2.

Cook, Lindsey Rogers. 2019. "How We Helped Our Reporters Learn to Love Spreadsheets." *NYT Open* (blog), *Medium*, June 12, 2019. https://open.nytimes.com/how-we-helped-our-reporters-learn-to-love-spreadsheets-adc43a93b919.

Cosentino, Dom. 2021. "How to Think About Football and Analytics Like a Coach, Not a Fuddy-Duddy." *TheScore.com*, December 2021. https://www.thescore.com/nfl/news/2250074.

Cowgill, Bo, and Catherine Tucker. 2017. "Algorithmic Bias: A Counterfactual Perspective." *Bitlab*. https://bitlab.cas.msu.edu/trustworthy-algorithms/whitepapers/Bo%20Cowgill.pdf.

Cramer, Maria. 2021. "A.I. Drone May Have Acted on Its Own in Attacking Fighters, U.N. Says." *New York Times*, June 3, 2021, sec. World.

Crawford, Kate, and Vladan Joler. 2019. "Anatomy of an AI System." *Virtual Creativity* 9, no. 1 (December 1, 2019): 117–120. https://doi.org/10.1386/vcr_00008_7.

Crawford, Susan P., and Laura Adler. 2016. "Culture Change and Digital Technology: The NYPD Under Commissioner William Bratton, 2014–2016." SSRN Scholarly Paper ID 2839004. Rochester, NY: Social Science Research Network. https://papers.ssrn.com/abstract=2839004.

Crosman, Penny. 2012. "ZestFinance Aims to Fix Underwriting for the Underbanked." *American Banker*, November 19, 2012. https://www.americanbanker.com/news/zestfinance-aims-to-fix-underwriting-for-the-underbanked.

Cunningham, Charles. 2021. Interviews by Zachary Tumin.

Cunningham, Nicholas. 2019. "Fuel Feuds in Mexico: Reviving Mexico's Oil Industry Will Require a Crackdown on Fuel Theft in Mexico. What Are the Implications for López Obrador's Progressive Mandate?" *NACLA Report on the Americas* 51, no. 2 (2019): 119–122. https://doi.org/10.1080/10714839.2019.1617466.

Cwalina, Alex, Christine Griffin, Jacob Haan, Nathan Kahn, Shehnaz Mannan, Alessandra Sauro, Ellen Shepard, Sterling Wiggins, Courtney Radcliff Gill, Timothy Reuter, and Harrison Wolf. 2021. *Medicine from the Sky: Opportunities and Lessons from Drones in Africa*. Insight Report, March 2021. Geneva: World Economic Forum. https://www.updwg.org/wp-content/uploads/2021/04/WEF_Medicine_from_the_Sky_2021.pdf.

Danaher, Brett. 2021. Interview by Zachary Tumin and Madeleine Want.

Darius, Sebastien. 2021. "Basketball Analytics, Part 2: Shot Quality." *Medium*, July 16, 2021. https://towardsdatascience.com/part-2-shot-quality-5ab27fd63f5e.

Datta, Shantanu, and Shibayan Sarkar. 2016. "A Review on Different Pipeline Fault Detection Methods." *Journal of Loss Prevention in the Process Industries* 41 (May 2016): 97–106. https://doi.org/10.1016/j.jlp.2016.03.010.

Davenport, Thomas H. 2011. "The Need for Analytical Service Lines." *International Institute for Analytics* (blog), October 28, 2011. Archived at the Wayback Machine. https://web.archive.org/web/20160623030702/http://iianalytics.com/research/the-need-for-analytical-service-lines.

Davenport, Thomas H. 2016. "Analytics and IT: New Opportunity for CIOs." *Harvard Business Review*, October 14, 2016. https://hbr.org/webinar/2016/09/analytics-and-it-new-opportunity-for-cios.

Davenport, Thomas H. 2018. *The AI Advantage: How to Put the Artificial Intelligence Revolution to Work*. Cambridge, MA: MIT Press.

Davenport, Thomas H., and David Brain. 2018. "Before Automating Your Company's Processes, Find Ways to Improve Them." *Harvard Business Review*, June 13, 2018. https://hbr.org/2018/06/before-automating-your-companys-processes-find-ways-to-improve-them.

Davenport, Thomas H., Abhijit Guha, and Dhruv Grewal. 2021. "How to Design an AI Marketing Strategy." *Harvard Business Review* (July–August 2021).

Davenport, Thomas H., Jeanne Harris, and Jeremy Shapiro. 2010. "Competing on Talent Analytics." *Harvard Business Review*, October 1, 2010. https://hbr.org/2010/10/competing-on-talent-analytics.

Davenport, Thomas H., and Rajeev Ronanki. 2018. "Artificial Intelligence for the Real World." *Harvard Business Review*, February 12, 2018. https://hbr.org/2018/01/artificial-intelligence-for-the-real-world.

Davis, Ben. 2017. "Ask the Experts: What's the Best Way to Target Programmatic Ads?" *Econsultancy* (blog), October 31, 2017. https://econsultancy.com/ask-the-experts-what-s-the-best-way-to-target-programmatic-ads/.

Davis, Lynn E., Michael J. McNerney, and Michael D. Greenberg. 2016. *Clarifying the Rules for Targeted Killing: An Analytical Framework for Policies Involving Long-Range Armed Drones*. Santa Monica, CA: RAND. https://www.rand.org/pubs/research_reports/RR1610.html.

DCVC. 2017. "John Deere Acquires Blue River Technology for $305 Million, Bringing Full Stack AI to Agriculture." *Medium* (blog), September 6, 2017. https://medium.com/@dcvc/john-deere-acquires-blue-river-technology-for-305-million-bringing-full-stack-ai-to-agriculture-7ca8c25a5fe1.

Debenedetti, Gabriel. 2020. "Election Night with Biden's Data Guru." *New York Magazine Intelligencer*, November 18, 2020. https://nymag.com/intelligencer/2020/11/election-night-with-bidens-data-guru.html.

Dehaye, Paul-Olivier. 2020. "Inferring Distance from Bluetooth Signal Strength: A Deep Dive." *Medium* (blog), May 19, 2020. https://medium.com/personaldata-io/inferring-distance-from-bluetooth-signal-strength-a-deep-dive-fe7badc2bb6d.

Del Beccaro, Mark A., et al. 2006. "Computerized Provider Order Entry Implementation: No Association with Increased Mortality Rates in an Intensive Care Unit." *Pediatrics* 118, no. 1 (2006): 290. https://doi.org/10.1542/peds.2006-0367.

Deldjoo, Yashar, Maurizio Ferrari Dacrema, Mihai Gabriel Constantin, et al. 2019. "Movie Genome: Alleviating New Item Cold Start in Movie Recommendation." *User Modeling and User-Adapted Interaction* 29, no. 2 (2019): 291–343. https://doi.org/10.1007/s11257-019-09221-y.

Delport, Jenna. 2020. "Personalisation: Differentiation Done Right." *ITNewsAfrica.com* (blog), February 1, 2020. https://www.itnewsafrica.com/2020/02/personalisation-differentiation-done-right/.

DeLuca, Joel R. 1999. *Political Savvy: Systematic Approaches to Leadership Behind the Scenes.* 2nd ed. Berwyn, PA: Evergreen Business Group.

De Martini, Garey. 2020. "Confessions of a Dem Dialer." *Marina Times*, November 10, 2020. https://www.marinatimes.com/2020/10/confessions-of-a-dem-dialer.

DeStefano, Anthony M. 2021. "Report: Stop and Frisks Fell 93 Percent After '13 Settlement.'" *Newsday*, September 1, 2021. https://www.newsday.com/long-island/crime/nypd-stop-and-frisk-peter-zimroth-1.50349874.

Detsch, Jack. 2022. "Drones Have Come of Age in Russia-Ukraine War." *Foreign Policy* (blog), April 27, 2022. https://foreignpolicy.com/2022/04/27/drones-russia-ukraine-war-donbas/.

Dewan, Pooja. 2019. "2018 IAAA Finalist: Automatic Train Identification." Presentation at the INFORMS Innovative Applications in Analytics Award. Published January 8, 2019. *YouTube*. https://youtu.be/cyQx66Of73s.

Dewey, John. 1954. *The Public and Its Problems.* Athens, GA: Swallow.

Dichter, Alex. 2021. Interviews by Zachary Tumin.

"Difference Between Direct-Sold and Programmatic Advertising." 2017. *DV Publisher Insights* (blog), November 6, 2017. https://pub.doubleverify.com/blog/differences-between-direct-sold-and-programmatic-advertising/.

Dighe, Amy, Lorenzo Cattarino, et al. 2020. "Response to COVID-19 in South Korea and Implications for Lifting Stringent Interventions." *BMC Medicine* 18 (1): 321. https://doi.org/10.1186/s12916-020-01791–8.

"Digital Disruption on the Farm." 2014. *The Economist/Schumpeter* (blog), May 24, 2014. https://www.economist.com/business/2014/05/24/digital-disruption-on-the-farm.

di Ieva, Antonio. 2019. "AI-Augmented Multidisciplinary Teams: Hype or Hope?" *The Lancet* 394, no. 10211 (2019): 1801. https://doi.org/10.1016/S0140-6736(19)32626-1.

Dijkstra, Andrea. 2021. "Africa's Flying Chemists." *NewsAfrica.net*, March 23, 2021. https://www.newsafrica.net/sections/business-economy/africa-s-flying-chemists.

Dillon, Emily. 2020. "DNC Tech Team Guide to Remote Work." *DNC Tech Team* (blog), *Medium*, March 16, 2020. https://medium.com/democratictech/dnc-tech-team-guide-to-remote-work-723507ff2e44.

Doctor, Ken. 2017. "Inside *The New York Times*' Digital Subscription Machine." *TheStreet* (blog), December 15, 2017. https://www.thestreet.com/investing/stocks/inside-the-new-york-times-digital-subscription-machine-14419255.

Doctor, Ken. 2019. "Newsonomics: CEO Mark Thompson on Offering More and More New York Times (and Charging More for It)." *Nieman Lab* (blog), November 13, 2019. https://www.niemanlab.org/2019/11/newsonomics-ceo-mark-thompson-on-offering-more-and-more-new-york-times-and-charging-more-for-it.

Doeland, Denis. 2019. "*The New York Times* Leader in the Digital Transformation." *Medium* (blog), September 13, 2019. https://medium.com/digital-assets-the-power-force-and-potential-of/the-new-york-times-leader-in-the-digital-transformation-611577f855a7.

Doerr, John, and Larry Page. 2018. *Measure What Matters: How Google, Bono, and the Gates Foundation Rock the World with OKRs*. New York: Portfolio.

Dolias, Kriton, and Vinessa Wan. 2019. "How We Prepared New York Times Engineering for the Midterm Elections." *NYT Open* (blog), *Medium*, February 15, 2019. https://open.nytimes.com/how-we-prepared-new-york-times-engineering-for-the-midterm-elections-2a615fe4196e.

Donahue, John D., and Mark H. Moore, eds. 2012. *Ports in a Storm: Public Management in a Turbulent World*. Washington, DC: Brookings Institution Press/Ash Center.

Donoho, David. 2017. "50 Years of Data Science." *Journal of Computational and Graphical Statistics* 26, no. 4 (2017): 745–766. https://doi.org/10.1080/10618600.2017.1384734.

Donohoe, Doug. 2020. "Published Assets? We Had a Few." *NYT Open* (blog), *Medium*, July 31, 2020. https://open.nytimes.com/publishing-assets-we-had-a-few-c3a844e98bac.

Dorn, Emma, Bryan Hancock, Jimmy Sarakatsannis, and Ellen Viruleg. 2021. "COVID-19 and Education: The Lingering Effects of Unfinished Learning." Last modified July 27, 2021. https://www.mckinsey.com/industries/public-and-social-sector/our-insights/covid-19-and-education-the-lingering-effects-of-unfinished-learning.

Dorn, Sara, and Ruth Weissmann. 2019. "Spiraling Morningside Park Crime Stats Show a Neighborhood Gripped by Violence." *New York Post* (blog), December 14, 2019. https://nypost.com/2019/12/14/spiraling-morningside-park-crime-stats-show-a-neighborhood-gripped-by-violence.

Dorner, Dietrich. 1997. *The Logic of Failure: Recognizing and Avoiding Error in Complex Situations*. Rev. ed. Cambridge, MA: Basic Books.

Douglas, Matt, Kristen Dudish, Elena Gianni, et al. 2018. "A Faster and More Flexible Home Page That Delivers the News Readers Want." *NYT Open* (blog), *Medium*, August 8, 2018. https://open.nytimes.com/a-faster-and-more-flexible-home-page-that-delivers-the-news-readers-want-1522ff64aa86.

DroneBlogger. 2021. "On Our Way: Zipline's Quest to Be the Medical Drone Logistics Company of Choice." Last modified January 16, 2021. https://dronenews.africa/on-our-way-ziplines-quest-to-be-the-medical-drone-logistics-company-of-choice.

Dubow, Josh. 2021a. "The Analytics Behind Fourth Down Play Calls in the NFL." *Berkshire Eagle*. Accessed November 28, 2021. https://www.berkshireeagle.com/the-analytics-behind-fourth-down-play-calls-in-the-nfl/article_c453b528-5c2a-11eb-93c8-c7b4329d6db0.html.

Dubow, Josh. 2021b. "4th Down Aggressiveness Increasing Rapidly Across NFL." *AP News*, November 12, 2021. https://apnews.com/article/nfl-sports-los-angeles-chargers-joe-lombardi-brandon-staley-a78f2cde1b8e1daac3ff4f78b9ef2d1f.

Duda, Richard O., Peter E. Hart, and David G. Stork. 2012. *Pattern Classification*. New York: John Wiley.

Duggan, William. 2004. *Napoleon's Glance: The Secret of Strategy*. 2nd ed. New York: Bold Type.

Duhalt, Adrian. 2017. "Issue Brief: Looting Fuel Pipelines in Mexico." Houston, TX: Rice University's Baker Institute for Public Policy.

Dullaghan, Ryan, Andrew Biga, Angela Duckworth, Amy Wrzesniewski, and Maryellen Reilly Lamb. 2015. "Case Studies: Predicting Employee Performance." Presentation at the Wharton People Analytics Conference in Philadelphia, PA. Published December 13, 2015. *YouTube*. https://youtu.be/Ob7J9T5QBnM.

Dunlap, Charles. 2003. "It Ain't No TV Show: JAGs and Modern Military Operations." *Chicago Journal of International Law* 4, no. 2 (Fall 2003): 479–491. https://chicagounbound.uchicago.edu/cjil/vol4/iss2/15.

Eager, Eric. 2021. Interviews by Zachary Tumin.

Ebersweiler, Cyril, and Benjamin Joffe. 2018. "10 Key Lessons About Tech Mergers and Acquisitions from Cisco's John Chambers." *TechCrunch* (blog), December 23, 2018. https://techcrunch.com/2018/12/23/twelve-key-lessons-about-tech-mergers-and-acquisitions-from-ciscos-john-chambers.

Editorial Board of the *Washington Post*. 2022. "Opinion: As the Pandemic Exploded, a Researcher Saw the Danger. China's Leaders Kept Silent." *Washington Post*, April 22, 2022. https://www.washingtonpost.com/opinions/interactive/2022/china-researcher-covid-19-coverup/.

EdjSports (@edjsports). 2021. "Brandon Staley and the #Chargers Getting in on the Action and Put Together an Outstanding TD Drive . . ." *Twitter*, October 10, 2021. https://twitter.com/edjsports/status/1447323641100869635.

Edmonds, Rick. 2019a. "*The New York Times* Sells Premium Ads Based on How an Article Makes You Feel." *Poynter* (blog), April 10, 2019. https://www.poynter.org/business-work/2019/the-new-york-times-sells-premium-ads-based-on-how-an-article-makes-you-feel.

Edmonds, Rick. 2019b. "Profitable New York Times Co. Readies Big Expansion Plans as Most Newspaper Companies Continue to Contract." *Poynter* (blog), February 6, 2019. https://www.poynter.org/business-work/2019/profitable-new-york-times-co-readies-big-expansion-plans-as-most-newspaper-companies-continue-to-contract.

*EdSurge*. 2016. "Teach to One: Math' Model Expands to 10 States." August 17, 2016. https://www.edsurge.com/news/2016-08-17-teach-to-one-math-model-expands-to-10-states.

Eisenmann, Thomas, Geoffrey Parker, and Marshall Van Alstyne. 2011. "Platform Envelopment." *Strategic Management Journal* 32, no. 12 (2011): 1270–1285.

Eisenmann, Thomas, Eric Ries, and Sarah Dillard. "Hypothesis-Driven Entrepreneurship: The Lean Startup." Harvard Business School Background Note 812–095, December 2011. Revised July 2013.

Ellis, Sean. 2017. "Growth Needs a North Star Metric." *Growth Hackers* (blog), June 5, 2017. https://blog.growthhackers.com/what-is-a-north-star-metric-b31a8512923f.

Elluri, Lavanya, Varun Mandalapu, and Nirmalya Roy. 2019. "Developing Machine Learning Based Predictive Models for Smart Policing." https://doi.org/10.1109/SMARTCOMP.2019.00053.

Emerson. 2016. "Best Practices in Leak and Theft Detection." https://www.emerson.com/documents/automation/white-paper-best-practices-in-leak-theft-detection-en-68294.pdf.

Epstein, Reid J. 2020. "Democrats Belatedly Launch Operation to Share Information on Voters." *New York Times*, September 6, 2020. Updated September 11, 2020. https://www.nytimes.com/2020/09/06/us/politics/Presidential-election-voting-Democrats.html.

Esquivel, Paloma. 2021. "Faced with Soaring Ds and Fs, Schools Are Ditching the Old Way of Grading." *Los Angeles Times*, November 8, 2021. https://www.latimes.com/california/story/2021-11-08/as-ds-and-fs-soar-schools-ditch-inequitable-grade-systems.

Evans, Benedict. 2022. "TV, Merchant Media and the Unbundling of Advertising." *Benedict Evans* (blog), March 18, 2022. https://www.ben-evans.com/benedictevans/2022/3/18/unbundling-advertising.

Evans, Laura. 2021. Interview by Zachary Tumin and Madeleine Want.

Fahim, Kareem. 2020. "Turkey's Military Campaign Beyond Its Borders Is Powered by Homemade Armed Drones." *Washington Post*, November 29, 2020. https://www.washingtonpost.com/world/middle_east/turkey-drones-libya-nagorno-karabakh/2020/11/29/d8c98b96-29de-11eb-9c21-3cc501d0981f_story.html.

Farley, Amy. 2020. "Zipline Mastered Medical Drone Delivery in Africa—Now It's Coming to the U.S." *Fast Company*, March 10, 2020. https://www.fastcompany.com/90457727/zipline-most-innovative-companies-2020.

Fehr, Tiff. 2019. “How We Sped Through 900 Pages of Cohen Documents in Under 10 Minutes.” *New York Times*, March 26, 2019. https://www.nytimes.com/2019/03/26/reader-center/times-documents-reporters-cohen.html.

Fehr, Tiff, and Josh Williams. 2021. “10 Million Data Requests: How Our COVID Team Tracked the Pandemic.” *New York Times*, June 24, 2021. https://www.nytimes.com/2021/06/24/insider/covid-tracking-data.html.

Feinberg, N. K. 2019. “To Design Better Products, Write Better UX Copy.” *NYT Open* (blog), *Medium*, October 10, 2019. https://open.nytimes.com/to-design-better-products-consider-the-language-f17b923f8bae.

Feloni, Richard. 2017. “Consumer-Goods Giant Unilever Has Been Hiring Employees Using Brain Games and Artificial Intelligence—and It’s a Huge Success.” *Business Insider*, June 28, 2017. https://www.businessinsider.com/unilever-artificial-intelligence-hiring-process-2017-6.

Fendos, Justin. 2020. “How Surveillance Technology Powered South Korea’s COVID-19 Response.” *Brookings TechStream* (blog), April 29, 2020. https://www.brookings.edu/techstream/how-surveillance-technology-powered-south-koreas-covid-19-response.

Ferrazzi, Keith. 2014. “Getting Virtual Teams Right.” *Harvard Business Review*, December 1, 2014. https://hbr.org/2014/12/getting-virtual-teams-right.

Fiedler, Jonathan. 2016. “An Overview of Pipeline Leak Detection Technologies.” American School of Gas Measurement Technology. https://asgmt.com/wp-content/uploads/2016/02/004.pdf.

Field, Hayden. 2021. “Nine Experts on the Single Biggest Obstacle Facing AI and Algorithms in the Next Five Years.” *Emerging Tech Brew* (blog), *Morning Brew*, January 22, 2021. https://www.morningbrew.com/emerging-tech/stories/2021/01/22/nine-experts-single-biggest-obstacle-facing-ai-algorithms-next-five-years.

Fink, Olga, Enrico Zio, and Ulrich Weidmann. 2013. “Predicting Time Series of Railway Speed Restrictions with Time-Dependent Machine Learning Techniques.” *Expert Systems with Applications* 40, no. 15 (2013): 6033–6040. https://doi.org/10.1016/j.eswa.2013.04.038.

Finnie, William, and Robert Randall. 2002. “Loyalty as a Philosophy and Strategy: An Interview with Frederick F. Reichheld.” *Strategy & Leadership* 30 (April 2002): 25–31. https://doi.org/10.1108/10878570210422120.

Fischer, Sara. 2021. “Trump Era Pushes *New York Times* to New Heights in Digital Subscriptions.” *Axios*, February 5, 2021. https://www.axios.com/new-york-times-digital-subscriptions-9be49fa1-3310-41d9-becb-2e77b18cd77e.html.

Fisher, Max, and Choe Sang-Hun. 2020. “How South Korea Flattened the Curve.” *New York Times*, March 23, 2020. https://www.nytimes.com/2020/03/23/world/asia/coronavirus-south-korea-flatten-curve.html.

Fisher, Roger, William L. Ury, and Bruce Patton. 2011. *Getting to Yes: Negotiating Agreement Without Giving In*. 3rd rev. ed. New York: Penguin.

Fleisher, Lisa. 2014. “Big Data Enters the Classroom.” *Wall Street Journal*, March 23, 2014. https://online.wsj.com/article/SB10001424052702304756104579451241225610478.html.

Flix, Nicolas. 2018. “Acquisition, Processing & Storage of RS CBM Data.” Presentation at the Big Data in Railway Maintenance Planning Conference, University of Delaware, December 2018.

Flyzipline.com. n.d. “Protecting Ghana’s Election: Instant Agility with Zipline’s Autonomous Delivery Network.” https://assets.ctfassets.net/pbn2i2zbvp41/3yrQaMNdJ1u1J2aSEucjzt/4412ea5d12896d15b7eb41a2212d0295/Zipline_Ghana_PPE_Global_Healthcare_Feb-2021.pdf.

Ford, Paul. 2020. “ ‘Real’ Programming Is an Elitist Myth.” *Wired*, August 18, 2020. https://www.wired.com/story/databases-coding-real-programming-myth.

Fortier, Sam. 2020. "The NFL's Analytics Movement Has Finally Reached the Sport's Mainstream." *Washington Post*, January 16, 2020. https://www.washingtonpost.com/sports/2020/01/16/nfls-analytics-movement-has-finally-reached-sports-mainstream/.

Fountaine, Tim, Brian McCarthy, and Tamin Saleh. 2019. "Building the AI-Powered Organization." *Harvard Business Review*, July 1, 2019.

Fournier, Camille. 2019. "Camille Fournier on Managing Technical Teams." Interview by Craig Cannon. Published July 17, 2019. *YouTube*. https://youtu.be/oxgfehnJ7GE.

Fox, Liam. 2021. "How the NFL Uses Analytics, According to the Lead Analyst of a Super Bowl Champion." *Forbes*, August 12, 2021. https://www.forbes.com/sites/liamfox/2021/08/12/how-the-nfl-uses-analytics-according-to-the-lead-analyst-of-a-super-bowl-champion/.

Franks, Bill. 2020. "Pitfalls When Measuring the Success of Analytics Programs." *LinkedIn*, June 12, 2020. https://www.linkedin.com/pulse/pitfalls-when-measuring-success-analytics-programs-bill-franks.

Free Management Ebooks. n.d. "Richard Hackman's 'Five Factor Model.'" Free Online Library for Managers. Accessed November 23, 2021. http://www.free-management-ebooks.com/faqld/development-03.htm.

Fretty, Peter. 2020. "This Is Not Your Pappy's Farm." *IndustryWeek*, August 17, 2020. https://www.industryweek.com/technology-and-iiot/article/21139221/this-is-not-your-pappys-farm.

Fuscaldo, Donna. 2019. "ZestFinance Using AI to Bring Fairness to Mortgage Lending." *Forbes*, March 19, 2019. https://www.forbes.com/sites/donnafuscaldo/2019/03/19/zestfinance-using-ai-to-bring-fairness-to-mortgage-lending.

Gallo, Amy. 2017. "A Refresher on A/B Testing." *Harvard Business Review*, June 28, 2017. https://hbr.org/2017/06/a-refresher-on-ab-testing.

Garcia, Cardiff, and Stacey Vanek Smith. September 11, 2020. "The Science of Hoops." *The Indicator from Planet Money*. https://www.npr.org/2020/09/11/911898347/the-science-of-hoops.

Garcia, Mike. 2018. "2018 FVCC Honors Symposium—Mike Garcia: 'The Internet of Things at BNSF Railway.'" Presentation at Flathead Valley Community College, February 27, 2018. Published May 1, 2018. *YouTube*. https://youtu.be/-HpD-pHMH7A.

Garey, Lorna. 2013. "New England Patriots' Winning Technology Plan." *InformationWeek* (blog), January 17, 2013. http://www.informationweek.com/mobility/muni-wireless/new-england-patriots-winning-technology/240146529.

Garvin, David A., and Alison Berkley Wagonfeld. 2013. "Google's Project Oxygen: Do Managers Matter?" Boston: Harvard Business School Publishing. https://hbsp.harvard.edu/product/313110-PDF-ENG.

Gayed, Jeremy, Said Ketchman, Oleksii Khliupin, et al. 2019. "We Re-Launched *The New York Times* Paywall and No One Noticed." *NYT Open* (blog), *Medium*, August 29, 2019. https://open.nytimes.com/we-re-launched-the-new-york-times-paywall-and-no-one-noticed-5cd1f795f76b.

Gee, Kelsey. 2017. "In Unilever's Radical Hiring Experiment, Resumes Are Out, Algorithms Are In." *Wall Street Journal*, June 26, 2017. https://www.wsj.com/articles/in-unilevers-radical-hiring-experiment-resumes-are-out-algorithms-are-in-1498478400.

Geiger, Gerhard, Thomas Werner, and Drago Matko. 2015. "Leak Detection and Locating: A Survey." Pipeline Simulation Interest Group. https://web.archive.org/web/20151129074144/http://fhge.opus.hbz-nrw.de/volltexte/2003/8/pdf/Paper_0301_for_FH.pdf.

Ghanchi, Asim. 2019. "BNSF Women in Stem." Presentation at the DisruptWell summit, September 2019. Published September 8, 2019. *Vimeo*. https://vimeo.com/455861938.

"Gigabytes Fly at Gillette Stadium." 2014. NECN. https://www.necn.com/news/business/gigabytes-fly-at-gillette-stadium_necn/43538/.

Gladwell, Malcolm. 1996. "The Tipping Point: Why Is the City Suddenly So Much Safer?" *New Yorker*, June 3, 1996, 32–38.

Glassdoor. n.d. "*The New York Times* Deputy Editor for Homescreen Personalization Job in New York, NY." Accessed July 19, 2021. https://www.glassdoor.com/job-listing/deputy-editor-for-homescreen-personalization-the-new-york-times-JV_IC1132348_KO0,44_KE45,63.htm?jl=4043959453.

Global Guerrillas. "DRONENET How to Build It." Accessed November 5, 2021. https://globalguerrillas.typepad.com/globalguerrillas/2013/01/dronenet-how-to-build-it.html.

Goh, Alison. 2020. "Zoom Video Meeting Best Practices." *DNC Tech Team* (blog), *Medium*, April 16, 2020. https://medium.com/democratictech/zoom-video-meeting-best-practices-fd255120bd0.

Goldsmith, Kevin. 2016. "Infoshare 2016: How Spotify Builds Products." Speech given May 19, 2016, in Gdańsk, Poland. Published June 29, 2016. *YouTube*. https://youtu.be/7B-qKT2sHD0.

Goldsmith, Kevin. 2017. "The Right Ingredients for Your Perfect Team." Presentation at the Lead Developer, New York, March 17, 2017. https://www.youtube.com/watch?v=JnCYxM9z1dE.

Goldsmith, Stephen, and William D. Eggers. 2004. *Governing by Network: The New Shape of the Public Sector*. Illustrated ed. Washington, DC: Brookings Institution Press/Ash Center.

Goldsmith, Stephen, and Donald F. Kettl, eds. 2009. *Unlocking the Power of Networks: Keys to High-Performance Government*. Washington, DC: Brookings Institution Press/Ash Center.

Goleman, Daniel. 2004. "What Makes a Leader?" *Harvard Business Review*, January 1, 2004. https://hbr.org/2004/01/what-makes-a-leader.

Goleman, Daniel. 2005. *Emotional Intelligence: Why It Can Matter More Than IQ*. New York: Random House.

Goodwin, Tom. 2015. "The Battle Is for the Customer Interface." *TechCrunch* (blog), March 3, 2015. https://social.techcrunch.com/2015/03/03/in-the-age-of-disintermediation-the-battle-is-all-for-the-customer-interface/.

Google Developers. "Recommendations: What and Why? | Machine Learning." 2022. August 5, 2022. https://developers.google.com/machine-learning/recommendation/overview.

Goswami, Debjani. 2021. "How to Do Usability Testing for Your AI Assistant?" *Qualitest* (blog), February 9, 2021. https://medium.com/qualitest/how-to-do-usability-testing-for-your-ai-assistant-a36887e1bbcb.

Gottlieb, Scott. 2021. *Uncontrolled Spread: Why COVID-19 Crushed Us and How We Can Defeat the Next Pandemic*. New York: Harper.

Gourville, John T. 2003. "Why Consumers Don't Buy: The Psychology of New Product Adoption." Harvard Business School Background Note 504-056, November 2003. Revised April 2004. https://www.hbs.edu/faculty/Pages/item.aspx?num=30597.

Gourville, John T. 2005. "Note on Innovation Diffusion: Rogers's Five Factors." Harvard Business School Background Note 505-075, May 2005. Revised April 2006. https://www.hbs.edu/faculty/Pages/item.aspx?num=32314.

Gourville, John T. 2006. "Eager Sellers and Stony Buyers: Understanding the Psychology of New-Product Adoption." *Harvard Business Review*, June 2006, 98–106. https://hbr.org/2006/06/eager-sellers-and-stony-buyers-understanding-the-psychology-of-new-product-adoption.

Government of the Republic of Korea. 2020. "Flattening the Curve on COVID-19: How Korea Responded to a Pandemic Using ICT." Last modified July 17, 2020. http://www.undp.org/content/seoul_policy_center/en/home/presscenter/articles/2019/flattening-the-curve-on-covid-19.html.

GovTech Singapore. 2020a. "Before You #TraceTogether, the Team First #TestsTogether." Last modified October 15, 2020. https://www.tech.gov.sg/media/technews/20201015-before-tracetogether-you-have-to-testtogether.

GovTech Singapore. 2020b. "Improving TraceTogether Through Community Engagement." Last modified July 6, 2020. https://www.tech.gov.sg/media/technews/2020-07-06-tracetogether-token-teardown.

GovTech Singapore. 2020c. "9 Geeky Myth-Busting Facts You Need to Know About TraceTogether." Last modified March 21, 2020. https://www.tech.gov.sg/media/technews/geeky-myth-busting-facts-you-need-to-know-about-tracetogether.

GovTech Singapore. 2020d. "6 Things About OpenTrace, the Open-Source Code Published by the TraceTogether Team." Last modified April 9, 2020. https://www.tech.gov.sg/media/technews/six-things-about-opentrace.

GovTech Singapore. 2020e. "TraceTogether—Behind the Scenes Look at Its Development Process." Last modified March 25, 2020. https://www.tech.gov.sg/media/technews/tracetogether-behind-the-scenes-look-at-its-development-process.

GovTech Singapore. 2020f. "Two Reasons Why Singapore Is Sticking with TraceTogether's Protocol." Last modified June 29, 2020. https://www.tech.gov.sg/media/technews/two-reasons-why-singapore-sticking-with-tracetogether-protocol.

Granville, Vincent. 2015. "Building Blocks of Data Science." *Vincent Granville's Blog* (blog), February 27, 2015. https://www.datasciencecentral.com/profiles/blogs/building-blocks-of-data-science.

Granville, Vincent. 2016. "40 Techniques Used by Data Scientists." *Vincent Granville's Blog* (blog), July 4, 2016. https://www.datasciencecentral.com/profiles/blogs/40-techniques-used-by-data-scientists.

Granville, Vincent. 2017. "Difference Between Machine Learning, Data Science, AI, Deep Learning, and Statistics." *Vincent Granville's Blog* (blog), January 2, 2017. https://www.datasciencecentral.com/profiles/blogs/difference-between-machine-learning-data-science-ai-deep-learning.

Green, Donald P., and Alan S. Gerber. 2019. *Get Out the Vote: How to Increase Voter Turnout*. 4th ed. Washington, DC: Brookings Institution Press.

Green, Donald P., Alan S. Gerber, and David W. Nickerson. 2003. "Getting Out the Vote in Local Elections: Results from Six Door-to-Door Canvassing Experiments." *Journal of Politics* 65, no. 4 (November 2003): 1083–1096. https://doi.org/10.1111/1468-2508.t01-1-00126.

Green, Donald P., Mary C. McGrath, and Peter M. Aronow. 2013. "Field Experiments and the Study of Voter Turnout." *Journal of Elections, Public Opinion and Parties* 23, no. 1 (February 2013): 27–48. https://doi.org/10.1080/17457289.2012.728223.

Greulich, William Walter, and S. Idell Pyle. 1959. *Radiographic Atlas of Skeletal Development of the Hand and Wrist*. Stanford, CA: Stanford University Press.

Greve, Ashley, Scott Dubin, and Ryan Triche. n.d. "Assessing Feasibility and Readiness for Cargo Drones in Health Supply Chains." United States Agency for International Development. https://www.updwg.org/wp-content/uploads/2021/06/Assessing_Feasibility_and_Readiness_for_Cargo_Drones_in_Health_Supply_Chains.pdf.

Grey Owl Engineering Ltd. 2018. "Best Management Practice: Pipeline Leak Detection Programs." Canadian Association of Petroleum Producers. https://www.capp.ca/wp-content/uploads/2019/12/Best_Management_Practice__Pipeline_Leak_Detection_Programs-310502.pdf.

Grotta, Jacob. 2019. "Rev 1 'Data Science in the Banking World.'" July 29, 2019. https://www.youtube.com/watch?v=iXxwlPzIBrg.

Grove, Andy. 2003. "Churning Things Up." *Fortune Magazine*, August 11, 2003, 115–118.

Gruen, David and Jonathan Messinger. 2021. "How to Bridge the Imaging AI Adoption Gap with an Ecosystem Approach—3 Learnings." Webinar, December 1, 2021. https://www.beckershospitalreview.com/healthcare-information-technology/how-to-bridge-the-imaging-ai-adoption-gap-with-an-ecosystem-approach-3-learnings.html.

Guerin, David, Manon Taylor, Charles Matemba, et al. n.d. "Drone Evidence Generation Toolkit." UAV for Payload Delivery Working Group. Accessed November 27, 2021. https://www.updwg.org/wp-content/uploads/Drone-Evidence-Generation-Toolkit.pdf.

Guo, Eileen, and Karen Hao. 2020. "This Is the Stanford Vaccine Algorithm That Left Out Frontline Doctors." *MIT Technology Review*, December 21, 2020. https://www.technologyreview.com/2020/12/21/1015303/stanford-vaccine-algorithm.

Guszcza, James, Iyad Rahwan, et al. 2018. "Why We Need to Audit Algorithms." *Harvard Business Review*, November 28, 2018. https://hbr.org/2018/11/why-we-need-to-audit-algorithms.

Guzman, Zack. 2018. "Bill Bratton Reveals What His 'Biggest Mistake' Taught Him About Ambition." *CNBC*, July 13, 2018. https://www.cnbc.com/2018/07/12/bill-bratton-reveals-what-his-biggest-mistake-taught-him-about-manag.html.

Ha, Richard, Christine Chin, and Jenika Karcich. 2019. "Prior to Initiation of Chemotherapy, Can We Predict Breast Tumor Response? Deep Learning Convolutional Neural Networks Approach Using a Breast MRI Tumor Dataset." *Journal of Digital Imaging* 32, no. 5 (2019): 693–701.

Ha, Richard, Simukayi Mutasa, et al. 2019. "Accuracy of Distinguishing Atypical Ductal Hyperplasia from Ductal Carcinoma in Situ with Convolutional Neural Network–Based Machine Learning Approach Using Mammographic Image Data." *American Journal of Roentgenology* 212, no. 5 (2019): 1166–1171. https://doi.org/10.2214/AJR.18.20250.

Haas, Martine, and Mark Mortensen. 2016. "The Secrets of Great Teamwork." *Harvard Business Review*, June 1, 2016. https://hbr.org/2016/06/the-secrets-of-great-teamwork.

Haass, Richard N. 1999. *The Bureaucratic Entrepreneur: How to Be Effective in Any Unruly Organization*. Washington, DC: Brookings Institution Press.

Hackman, J. Richard. 2002. *Leading Teams: Setting the Stage for Great Performances*. Boston: Harvard Business Review Press.

Hackman, J. Richard. 2004. "What Makes for a Great Team?" Psychological Science Agenda. https://www.apa.org/science/about/psa/2004/06/hackman.

Hackman, J. Richard, and Richard E. Walton. 1986. "Leading Groups in Organizations." In *Designing Effective Work Groups*, ed. Paul S. Goodman, 72–119. San Francisco: Jossey-Bass.

Haimson, Leonie. 2019. "NYC Public School Parents: The Reality vs the Hype of Teach to One." *NYC Public School Parents* (blog), February 28, 2019. https://nycpublicschoolparents.blogspot.com/2019/02/the-reality-vs-hype-of-teach-to-one.html.

Hall, Madison. 2021. "The DOJ Is Mapping Cell Phone Location Data from Capitol Rioters." *Business Insider France*, March 24, 2021. https://www.businessinsider.com/doj-is-mapping-cell-phone-location-data-from-capitol-rioters-2021-3.

Hamman, Brian. 2018a. "How We Hire Front-End Engineers at *The New York Times*." *NYT Open* (blog), *Medium*, August 10, 2018. https://open.nytimes.com/how-we-hire-front-end-engineers-at-the-new-york-times-e1294ea8e3f8.

Hamman, Brian. 2018b. "Part I: An Interview with Brian Hamman, VP of Engineering for News Products at *The New York Times*." *Medium* (blog), June 3, 2018. https://medium.com/enigma-engineering/part-i-an-interview-with-brian-hamman-vp-of-engineering-for-news-products-at-the-new-york-times-f3f59fe18d69.

Hamman, Brian. 2018c. "Part II: An Interview with Brian Hamman, VP of Engineering for News Products at *The New York Times*." *Medium* (blog). June 5, 2018. https://medium.com/enigma-engineering/part-ii-an-interview-with-brian-hamman-vp-of-engineering-for-news-products-at-the-new-york-times-db14bb89641c.

Hamman, Brian. 2019. "From Designing Boxes to Designing Algorithms: How Programming the News Has Evolved at *The New York Times*." Summary of talk presented at the Computation + Journalism Symposium, Miami, FL, February 2, 2019. http://cplusj.org/keynote-speaker-brian-hamman.

Han, Yong Y., Joseph A. Carcillo, Shekhar T. Venkataraman, et al. 2005. "Unexpected Increased Mortality After Implementation of a Commercially Sold Computerized Physician Order Entry System." *Pediatrics* 116 (2005):1506–1512.

Hansen, Morton T. 2011. "IDEO CEO Tim Brown: T-Shaped Stars: The Backbone of IDEO's Collaborative Culture." ChiefExecutive.net, March 29, 2011. https://web.archive.org/web/20110329003842/http://www.chiefexecutive.net/ME2/dirmod.asp?sid=&nm=&type=Publishing&mod=Publications::Article&mid=8F3A7027421841978F18BE895F87F791&tier=4&id=F42A23CB49174C5E9426C43CB0A0BC46.

Harlan, Chico. 2021. "Highly Vaccinated Countries Thought They Were over the Worst. Denmark Says the Pandemic's Toughest Month Is Just Beginning." *Washington Post*, December 18, 2021. https://www.washingtonpost.com/world/2021/12/18/omicron-variant-denmark/.

Hartley, Jon. 2020. "The Baltimore Ravens Fell Short of the Super Bowl, but Their NFL Analytics Revolution Is Just Beginning." *National Review* (blog), February 2, 2020. https://www.nationalreview.com/2020/02/baltimore-ravens-nfl-analytics-data-driven-strategies-transform-football.

Harwell, Drew. 2020. "Algorithms Are Deciding Who Gets the First Vaccines. Should We Trust Them?" *Washington Post*, December 23, 2020. https://www.washingtonpost.com/technology/2020/12/23/covid-vaccine-algorithm-failure.

Hassan, Norel. 2018. "Announcing a *New York Times* iOS Feature That Helps Readers Find Stories Relevant to Them." *NYT Open* (blog), *Medium*, August 3, 2018. https://open.nytimes.com/announcing-a-new-ios-feature-that-helps-readers-find-stories-relevant-to-them-a8273f8fcca4.

Hawkins, Beth. 2021. "A Better Equation: New Pandemic Data Supports Acceleration Rather Than Remediation to Make Up for COVID Learning Loss." *The 74*, May 24, 2021. https://www.the74million.org/a-better-equation-new-pandemic-data-supports-acceleration-rather-than-remediation-to-make-up-for-covid-learning-loss.

HBS Online. 2021. "What Is Predictive Analytics? 5 Examples." *Business Insights Blog* (blog). October 26, 2021. https://online.hbs.edu/blog/post/predictive-analytics.

Heaven, Will Douglas. 2020. "Predictive Policing Algorithms Are Racist. They Need to Be Dismantled." *MIT Technology Review*. Accessed February 22, 2021. https://www.technologyreview.com/2020/07/17/1005396/predictive-policing-algorithms-racist-dismantled-machine-learning-bias-criminal-justice/.

Hebert, Adam J. 2003. "Compressing the Kill Chain." *Air Force Magazine*, March 1, 2003. https://www.airforcemag.com/article/0303killchain.

Heideman, Justin. 2017. "Quick and Statistically Useful Validation of Page Performance Tweaks." *NYT Open* (blog), *Medium*, February 6, 2017. https://open.nytimes.com/quick-and-statistically-useful-validation-of-page-performance-tweaks-39ecce4328d7.

Henderson, Rebecca M., and Kim B. Clark. 1990. "Architectural Innovation: The Reconfiguration of Existing Product Technologies and the Failure of Established Firms." *Administrative Science Quarterly* 35, no. 1 (1990): 9–30. https://doi.org/10.2307/2393549.

Henry, Vincent E. 2005. "Compstat Management in the NYPD: Reducing Crime and Improving Quality of Life in New York City." Paper presented at the United Nations Asia and Far East Institute for the Prevention of Crime and Treatment of Offenders 129th International Seniors Seminar, Tokyo, Japan, January–February 2005. *Resource Material Series* no. 68 (2005), 100–116.

Hensley, Jamison. 2019. "Ravens' John Harbaugh Being Lauded as 'Prince' of Football Analytics." *ESPN* (blog), September 25, 2019. https://www.espn.com/blog/baltimore-ravens/post/_/id/50798/ravens-john-harbaugh-being-lauded-as-prince-of-football-analytics.

Hensley, Jamison. 2020. "Baltimore Ravens DC Unhappy with Late Cincinnati Bengals Field Goal to Avoid Shutout." *ESPN*, October 15, 2020. https://www.espn.com/nfl/story/_/id/30121047.

Hensley, Jamison. 2021. "Ravens' John Harbaugh on Decision to Go for 2 Instead of Game-Tying PAT in Loss to Steelers—'We Were Pretty Much out of Corners.'" ESPN.com, December 6, 2021. https://www.espn.com/nfl/story/_/id/32800759/ravens-john-harbaugh-decision-go-2-game-tying-pat-loss-steelers-were-pretty-much-corners.

Heraud, Jorge. 2017. "How an Artificial Intelligence & Robotics Startup Got Acquired by John Deere." Interview by Louisa Burwood-Taylor. Published October 3, 2017. Simplecast podcast. https://agfundernews.com/artificial-intelligence-robotics-startup-got-acquired-john-deere.html.

Hess, Rick. 2017. "Straight Up Conversation: Teach to One CEO Joel Rose." *Education Week*, September 18, 2017. https://www.edweek.org/education/opinion-straight-up-conversation-teach-to-one-ceo-joel-rose/2017/09.

Heymann, Phillip B. 1987. *The Politics of Public Management*. New Haven, CT: Yale University Press.

Hickins, Michael. 2012. "Union Pacific Using Predictive Software to Reduce Train Derailments." *Wall Street Journal*, April 2, 2012. https://www.wsj.com/articles/BL-CIOB-102.

Hill, Carolyn J., and Laurence E. Lynn. 2015. *Public Management: Thinking and Acting in Three Dimensions*. 2nd ed. Los Angeles: CQ Press.

Hill, Kashmir. 2021. "What Happens When Our Faces Are Tracked Everywhere We Go?" *New York Times Magazine*, March 18, 2021. https://www.nytimes.com/interactive/2021/03/18/magazine/facial-recognition-clearview-ai.html.

Hiller, Jennifer. 2022. "While Electric Vehicles Proliferate, Charging Stations Lag Behind." *Wall Street Journal*. May 30, 2022. https://www.wsj.com/articles/electric-vehicles-proliferate-while-charging-stations-lag-behind-11653903180.

HireVue. n.d. "Bias, AI Ethics and the HireVue Approach." HireVue. Accessed November 13, 2021. https://www.hirevue.com/why-hirevue/ai-ethics.

HireVue. 2020. "Unilever Finds Top Talent Faster with HireVue Assessments." Unilever + HireVue. https://webapi.hirevue.com/wp-content/uploads/2020/09/Unilever-Success-Story-PDF.pdf?_ga=2.14249143.1021990485.1637083562–1312617268.1637083562.

Hodges, Matt (@hodgesmr). 2020. "2012 was known for reimagining what campaign tech and data could do. 2016 was known for an armada in-house team. 2020 should be known for the sweeping role tech vendors played across the ecosystem. The future of Democratic tech is distributed. That's investing in infrastructure." *Twitter*, December 7, 2020. https://twitter.com/hodgesmr/status/1336012756563050496.

Hofstadter, Douglas. 2022. "Artificial Neural Networks Today Are Not Conscious, According to Douglas Hofstadter." *The Economist*, June 9, 2022. https://www.economist.com/by-invitation/2022/06/09/artificial-neural-networks-today-are-not-conscious-according-to-douglas-hofstadter.

Holdman, Jessica. 2019. "BNSF Explains New Cameras Able to Detect Rail Flaws at 70 mph at ND Capitol Exhibit." *Jamestown Sun*, January 23, 2019. https://www.jamestownsun.com/business/technology/956943-BNSF-explains-new-cameras-able-to-detect-rail-flaws-at-70-mph-at-ND-Capitol-exhibit.

Holguín-Veras, Jose, Saif Benjaafar, Pooja Dewan, et al. 2018. "Transportation Panel at INFORMS 2018 Government & Analytics Summit." Panel discussion. Published May 24, 2018. *YouTube*. https://youtu.be/7Daebc9B4wo.

Holmes, Aaron. 2020. "South Korea Is Relying on Technology to Contain COVID-19, Including Measures That Would Break Privacy Laws in the US—And So Far, It's Working." *Business Insider*, May 2, 2020. https://www.businessinsider.com/coronavirus-south-korea-tech-contact-tracing-testing-fight-covid-19-2020-5.

Honan, James P., Stacey Childress, and Caroline King. 2004. "Aligning Resources to Improve Student Achievement: San Diego City Schools." Boston: Harvard Business Publishing, September 13, 2004. https://hbsp.harvard.edu/product/PEL003-PDF-ENG?Ntt=san%20 diego*.

Hoops Geek. 2021. "The History and Evolution of the Three-Point Shot." *The Hoops Geek* (blog), February 10, 2021. https://www.thehoopsgeek.com/history-three-pointer/.

Hoppe, Anisha. 2021. "'Fire, Forget, and Find': Kamikaze Drones Are Already a Thing in Libya." TAG24. https://www.tag24.com/tech/fire-forget-and-find-kamikaze-drones-are-already-a-thing-in-libya-1988433. June 3, 2021.

Hu, Nick, 2021. "Zipline, on the Future of Healthcare Supply Chains." Interview with Dandi Zhu. In *Digital Health Forward*. Published January 13, 2021. Spotify podcast. https://open.spotify.com/episode/3rZtHOBfBcgsKa3APYOm1U.

Huang, Gregory T. 2013. "Xconomy: Enterasys, Boston Sports Teams Talk Future of Stadium Tech at Gillette." Xconomy. January 16, 2013. https://xconomy.com/boston/2013/01/16/enterasys-and-boston-sports-teams-talk-future-of-stadium-tech/.

Huawei. n.d. "Digital PEMEX Raises Efficiency and Profits." Huawei Enterprise. Accessed June 19, 2020. https://e.huawei.com/mx/case-studies/leading-new-ict/2018/201810171521.

Hubert, Anne. 2015. "Millennial Disruption Index." Presentation at the 2015 Yale Customer Insights Conference, Center for Customer Insights at Yale SOM, May 19, 2015. https://www.youtube.com/watch?v=85kIjWp74NI.

Human Resources Online. 2014. "Case Study: Unilever." Humanresourcesonline.net. August 27, 2014. https://www.humanresourcesonline.net/case-study-unilever.

Huntsberry, William. 2015. "Meet the Classroom of the Future." *NPR*, January 12, 2015. https://www.npr.org/sections/ed/2015/01/12/370966699/meet-the-classroom-of-the-future.

Iliakopoulou, Katerina. 2019. "Architecting for News Recommendations: The Many Hidden Colors of the Gray Lady." Presentation at the O'Reilly Software Architecture Conference 2019, New York, February 3, 2019. Video. https://www.oreilly.com/library/view/oreilly-software-architecture/9781492050506/video323974.html.

Iliakopoulou, Katerina. 2020. "Building for Rapid Scale: A Deep Dive into *The New York Times*' Messaging Platform." Presentation at the O'Reilly Software Architecture Conference 2020, New York, February 23, 2020. Video. https://www.oreilly.com/library/view/oreilly-software-architecture/0636920333777/video329403.html.

Ingram, David, and Jacob Ward. 2020. "Behind the Global Efforts to Make a Privacy-First Coronavirus Tracking App." *NBC News*, April 7, 2020. https://www.nbcnews.com/tech/tech-news/behind-global-efforts-make-privacy-first-coronavirus-tracking-app-n1177871.

Institution for Social and Policy Studies. 2021. "Lessons from GOTV Experiments." 2021. https://isps.yale.edu/node/16698.

Intersoft Consulting. n.d. "GDPR: Privacy by Design." Key Issues. Accessed October 22, 2021. https://gdpr-info.eu/issues/privacy-by-design.

Issenberg, Sasha. 2013. *The Victory Lab: The Secret Science of Winning Campaigns*. Reprint ed. New York: Crown.

Issenberg, Sasha. 2016. "The Meticulously Engineered Grassroots Network Behind the Bernie Sanders Revolution." *Bloomberg*, February 24, 2016. https://www.bloomberg.com/news/features/2016-02-24/behind-bernie-sanders-revolution-lies-a-meticulously-engineered-grassroots-network.

Jacob, Mark. 2019. "Building Habit—Not Pageviews—Matters Most for Keeping Subscribers, Data Analysis Finds." *Poynter* (blog), February 5, 2019. https://www.poynter.org/business-work/2019/building-habit-not-pageviews-matters-most-for-keeping-subscribers-data-analysis-finds/.

James, Sarah. 2020. "User Testing for a Niche Product." *DNC Tech Team* (blog), *Medium*, May 19, 2020. https://medium.com/democratictech/user-testing-for-a-niche-product-12d53dd1bca1.

*Japan Times*. 2020. "South Korea Crushed a Huge Coronavirus Outbreak. Can It Beat a Second Wave?" June 15, 2020. https://www.japantimes.co.jp/news/2020/06/15/asia-pacific/south-korea-coronavirus-second-wave.

Jardine, Andrew K. S., Daming Lin, and Dragan Banjevic. 2006. "A Review on Machinery Diagnostics and Prognostics Implementing Condition-Based Maintenance." *Mechanical Systems and Signal Processing* 20, no. 7 (2006): 1483–1510. https://doi.org/10.1016/j.ymssp.2005.09.012.

Jimmy (@JimmySecUK). 2022. "According to Ukrainian Sources the Russians Once Again Attempted to Ford the Siverskyi Donets River, near Bilohorivka, Earlier This Afternoon. It Went about as Well as Their Previous Attempts." *Twitter*. https://twitter.com/JimmySecUK/status/1524783663908634624.

Johnson, Lynne D., Lisa Howard, and Eden Sandlin. 2021. "Building a Successful Advertising Business from a Subscription-First Model." Presentation at the AdMonsters' Publisher Forum Virtual, 2021. Published May 7, 2021. *YouTube*. https://youtu.be/NIi-Xdd-sx8.

Johnson, Sydney. 2017. "Pulling the Plug on a Personalized Learning Pilot." *EdSurge*, January 19, 2017. https://www.edsurge.com/news/2017-01-19-pulling-the-plug-on-a-personalized-learning-pilot.

Jones, Nathan P., and John Sullivan. 2019. "Huachicoleros: Criminal Cartels, Fuel Theft, and Violence in Mexico." *Journal of Strategic Security* 12, no. 4 (2019): 1–24. https://doi.org/10.5038/1944-0472.12.4.1742.

Joyce, Greg. 2019. "Dave Gettleman's New Giants Promises: 'Computer Folks' and Eventual Success." *New York Post* (blog), December 31, 2019. https://nypost.com/2019/12/31/dave-gettlemans-new-giants-promises-computer-folks-and-eventual-success/.

Jumper, John. (2012–2021). Interviews with Zachary Tumin.

Kahneman, Daniel, Jack L. Knetsch, and Richard H. Thaler. 1990. "Experimental Test of the Endowment Effect and the Coase Theorem." *Journal of Political Economy* 98, no. 6 (December 1990): 1325–1348.

Kahneman, Daniel, and Amos Tversky. 1979. "Prospect Theory: An Analysis of Decision Under Risk." *Econometrica* 47, no. 2 (March): 263–291.

Kalla, Joshua L., and David E. Broockman. 2018. "The Minimal Persuasive Effects of Campaign Contact in General Elections: Evidence from 49 Field Experiments." *American Political Science Review* 112, no. 1 (February 2018): 148–166. https://doi.org/10.1017/S0003055417000363.

Kanter, Rosabeth Moss. 1985. *Change Masters*. New York: Free Press.

Kanter, Rosabeth Moss, and Tuna Cem Hayirli. 2022. "Creating High Impact Coalitions." *Harvard Business Review*, March-April 2022.

Kanter, Rosabeth Moss, Barry A. Stein, and Todd D. Jick. 1992. *The Challenge of Organizational Change: How Companies Experience It and Leaders Guide It*. New York: Free Press.

Kapadia, Sheil. 2019. "Analytical Edge: How John Harbaugh and Ravens Have Gained an Advantage with Fourth Down Aggressiveness." *The Athletic*, November 22, 2019. https://theathletic.com/1396091/2019/11/22/analytical-edge-how-john-harbaugh-and-ravens-have-gained-an-advantage-with-fourth-down-aggressiveness.

Kaplan, Evan. 2018. "The Age of Instrumentation Enables Better Business Outcomes." *Upside* (blog), *Transforming Data with Intelligence*, May 30, 2018. https://tdwi.org/articles/2018/05/30/arch-all-age-of-instrumentation-better-business-outcomes.aspx.

Kaplan, Robert E., and Mignon Mazique. 1987. *Trade Routes: The Manager's Network of Relationships*. Greensboro, NC: Center for Creative Leadership.

Kasinitz, Aaron. 2019. "Baltimore Ravens Make Hires for Analytics Department." *PennLive*, June 19, 2019. https://www.pennlive.com/baltimore-ravens/2019/06/baltimore-ravens-make-hires-for-analytics-department.html.

Kaste, Martin. 2020. "NYPD Study: Implicit Bias Training Changes Minds, Not Necessarily Behavior." *NPR*, September 10, 2020, sec. Hidden Brain. https://www.npr.org/2020/09/10/909380525/nypd-study-implicit-bias-training-changes-minds-not-necessarily-behavior.

Katzenbach, Jon R., and Douglas K. Smith. 2009. *The Discipline of Teams*. Boston: Harvard Business Review Press.

Katzenbach, Jon R., and Douglas K. Smith. 2015. *The Wisdom of Teams: Creating the High-Performance Organization*. Reprint ed. Boston: Harvard Business Review Press.

Kawaguchi, Akihito, Masashi Miwa, and Koichiro Terada. 2005. "Actual Data Analysis of Alignment Irregularity Growth and Its Prediction Model." *Quarterly Report of Railway Technical Research Institute* 46, no. 4 (2005): 262–268. https://doi.org/10.2219/rtriqr.46.262.

Kaye, Kate. 2022. "Big Data Is Dead. Small AI Is Here." Protocol, January 13, 2022. https://www.protocol.com/newsletters/protocol-enterprise/manufacturing-ai-salesforce-shay-banon.

Keane, Pearse A., and Eric J. Topol. 2018. "With an Eye to AI and Autonomous Diagnosis." *NPJ Digital Medicine* 1, no. 1 (2018): 1–3. https://doi.org/10.1038/s41746-018-0048-y.

Kedet. 2021. "Programmatic Direct vs Real-Time Bidding: What's the Difference?" *War Room Inc* (blog), May 10, 2021. https://www.warroominc.com/institute-library/blog/programmatic-direct-vs-real-time-bidding/.

Keller, Kevin Lane. 2001. "Building Customer-Based Brand Equity: A Blueprint for Creating Strong Brands." Marketing Science Institute Working Paper Series, Report 01–107, Cambridge, MA, January 1, 2001. https://www.msi.org/wp-content/uploads/2020/06/MSI_Report_01-107.pdf.

Kelman, Steven. 2005. *Unleashing Change: A Study of Organizational Renewal in Government*. Washington, DC: Brookings Institution Press.

Kendrick, Tom. 2012. *Results Without Authority: Controlling a Project When the Team Doesn't Report to You*. 2nd ed. New York: AMACOM.

Kennedy, David. 2017. "What Cops Need to Do If They Want the Public's Trust." *Oprah* (blog), May 2017. https://www.oprah.com/inspiration/what-cops-need-to-do-if-they-want-the-publics-trust.

Kerr, William R., Mark Roberge, and Paul A. Gompers. 2017. "The Entrepreneurial Manager, Module I: Defining and Developing the Business Model." Harvard Business School Module Note 817–108, February 2017. Revised December 2018.

Kerry, Cameron F. 2020. "Protecting Privacy in an AI-Driven World." Brookings. Last modified February 10, 2020. https://www.brookings.edu/research/protecting-privacy-in-an-ai-driven-world.

Khan, Zainub. n.d. "Teach to One Math with Ms. Khan." *Elevate Chicago* (blog). Accessed February 6, 2019. https://chartersforchange.org/teach-to-one-math-with-ms-khan.

Khanna, Harry. 2020. "Tools for Rapid Response on Election Day." *DNC Tech Team* (blog), *Medium*, July 20, 2020. https://medium.com/democratictech/tools-for-rapid-response-on-election-day-2feb7a38579f.

Khanna, Tarun. 2018. "When Technology Gets Ahead of Society." *Harvard Business Review* 96, no. 4 (July–August 2018), 86–95.

Khanna, Tarun, and George Gonzalez. 2020. "Zipline: The World's Largest Drone Delivery Network." Harvard Business School Case 721–366, November 2020. Revised January 2021. https://www.hbs.edu/faculty/Pages/item.aspx?num=59187.

Kilgannon, Corey. 2010. "With a Squeegee and a Smile, a Relic of an Unruly Past Wipes On." *New York Times*, October 3, 2010. https://www.nytimes.com/2010/10/04/nyregion/04squeegee.html.

Kilgannon, Corey. 2019. "A Park Shed Its Reputation. Then Came the Tessa Majors Murder." *New York Times*, December 14, 2019. https://www.nytimes.com/2019/12/14/nyregion/tessa-majors-columbia-morningside.html.

Kilner, James. 2022. "'Invincible' Russian Tank Equipped with Exploding Armor Destroyed by Ukrainian Troops." *Telegraph*, May 7, 2022. https://www.telegraph.co.uk/world-news/2022/05/07/invincible-russian-tank-equipped-exploding-armour-destroyed/.

Kim, Eugene. 2015. "Google's Former CIO Was Deaf and Dyslexic as a Kid—Now He's Helping Millions of Americans Get Emergency Loans." *Business Insider*, July 18, 2015. https://www.businessinsider.com/zest-finance-douglas-merrill-2015-7.

Kim, June-Ho, Julia Ah-Reum An, SeungJu Jackie Oh, Juhwan Oh, and Jong-Koo Lee. 2020. "Emerging COVID-19 Success Story: South Korea Learned the Lessons of MERS." Our World in Data. Last modified June 30, 2020. https://ourworldindata.org/covid-exemplar-south-korea.

Kingsmill, Sylvia, and Ann Cavoukian. n.d. *Privacy by Design: Setting a New Standard for Privacy Certification.* Toronto: Deloitte Canada and Ryerson University. https://www2.deloitte.com/content/dam/Deloitte/ca/Documents/risk/ca-en-ers-privacy-by-design-brochure.PDF.

Kite-Powell, Jennifer. 2019. "How to Transform 60,000 Irrigation Pivots into Autonomous Growing Machines." *Forbes*, March 19, 2019. https://www.forbes.com/sites/jenniferhicks/2019/03/19/how-to-transform-60000-irrigation-pivots-into-autonomous-growing-machines/.

Kizilcec, Rene F., Justin Reich, Michael Yeomans, Christoph Dann, Emma Brunskill, Glenn Lopez, Selen Turkay, Joseph Jay Williams, and Dustin Tingley. 2020. "Scaling Up Behavioral Science Interventions in Online Education." *Proceedings of the National Academy of Sciences* 117, no. 26 (June 2020): 14900–14905.

Klippenstein, Ken, and Eric Lichtblau. 2021. "FBI Seized Congressional Cellphone Records Related to Capitol Attack." *Intercept*, February 22 2021. https://theintercept.com/2021/02/22/capitol-riot-fbi-cellphone-records.

Knetsch, Jack L. 1989. "The Endowment Effect and Evidence of Nonreversible Indifference Curves." *American Economic Review* 79, no. 5 (December 1989): 1277–1288.

Kniberg, Henrik. 2016. "Making Sense of MVP (Minimum Viable Product)—and Why I Prefer Earliest Testable/Usable/Lovable." *Crisp's Blog* (blog), January 25, 2016. https://blog.crisp.se/2016/01/25/henrikkniberg/making-sense-of-mvp.

Kohavi, Ron, Thomas Crook, and Roger Longbotham. 2009. "Online Experimentation at Microsoft." Microsoft. https://doi.org/10.1.1.147.9056.

Kohavi, Ron, Diane Tang, and Ya Xu. 2020. *Trustworthy Online Controlled Experiments: A Practical Guide to A/B Testing.* Cambridge: Cambridge University Press.

Koppel, Ross, et al. 2005. "Role of Computerized Physician Order Entry Systems in Facilitating Medication Errors." *JAMA* 293, no. 10 (2005): 1197–1203. https://doi.org/10.1001/jama.293.10.1197.

Korea Centers for Disease Control and Prevention. 2020. "Coronavirus Disease-19: Summary of 2,370 Contact Investigations of the First 30 Cases in the Republic of Korea." *Osong Public Health and Research Perspectives* 11, no. 2 (2020): 81–84. https://doi.org/10.24171/j.phrp.2020.11.2.04.

Kornhauser, Wilhem. 2021. "Measuring Success of Machine Learning Products." *Towards Data Science* (blog), *Medium*, January 4, 2021. https://towardsdatascience.com/measuring-success-ef3aff9c28e4.

Kosslyn, Stephen M. 2019. "Integrating the Science of How We Learn into Education Technology." *Harvard Business Review*, October 11, 2019. https://hbr.org/2019/10/integrating-the-science-of-how-we-learn-into-education-technology.

Kotter, John P. 2012. *Leading Change, with a New Preface by the Author*. Boston: Harvard Business Review Press.

Kramer, Andrew E. and David Guttenfelder. 2022. "From the Workshop to the War: Creative Use of Drones Lifts Ukraine." *The New York Times*, August 10, 2022, sec. World. https://www.nytimes.com/2022/08/10/world/europe/ukraine-drones.html.

Krasadakis, George. 2020. "Data Quality in the Era of A.I." Innovation Machine, October 11, 2020. https://medium.com/innovation-machine/data-quality-in-the-era-of-a-i-d8e398a91bef.

Krasno, Jonathan S., and Donald P. Green. 2008. "Do Televised Presidential Ads Increase Voter Turnout? Evidence from a Natural Experiment." *Journal of Politics* 70, no. 1 (January 2008): 245–261. https://www.jstor.org/stable/10.1017/s0022381607080176.

Kravitz, Joshua. 2021. "Using Tech and Data to Supercharge Relational: Principles, Highlights and Challenges." Google Docs. March 21, 2021.

Krensky, Peter, Carlie Idoine, and Erick Brethenoux. 2021. "Magic Quadrant for Data Science and Machine Learning Platforms." Gartner, March 1, 2021. https://www.gartner.com/en/documents/3998753/magic-quadrant-for-data-science-and-machine-learning-pla.

Krikorian, Raffi. 2017a. "The Democrats' Tech Team Is Hiring." *DNC Tech Team* (blog), *Medium*, July 11, 2017. https://medium.com/democratictech/the-democrats-tech-team-is-hiring-f5a66f0a714f.

Krikorian, Raffi. 2017b. "Want to Change the World?: Here's How." *DNC Tech Team* (blog), *Medium*, July 31, 2017. https://medium.com/democratictech/want-to-change-the-world-heres-how-e486f0992d69.

Krikorian, Raffi. 2017c. "We Need Everyone to Pitch In; We've Got Elections to Win." *DNC Tech Team* (blog), *Medium*, November 9, 2017. https://medium.com/democratictech/we-need-everyone-to-pitch-in-weve-got-elections-to-win-2ed8ace719df.

Kroll, Andy. 2020. "The Best Way to Beat Trumpism?: Talk Less, Listen More." *Rolling Stone*, September 15, 2020. https://www.rollingstone.com/politics/politics-news/2020-presidential-campaign-tactic-deep-canvassing-1059531.

Kushner, Theresa. 2012. "Connecting the Stars: Applying Social Media Understanding to a Structured Marketing Data Environment." Presentation at the INFORMS Conference on Business Analytics and Operations Research, April 2012. Published September 10, 2012. *YouTube*. https://youtu.be/6uX2yVVT54c.

Kushner, Theresa. 2012–2020. Interviews by Zachary Tumin and Madeleine Want.

Kushner, Theresa. 2021. "The Marketing Database Is Dead, So, Now What?" *Medium* (blog), May 19, 2021. https://theresakushner.medium.com/the-marketing-database-is-dead-so-now-what-2e630eb2e3da.

LaForme, Ren. 2018. "*The New York Times* Homepage Is Far from Dead: Its Growing." *Poynter* (blog), August 23, 2018. https://www.poynter.org/tech-tools/2018/the-new-york-times-homepage-is-far-from-dead-its-growing.

Lampen, Amanda Arnold, Claire. 2020. "Everything We Know About the Tessa Majors Murder Case." The Cut, February 19, 2020. https://www.thecut.com/2020/02/tessa-majors-barnard-student-death.html.

Landau, Susan. 2020. "Location Surveillance to Counter COVID-19: Efficacy Is What Matters." *Lawfare* (blog), March 25, 2020. https://www.lawfareblog.com/location-surveillance-counter-covid-19-efficacy-what-matters.

Lardner, James, and Thomas Reppetto. 2001. *NYPD: A City and Its Police*. New York: Holt Paperbacks.

Larionov, Daniil, Natalia Romantsova, and Roman Shalymov. 2019. "Multiphysical System of Operational Monitoring of the Condition of the Railway Track." Paper presented at the Fourth International Conference on Intelligent Transportation Engineering, Singapore, September 2019. https://doi.org/10.1109/ICITE.2019.8880225.

Lariviere, Marty. 2013. "The Process of Orange Juice." Operations Room, February 4, 2013. https://operationsroom.wordpress.com/2013/02/04/the-process-of-orange-juiced/.

Larrier, Travis. 2021. "Stop Learning Loss in Its Tracks: Part 1 with Joel Rose & Travis Larrier." Webinar presentation May 1, 2021. Transcript included. Published May 14, 2021. *YouTube*. https://newclassrooms.org/2021/05/14/webinar-part-1-stop-learning-loss-in-its-tracks-with-joel-rose-travis-larrier.

Larson, David B., Matthew C. Chen, Matthew P. Lungren, et al. 2018. "Performance of a Deep-Learning Neural Network Model in Assessing Skeletal Maturity on Pediatric Hand Radiographs." *Radiology* 287, no. 1 (2018): 313–22. https://doi.org/10.1148/radiol.2017170236.

Leape, Lucian L., et al. 1995. "Systems Analysis of Adverse Drug Events." *JAMA* 274, no. 1 (1995): 35–43.

Lederman, George, Siheng Chen, James H. Garrett, Jelena Kovačević, Hae Young Noh, and Jacobo Bielak. 2017. "A Data Fusion Approach for Track Monitoring from Multiple In-Service Trains." *Mechanical Systems and Signal Processing* 95 (October 2017): 363–379. https://doi.org/10.1016/j.ymssp.2017.03.023.

Lee, Edmund. 2012. "Newspapers Lose $10 in Print for Every Digital $1." Bloomberg.com, March 19, 2012. https://www.bloomberg.com/news/articles/2012-03-19/newspapers-lose-10-dollars-in-print-for-every-digital-1.

Lee, Edmund. 2020. "New York Times Hits 7 Million Subscribers as Digital Revenue Rises." *New York Times*, November 5, 2020. https://www.nytimes.com/2020/11/05/business/media/new-york-times-q3-2020-earnings-nyt.html.

Lee, Wan-Jui. 2020. "Anomaly Detection and Severity Prediction of Air Leakage in Train Braking Pipes." *International Journal of Prognostics and Health Management* 8, no. 3 (November 2020): 1–12. https://doi.org/10.36001/ijphm.2017.v8i3.2662.

Lee, Yoolim. 2020. "Singapore App Halves Contact-Tracing Time, Top Engineer Says." *Bloomberg*, December 8, 2020. https://www.bloomberg.com/news/articles/2020-12-08/singapore-app-halves-contact-tracing-time-leading-engineer-says.

Leighninger, Harrison J. 2020. "Evaluating the Effectiveness of Relational Organizing: The Buttigieg 2020 Campaign in New Hampshire." Senior diss., Haverford College, 2020. http://hdl.handle.net/10066/22744.

Lemouche, Patrick. 2021. "Unilever HireVue Interview: Questions and Strategy." *Voomer Blog* (blog), July 9, 2021. https://blog.tryvoomer.com/unilever-hirevue-interview-questions-answers-and-strategy.

Leonard, Davis. 2022. "How We Built a Relational Network of 160k Voters in Less than a Month." *Medium* (blog). April 11, 2022. https://medium.com/@davisleonard/how-we-built-a-relational-network-of-160k-voters-in-less-than-a-month-92262926fdb0.

Leone, Dario. 2020. "An In-Depth Analysis of How Serbs Were Able to Shoot Down An F-117 Stealth Fighter during Operation Allied Force." The Aviation Geek Club. March 26, 2020. https://theaviationgeekclub.com/an-in-depth-analysis-of-how-serbs-were-able-to-shoot-down-an-f-117-stealth-fighter-during-operation-allied-force/.

Leslie, Mark, Russell Lewis Siegelman, and Austin Kiesseg. 2013. "Blue River Technology (A)." Stanford Graduate School of Business Case E480A, August 26, 2013. https://www.gsb.stanford.edu/faculty-research/case-studies/blue-river-technology.

Levitt, Leonard. 2010. *NYPD Confidential: Power and Corruption in the Country's Greatest Police Force*. New York: St. Martin's.

Li, Hongfei, Dhaivat Parikh, Qing He, Buyue Qian, Zhiguo Li, Dongping Fang, and Arun Hampapur. 2014. "Improving Rail Network Velocity: A Machine Learning Approach to Predictive Maintenance." *Transportation Research Part C: Emerging Technologies* 45 (August 2014): 17–26. https://doi.org/10.1016/j.trc.2014.04.013.

Li, Oscar. 2017. "Review: Artificial Intelligence Is the New Electricity—Andrew Ng." SyncedReview, April 28, 2017. https://medium.com/syncedreview/artificial-intelligence-is-the-new-electricity-andrew-ng-cc132ea6264.

Li, Yu-Sheng, Hong Chi, Xue-Yan Shao, Ming-Liang Qi, and Bao-Guang Xu. 2020. "A Novel Random Forest Approach for Imbalance Problem in Crime Linkage." *Knowledge-Based Systems* 195 (May 11, 2020): 105738. https://doi.org/10.1016/j.knosys.2020.105738.

Lillis, Katie Bo, and Natasha Bertrand. 2022. "US Intelligence Community Launches Review Following Ukraine and Afghanistan Intel Failings." *CNN*, May 13, 2022. https://www.cnn.com/2022/05/13/politics/us-intelligence-review-ukraine/index.html.

Lipetri, Michael. 2021. Interviews by Zachary Tumin.

Litow, Stanley S., Michael Casserly, Bruce MacLaury, and Joseph P. Viteritti. 1999. "Problems of Managing a Big-City School System." *Brookings Papers on Education Policy* no. 2 (1999): 185–230. http://www.jstor.org/stable/20067209.

Liu, Chris. 2019. "An Engineer's Perspective on Engineering and Data Science Collaboration for Data Products." Medium, May 20, 2019. https://medium.com/coursera-engineering/an-engineers-perspective-on-engineering-and-data-science-collaboration-for-data-products-84cf9b38cd52.

Liu, Cui-wei, Yu-xing Li, Yu-kun Yan, Jun-tao Fu, and Yu-qian Zhang. 2015. "A New Leak Location Method Based on Leakage Acoustic Waves for Oil and Gas Pipelines." *Journal of Loss Prevention in the Process Industries* 35 (May 1, 2015): 236–246. https://doi.org/10.1016/j.jlp.2015.05.006.

Locke, Julian, and Sameer More. 2021. "How We Rearchitected Mobile A/B Testing at *The New York Times*." *NYT Open* (blog), *Medium*, March 4, 2021. https://open.nytimes.com/how-we-rearchitected-mobile-a-b-testing-at-the-new-york-times-78eb428d9132.

Lores, Enrique. 2021. "Conversations with Mike Milken." Podcast. https://mikemilken.com/podcast/Conversations-with-Mike-Milken-Enrique-Lores-1-26–21.pdf.

Lorica, Ben. 2021. "Applications of Reinforcement Learning: Recent Examples from Large US Companies." Gradient Flow, May 19, 2021. https://gradientflow.com/applications-of-reinforcement-learning-recent-examples-from-large-us-companies/.

Louque, Jake. 2019. "Ravens Bolster Their Analytics Department with New Front Office Additions." Baltimore Beatdown, June 19, 2019. https://www.baltimorebeatdown.com/2019/6/19/18684490/ravens-to-beef-up-analytics-department-with-new-front-office-additions-eric-decosta-john-harbaugh.

Luckin, Rose. 2020. "Turing Lecture: Is Education AI-Ready?" Lecture for The Alan Turing Institute. Published August 6, 2020. *YouTube*. https://youtu.be/PqQWQr4V-JE.

Lundberg, Brandon. 2021. "The Ravens Aggressive Offensive Style Cultivated by Lamar Jackson, Analytics & Early Down Success." Football Scout 365, September 25, 2021. https://www.footballscout365.com/post/the-ravens-aggressive-offensive-style-cultivated-by-lamar-jackson-analytics-early-down-success.

Lupesko, Hagay. 2019. "Personalization at Scale: Challenges and Practical Techniques." Presentation at the O'Reilly Artificial Intelligence Conference 2019, San Jose, CA, September 9, 2019. Video. https://www.oreilly.com/library/view/personalization-at-scale/0636920371144/video329159.html.

Lyell, David, and Enrico Coiera. 2017. "Automation Bias and Verification Complexity: A Systematic Review." *Journal of the American Medical Informatics Association* 24, no. 2 (2017): 423–431.

Lynch, Matthew. 2019. "Chronicling the Biggest Edtech Failures of the Last Decade." Tech Edvocate. Last modified July 10, 2019. https://www.thetechedvocate.org/chronicling-the-biggest-edtech-failures-of-the-last-decade.

Lynn, Lawrence E. 1982. "Government Executives as Gamesmen: A Metaphor for Analyzing Managerial Behavior." *Journal of Policy Analysis and Management* 1, no. 4 (1983): 482–495.

Lyons, Keith. n.d. "Pattern Recognition." Sport Informatics and Analytics. Accessed December 19, 2021. https://sites.google.com/site/ucsportinformaticsandanalytics/pattern-recognition.

Macaulay, Thomas. 2022. "Meta's Free GPT-3 Replica Exposes the Business Benefits of AI Transparency." TNW Neural, May 4, 2022. https://thenextweb.com/news/meta-new-ai-model-opt-replicates-gpt-3-and-researchers-can-download-it-free.

Mahesh, C. K. 2021. "New Orleans Educators Share How Teach to One Helps Save Students from the Iceberg Problem." *RecentlyHeard* (blog), November 8, 2021. https://recentlyheard.com/2020/06/22/new-orleans-educators-share-how-teach-to-one-helps-save-students-from-the-iceberg-problem.

Mahnken, Kevin. 2021. "Study: Chicago Tutoring Program Delivered Huge Math Gains; Personalization May Be the Key." The 74 Million, March 8, 2021. https://www.the74million.org/study-chicago-tutoring-program-delivered-huge-math-gains-personalization-may-be-the-key.

Maier, Maximilian, František Bartoš, et al. 2022. "No Evidence for Nudging after Adjusting for Publication Bias." *Proceedings of the National Academy of Sciences* 119 (31): e2200300119. https://doi.org/10.1073/pnas.2200300119.

Malik, Momin. 2020 "Types of a Hierarchy of Limitations in Machine Learning: Data Biases." UOC 2020, Universitat Oberta De Catalunya, 2020, https://www.mominmalik.com/uoc2020.pdf.

Mallampati, Dasaradh. 2021. Interviews by Zachary Tumin.

Malone, Thomas, Daniela Rus, and Robert Laubacher. 2020. "Artificial Intelligence and the Future of Work." Research Brief RB 17–2020. MIT Work of the Future. MIT. https://workofthefuture.mit.edu/research-post/artificial-intelligence-and-the-future-of-work/.

Mamet, Matthew. 2016. "Directly Responsible Individuals." *Medium* (blog), November 13, 2016. https://medium.com/@mmamet/directly-responsible-individuals-f5009f465da4.

Man, Romeo. 2020. "Sean Ellis on Successful Growth Teams & The North Star Metric." *StartUs Magazine*, January 29, 2020. https://magazine.startus.cc/sean-ellis-on-the-backbone-of-successful-growth-teams-the-north-star-metric.

Mandel, Eugene. 2016. "Data Science for Product Managers (DS4PM)." *Medium* (blog), October 31, 2016. https://medium.com/@eugmandel/data-science-for-product-managers-fbd2036536a0.

Manyika, James, Jake Silberg, and Brittany Presten. 2019. "What Do We Do About the Biases in AI?" *Harvard Business Review*, October 25, 2019. https://hbr.org/2019/10/what-do-we-do-about-the-biases-in-ai.

Maple, Jack. 1999. *The Crime Fighter*. New York: Broadway Books.

Marin, Nina, and Asiya Yakhina. 2019. "What It Means to Design for Growth at *The New York Times*." Interview by Sarah Bures. *NYT Open* (blog), *Medium*, June 13, 2019. https://open.nytimes.com/what-it-means-to-design-for-growth-at-the-new-york-times-2041e0f5e64a.

Markus, M. Lynne, and Robert I Benjamin. 1997. "The Magic Bullet Theory in IT-Enabled Transformation." *Sloan Management Review* 38, no. 2 (Winter 1997): 55.

Marr, Bernard. 2019. "The Amazing Ways John Deere Uses AI And Machine Vision to Help Feed 10 Billion People." *Forbes*, March 15, 2019. https://www.forbes.com/sites/bernardmarr/2019/03/15/the-amazing-ways-john-deere-uses-ai-and-machine-vision-to-help-feed-10-billion-people/.

Marsh, Allison. 2018. "John Deere and the Birth of Precision Agriculture." *IEEE Spectrum*, February 28, 2018. https://spectrum.ieee.org/tech-history/silicon-revolution/john-deere-and-the-birth-of-precision-agriculture.

Marson, James, and Brett Forrest. 2021. "Armed Low-Cost Drones, Made by Turkey, Reshape Battlefields and Geopolitics." *Wall Street Journal*, June 3, 2021. https://www.wsj.com/articles/armed-low-cost-drones-made-by-turkey-reshape-battlefields-and-geopolitics-11622727370.

Martin, Scott. 2019. "Blue River Harvests AI to Reduce Herbicides." *NVIDIA Blog* (blog), May 2, 2019. https://blogs.nvidia.com/blog/2019/05/02/blue-river-john-deere-reduce-herbicide/.

Marx, Sally. 2017. "Calling All Data and Tech Innovators: We Have Elections to Win." *DNC Tech Team* (blog), *Medium*, November 27, 2017. https://medium.com/democratictech/calling-all-data-and-tech-innovators-we-have-elections-to-win-28893599ea21.

Mates, Stacey. 2020. "Why I Joined the DNC Tech Team." *DNC Tech Team* (blog), *Medium*, July 15, 2020. https://medium.com/democratictech/why-i-joined-the-dnc-tech-team-e1bfad994bb9.

Mathias, Craig. 2012. "A Really, Really Big Wireless LAN: Gillette Stadium and the New England Patriots." *Nearpoints* (blog), Network World, November 6, 2012. http://www.networkworld.com/community/blog/really-really-big-wireless-lan-gillette-stadium-and-new-england-patriots.

Mathias, Craig. 2013. "Returning to Gillette Stadium: Implications for the Enterprise." *Nearpoints* (blog), Network World, January 16, 2013. http://www.networkworld.com/community/node/82169.

Mathieu, John E., John R. Hollenbeck, Daan van Knippenberg, and Daniel R. Ilgen. 2017. "A Century of Work Teams in the Journal of Applied Psychology." *Journal of Applied Psychology* 102, no. 3 (2017): 452–467. https://doi.org/10.1037/apl0000128.

Maurer, Roy. 2021. "HireVue Discontinues Facial Analysis Screening." SHRM, February 3, 2021. https://www.shrm.org/resourcesandtools/hr-topics/talent-acquisition/pages/hirevue-discontinues-facial-analysis-screening.aspx.

Mayer, Jane. 2009. "The Predator War." *New Yorker*, October 19, 2009. http://www.newyorker.com/magazine/2009/10/26/the-predator-war.

Mayfield, Bob. 2016. "John Deere Hands-Free Guidance System Continues Its Evolution." March 29, 2016. https://johndeerejournal.com/2016/03/terry-picket-first-gps-unit/.

Mays, Robert. 2014. "What Is the NFL's 'Corner 3'?" *Grantland* (blog), August 21, 2014. https://grantland.com/features/nfl-corner-3-nba-offense-success-strategy-run-game-screen-pass-play-action/.

McAfee, Andrew. 2006. "The 9X Email Problem." *Andrew McAfee's Blog* (blog), September 29, 2006. Archived at the Wayback Machine. https://web.archive.org/web/20120606073022/http://andrewmcafee.org/2006/09/the_9x_email_problem.

McCaskill, Steve. 2021. "NFL Using Data Analytics to Monitor Excitement Level of Each Game." *SportsProMedia.Com* (blog). November 23, 2021. https://www.sportspromedia.com/news/nfl-data-analytics-excitement-engagement-paul-ballew/.

McChrystal, Gen. Stanley, Tantum Collins, David Silverman, and Chris Fussell. 2015. *Team of Teams: New Rules of Engagement for a Complex World*. Illustrated ed. New York: Portfolio.

McCord, Patty. 2014. "How Netflix Reinvented HR." *Harvard Business Review*, January 1, 2014. https://hbsp.harvard.edu/product/R1401E-PDF-ENG.

McCurry, Justin. 2020. "Test, Trace, Contain: How South Korea Flattened Its Coronavirus Curve." *Guardian*, April 23, 2020. https://www.theguardian.com/world/2020/apr/23/test-trace-contain-how-south-korea-flattened-its-coronavirus-curve.

McDonald, Drew. 2020a. "Introducing DNC Blueprint!" *DNC Tech Team* (blog), *Medium*, April 9, 2020. https://medium.com/democratictech/introducing-dnc-blueprint-3d0abc161c23.

McDonald, Drew. 2020b. "Our Data Science Clearinghouse." *DNC Tech Team* (blog), *Medium*, August 24, 2020. https://medium.com/democratictech/our-data-science-clearinghouse-e9f12fd4a86.

McDonnell, Patrick J. 2017. "Clashes Between Soldiers and Gasoline Smugglers Leave 10 Dead in Mexico—LA Times." *Los Angeles Times*, May 4, 2017. https://web.archive.org/web/20170505014014/https://www.latimes.com/world/mexico-americas/la-fg-mexico-gas-smugglers-20170504-story.html. https://www.latimes.com/world/mexico-americas/la-fg-mexico-gas-smugglers-20170504-story.html.

McGowan, Candace. 2019. "Police Investigate Robbery Pattern near Scene of Barnard Student's Murder." *ABC 7 New York*, December 18, 2019. https://abc7ny.com/5764879.

McGuinness, Julia E., Vicky Ro, et al. 2021. "Abstract PR-04: Effect of Breast Cancer Chemoprevention on a Convolutional Neural Network-Based Mammographic Evaluation Using a Mammographic Dataset of Women with Atypical Hyperplasia, Lobular or Ductal Carcinoma in Situ." *Clinical Cancer Research* 27, no. 5 (2021): PR-PR-04. https://doi.org/10.1158/1557-3265.ADI21-PR-04.

McKay, Daniel. 2018. "BNSF Combining Technology, Human Expertise to Improve Operations." Transport Topics, March 9, 2018. https://www.ttnews.com/articles/bnsf-combining-technology-human-expertise-improve-operations.

McLaughlin, Corey. 2019. "Meet the Ravens' 25-Year-Old, Number-Crunching Whiz Who Has John Harbaugh's Ear." *Baltimore Magazine*, December 30, 2019. https://www.baltimoremagazine.com/section/sports/meet-daniel-stern-the-ravens-25-year-old-number-crunching-whiz-who-has-john-harbaughs-ear.

Meadows, Donella H. 2008. In *Thinking in Systems: International Bestseller*, ed. Diana Wright. White River Junction, VT: Chelsea Green.

Megler, Veronika. 2019. "Managing Machine Learning Projects" Scribd, February 2019, *Amazon Web Services*. https://www.scribd.com/document/451227438/aws-managing-ml-projects.

Merten, Paxtyn. 2019. "Newsroom Automation, Election Forecasting and UX at *The New York Times*: Keynotes at C+J 2019." *Storybench* (blog), March 8, 2019. https://www.storybench.org/newsroom-automation-election-forecasting-and-u-x-at-the-new-york-times-keynotes-at-cj-2019.

Meyer, Christopher, and Julia Kirby. 2010. "Leadership in the Age of Transparency." *Harvard Business Review*, April 1, 2010.

Michel, Arthur Holland. 2015. "How Rogue Techies Armed the Predator, Almost Stopped 9/11, and Accidentally Invented Remote War." *Wired*, December 17, 2015. https://www.wired.com/2015/12/how-rogue-techies-armed-the-predator-almost-stopped-911-and-accidentally-invented-remote-war/.

Mill, Daniel. 2018. "Measuring What Makes Readers Subscribe to *The New York Times*." *NYT Open* (blog), *Medium*, November 15, 2018. https://open.nytimes.com/measuring-what-makes-readers-subscribe-to-the-time-fa31f00a3cdd.

Miller, Kerry. 2022. "The Newest Trend Taking over MLB in 2022 and Beyond." Bleacher Report, April 27, 2022. https://bleacherreport.com/articles/2955675-the-newest-trend-taking-over-mlb-in-2022-and-beyond.

Miller, Lisa. 2020. "The Stabbing in Morningside Park." *New York Magazine Intelligencer*, March 16, 2020. https://nymag.com/intelligencer/2020/03/tessa-majors-murder-morningside-park.html.

Miller, Myles. 2019. "NYPD Software Credited with Helping to Spot Crime Patterns." *Spectrum News NY1*, March 20, 2019. https://www.ny1.com/nyc/all-boroughs/news/2019/03/20/nypd-software-credited-with-helping-to-spot-crime-patterns.

Miller, Sean J. 2021. "New Studies Show the Impact Texting Can Have on Voter Turnout." *Campaigns & Elections* (blog), June 9, 2021. https://campaignsandelections.com/campaigntech/new-studies-show-the-impact-texting-can-have-on-voter-turnout/.

Miller, Steven M. 2018. "AI: Augmentation, More So Than Automation." *Asian Management Insights* 5, no. 1 (2018): 1–20. https://ink.library.smu.edu.sg/ami/83.

Miller, Zeke and Aamer Madhan. 2022. "Watching Al-Qaida Chief's 'pattern of Life' Key to His Death." AP NEWS. August 2, 2022. https://apnews.com/article/al-qaida-biden-ayman-zawahri-covid-health-595c6bda6d17fdd0c1c936137fe1e7c6.

Mishra, Ashwani. 2018. "How Gururaj Rao, CIO, Mahindra Finance Is Using Tech to Drive Business Outcomes." *ETCIO*, December 24, 2018. https://cio.economictimes.indiatimes.com/news/strategy-and-management/how-gururaj-rao-cio-mahindra-finance-is-using-tech-to-drive-business-outcomes/67218997.

Mistry, Ravi, and Richard Battle. 2021. "Analytics & Data Science in Sports Interview by Alan Jacobson." Webinar. https://pages.alteryx.com/analytics-and-data-science-in-sports-on-demand-ty.html?aliId=eyJpIjoiZXkybWJhblFLaotGek9KbyIsInQiOiJTNnJxRFFCQStRcDE3Vol3RTJEM3pnPToifQ%253D%253D.

Molteni, Megan. 2022. "For Patients, Seeing the Benefits of the New, Fully Sequenced Genome Could Take Years." *StatNews* (blog), April 8, 2022. https://www.statnews.com/2022/04/08/for-patients-seeing-the-benefits-of-the-new-fully-sequenced-genome-could-take-years/.

Monkeypoxtally [@Monkeypoxtally]. 2022. "Monkeypox Cases around the World F30D; Timeline. Tweet. *Twitter*. https://twitter.com/Monkeypoxtally/status/1553430250805133312. Retweeted by Carr, Kareem [@kareem_carr]. 2022. "One of the Most Fascinating Data Science Lessons I've Learned on Social Media . . . " Tweet. *Twitter*. https://twitter.com/kareem_carr/status/1553775608664031236.

Montal, Tal, and Zvi Reich. 2016. "I, Robot. You, Journalist. Who Is the Author?: Authorship, Bylines and Full Disclosure in Automated Journalism." *Digital Journalism* 5 (August 2016): 1–21. https://doi.org/10.1080/21670811.2016.1209083.

Moon, Grace. 2020. "South Korea's Return to Normal Interrupted by Uptick in Coronavirus Cases." *NBC News*, April 5, 2020. https://www.nbcnews.com/news/world/south-korea-s-return-normal-interrupted-uptick-coronavirus-cases-n1176021.

Moore, Geoffrey A. 2006. *Crossing the Chasm: Marketing and Selling High-Tech Products to Mainstream Customers*. Rev. ed. New York: HarperBusiness.

Moore, Mark H. 1995. *Creating Public Value: Strategic Management in Government*. Rev. ed. Cambridge, MA: Harvard University Press.

Moore, Mark H. 2013. *Recognizing Public Value*. Illustrated ed. Cambridge, MA: Harvard University Press.

Moore, Mark H., and Anthony A. Braga. 2003. "Measuring and Improving Police Performance: The Lessons of CompStat and Its Progeny." *Policing: An International Journal of Police Strategies & Management* 26, no. 3 (2003): 439–453. https://doi.org/10.1108/13639510310489485.

Moore, Mark H., and Archon Fong. 2012. "Calling Publics into Existence: The Political Arts of Public Management." In *Ports in a Storm: Public Management in a Turbulent World*, ed. John D. Donahue and Mark H. Moore, 180–210. Washington, DC: Brookings Institution Press/Ash Center.

Moore, Tina, Larry Celona, Olivia Bensimon, and Bruce Golding. 2019. "Barnard Student Tessa Majors' Suspected Killer in NYPD Custody." *New York Post* (blog), December 26, 2019. https://nypost.com/2019/12/26/barnard-student-tessa-majors-suspected-killer-located-by-nypd.

Moran, Max. 2019. "Freddie Mac Using Shady AI Company for Mortgage Loans." *American Prospect*, October 7, 2019. https://prospect.org/api/content/71c641a2-e6f3-11e9-9f97-12f1225286c6.

Morozov, Evgeny. 2013. *To Save Everything, Click Here: The Folly of Technological Solutionism.* New York: PublicAffairs.

Morris, Bob. 2013. "Scoble Interviews Marc Andreessen on What's Coming." Politics in the Zeros, February 22, 2013. https://polizeros.com/2013/02/22/scoble-interviews-marc-andreessen-on-whats-coming/.

Morrissey, Janet. 2020. "Using Technology to Tailor Lessons to Each Student." *New York Times*, September 29, 2020. Updated November 4, 2021. https://www.nytimes.com/2020/09/29/education/schools-technology-future-pandemic.html.

Morse, Ben. 2021. "How Steph Curry 'Revolutionized' the NBA." *CNN*, December 15, 2021. https://www.cnn.com/2021/12/14/sport/steph-curry-ray-allen-three-point-record-spt-intl/index.html.

Moses, Lucia. 2017. "The *Washington Post*'s Robot Reporter Has Published 850 Articles in the Past Year." *Digiday* (blog), September 14, 2017. https://digiday.com/media/washington-posts-robot-reporter-published-500-articles-last-year.

Moss, Sebastian. 2017. "For *The New York Times*, a Move to the Cloud." Data Center Dynamics. Last modified April 19, 2017. https://www.datacenterdynamics.com/en/news/for-the-new-york-times-a-move-to-the-cloud.

Motley Fool. 2021a. "New York Times Co (NYT) Q1 2021 Earnings Call Transcript." Motley Fool. Last modified May 5, 2021. https://www.fool.com/earnings/call-transcripts/2021/05/05/new-york-times-co-nyt-q1-2021-earnings-call-transc.

Motley Fool. 2021b. "*The New York Times* Company (NYT) Q2 2021 Earnings Call Transcript." Motley Fool. Last modified August 4, 2021. https://www.fool.com/earnings/call-transcripts/2021/08/04/the-new-york-times-company-nyt-q2-2021-earnings-ca.

Mozur, Paul, and Don Clark. 2020. "China's Surveillance State Sucks Up Data. US Tech Is Key to Sorting It." *New York Times*, November 22, 2020. Updated January 20, 2021. https://www.nytimes.com/2020/11/22/technology/china-intel-nvidia-xinjiang.html.

Mullen, Benjamin. 2018. "New York Times Adapts Data Science Tools for Advertisers." *Wall Street Journal*, February 15, 2018. https://www.wsj.com/articles/new-york-times-adapts-data-science-tools-for-advertisers-1518714077.

Mullin, Benjamin. 2016. "Meet Beta, the Team That Brings *The New York Times* to Your Smartphone." *Poynter* (blog), March 25, 2016. https://www.poynter.org/tech-tools/2016/meet-beta-the-team-that-brings-the-new-york-times-to-your-smartphone.

Mullin, Emily. 2019. "IVF Often Doesn't Work. Could an Algorithm Help?" WSJ.com, April 4, 2019. https://www.wsj.com/articles/ivf-often-doesnt-work-could-an-algorithm-help-11554386243.

Mulvenney, Nick. 2021. "Sailing: America's Cup Yachts Close on Speeds of 100 Kph." *Reuters*, March 9, 2021. https://www.reuters.com/article/us-sailing-americas-cup-idUSKBN2B105E.

Murphy, Mike. 2020. "The Future of Farming Is One Giant A/B Test on All the Crops in the World at Once." Protocol—The People, Power and Politics of Tech, August 11, 2020. https://www.protocol.com/the-future-of-farming-is-math.

Mushnick, Phil. 2021. "How a Tired Aaron Boone Tactic Doomed the Yankees." *New York Post*, October 9, 2021. https://nypost.com/2021/10/09/tired-aaron-boone-bullpen-tactic-dooms-yankees-yet-again/.

Mutasa, Simukayi. 2021, various dates. Interviews by Zachary Tumin and Madeleine Want.

Mutasa, Simukayi, Peter D. Chang, Carrie Ruzal-Shapiro and Rama Ayyala. 2018. "MABAL: A Novel Deep-Learning Architecture for Machine-Assisted Bone Age Labeling." *Journal of Digital Imaging* 31, no. 4 (February 2018): 513–519. https://doi.org/10.1007/s10278-018-0053-3.

Nagendran, Myura, and Hugh Harvey. 2020. "How Good Is the Evidence Supporting AI in Radiology?" Interview by Brian Casey and Philip Ward. https://www.auntminnie.com/index.aspx?sec=rca&sub=rsna_2020&pag=dis&ItemID=130976.

Nan, Wong Yuet, Jovina Ang, and Steven Miller. 2022. "Digital Product Management Under Extreme Uncertainty: The Singapore Tracetogether Story for COVID-19 Contact Tracing." Singapore Management University.

Narcisi, Gina. 2013. "Stadium vs. the Couch: High-Density Wireless Improves Fan Experience." *TechTarget* (blog), January 21, 2013. Archived at the Wayback Machine. https://web.archive.org/web/20190728194845/https://searchnetworking.techtarget.com/news/2240176377/Stadium-vs-the-couch-High-density-wireless-improves-fan-experience.

Narisetti, Raju and Yael Taqqu. 2020. "Building a Digital *New York Times*: CEO Mark Thompson." Interview by McKinsey & Company. Last modified August 10, 2020. https://www.mckinsey.com/industries/technology-media-and-telecommunications/our-insights/building-a-digital-new-york-times-ceo-mark-thompson.

National Museum of American History. n.d. "Precision Farming." American Enterprise. Accessed October 24, 2021. https://americanhistory.si.edu/american-enterprise/new-perspectives/precision-farming.

Navarro, Carlos. 2017. "Fuel Thefts Increase Significantly in the Triángulo Rojo Region of Puebla State." SourceMex, March 15, 2017. https://digitalrepository.unm.edu/sourcemex/6332.

Navarro, Carlos. 2015. "Mexican Consumers Pay More at the Pump Despite Sharp Drop in Global Oil Prices." SourceMex, January 7, 2015. https://digitalrepository.unm.edu/sourcemex/6129.

Nayak, Akhilesh. 2018. "Growing a Successful and Collaborative Team." *NYT Open* (blog), *Medium*, September 27, 2018. https://open.nytimes.com/growing-a-successful-and-collaborative-team-4e4c608ab2fc.

NEDARC. 2019. "Hypothesis Testing." August 5, 2019. https://aws.amazon.com/blogs/machine-learning/dynamic-a-b-testing-for-machine-learning-models-with-amazon-sagemaker-mlops-projects/.

Nesterak, Evan. 2014. "Google Re:Work: Shaping the Future of HR." *Behavioral Scientist*, December 2, 2014. https://behavioralscientist.org/google-rework-shaping-future-hr.

Netroots Nation. n.d. "Proven Persuasion and Inoculation: Breakthroughs in Deep Canvassing." Presented at Netroots Nation. Accessed May 11, 2021. https://www.netrootsnation.org/nn_events/nn19/proven-persuasion-and-inoculation-breakthroughs-in-deep-canvassing.

Neustadt, Richard E. 2000. *Preparing to Be President: The Memos of Richard E. Neustadt*, ed. Charles Jones. Washington, DC: AEI Press.

New Classrooms. 2019. "The Iceberg Problem: How Assessment and Accountability Policies Cause Learning Gaps in Math to Persist Below the Surface . . . and What to Do About It." Panel discussion. Published September 25, 2019. *YouTube*. https://youtu.be/re3RVNauC7M.

New Classrooms. 2021. "Webinar: 'Stop Learning Loss in Its Tracks': Part 2 with Teacher Panelists." New Classrooms, May 14, 2021. https://newclassrooms.org/2021/05/14/webinar-part-2-stop-learning-loss-in-its-tracks-with-teacher-panelists.

Newton, Casey. 2020. "Why Countries Keep Bowing to Apple and Google's Contact Tracing App Requirements." *Interface* (blog), *Verge*, May 8, 2020. https://www.theverge.com/interface/2020/5/8/21250744/apple-google-contact-tracing-england-germany-exposure-notification-india-privacy.

New York Civil Liberties Union. 2012. "Stop-and-Frisk Data." January 2, 2012. https://www.nyclu.org/en/stop-and-frisk-data.

*New York Times*. 2018. "Meet Our New Home Page." August 8, 2018. https://www.nytimes.com/2018/08/08/homepage/meet-our-new-home-page.html.

New York Times Company. 2018. "*The New York Times* Introduces First Personalized Editorial Newsletter, 'Your Weekly Edition.'" Last modified June 13, 2018. https://www.nytco.com/press/the-new-york-times-introduces-first-personalized-editorial-newsletter-your-weekly-edition.

Ng, Andrew. 2017. "Artificial Intelligence Is the New Electricity—YouTube." Presentation at the Stanford MSx Future Forum, Stanford Graduate School of Business, Stanford, CA, January 25. https://www.youtube.com/watch?v=21EiKfQYZXc&ab_channel=StanfordGraduateSchoolofBusiness.

Ng, Andrew. 2021. "AI Matches Patients to Drugs, Robots Crawl Sewers, New Voices for Atypical Speech, Graph Neural Networks Go Deep." The Batch, DeepLearning.AI, December 1, 2021. https://read.deeplearning.ai/the-batch/issue-120/?hss_channel=tw-992153930095251456.

Nielsen, Jakob. 2012. "How Many Test Users in a Usability Study?" Nielsen Norman Group, June 3, 2012. https://www.nngroup.com/articles/how-many-test-users/.

Nieman Lab. 2017. "*The New York Times* Is Experimenting with Personalization." *Outriders*, December 4, 2017. https://outride.rs/en/the-new-york-times-is-experimenting-with-personalization.

"1999 F-117A Shootdown." 2022. In *Wikipedia*. https://en.wikipedia.org/w/index.php?title=1999_F-117A_shootdown&oldid=1092041414.

Norfolk Southern Corp. 2019. "NS Uses Predictive Data Analytics to Improve Customer Service." Published March 29, 2019. *YouTube*. https://youtu.be/DBf7pM_c5MQ.

Norfolk Southern Corp. 2020. "Pioneering a New Way to Inspect Track That Enhances the Safety and Efficiency of Railroad Operations." Published March 5, 2020. *YouTube*. https://youtu.be/vpa85Vx9JTc.

Novet, Jordan. 2020. "No Emails Have Leaked from the 2020 Election Campaigns Yet—Tiny USB Sticks May Be One Reason Why." *CNBC*, December 23, 2020. https://www.cnbc.com/2020/12/23/physical-security-keys-protected-2016-election-campaigns-against-leaks.html.

Nowobilska, Aneta. 2018. "Three Key Tech Drivers to Monetise a Premium News App." Presentation at Scale18, November 2018. Published November 6, 2018. *YouTube*. https://youtu.be/Y6J55hQLMGc.

Nunez, Jose. 2020. "Grassroots Organizing During a Pandemic." *Medium* (blog), December 1, 2020. https://josenuneziv.medium.com/grassroots-organizing-during-a-pandemic-99847b7d6916.

Nunez, Jose. 2021. Interviews by Zachary Tumin.

Nye, Jr., Joseph S. 2010. *The Powers to Lead*. Reprint ed. Oxford: Oxford University Press.

NYPD Project Management Office. 2017a. *ReBoot 311: Current States and Discoveries*. New York: New York Police Department.

NYPD Project Management Office. 2017b. *ReBoot 311: Proposed TEA Proof of Concept*. New York: New York Police Department.

NYT Open Team. 2020. "Meeting . . . Katerina Iliakopoulou, Lead Software Engineer." *NYT Open* (blog), *Medium*, August 20, 2020. https://open.nytimes.com/meeting-katerina-iliakopoulou-lead-software-engineer-82b1dec8cc18.

NYT Open Team. 2021a. "Meeting . . . Alexandra Shaheen, Program Manager at *The New York Times*." *NYT Open* (blog), *Medium*, March 30, 2021. https://open.nytimes.com/meeting-alexandra-shaheen-program-manager-at-the-new-york-times-2450946576f6.

NYT Open Team. 2021b. "Meeting . . . Charity Garcia, Software Engineer at *The New York Times*." *NYT Open* (blog), *Medium*, March 22, 2021. https://open.nytimes.com/meeting-charity-garcia-software-engineer-at-the-new-york-times-a7f6defa0bde.

NYT Open Team. 2021c. "Meeting . . . Cindy Taibi, Chief Information Officer at *The New York Times*." *NYT Open* (blog), *Medium*, March 8, 2021. https://open.nytimes.com/meeting-cindy-taibi-chief-information-officer-at-the-new-york-times-795022c3428f.

NYT Open Team. 2021d. "Meeting . . . Kathleen Kincaid, Executive Director for Development & Engagement at *The New York Times*." *NYT Open* (blog), *Medium*, June 15, 2021. https://open.nytimes.com/meeting-kathleen-kincaid-executive-director-for-development-engagement-at-the-new-york-times-39ceb00c1137.

NYT Open Team. 2021e. "Meeting . . . Véronique Brossier, Lead Software Engineer at *The New York Times*." *NYT Open* (blog), *Medium*, March 29, 2021. https://open.nytimes.com/meeting-v%C3%A9ronique-brossier-lead-software-engineer-at-the-new-york-times-52088bb534ac.

NYT Open Team. 2021f. "Meeting . . . Vicki Crosson, Software Engineer at *The New York Times*." *NYT Open* (blog), *Medium*, March 26, 2021. https://open.nytimes.com/meeting-vicki-crosson-software-engineer-at-the-new-york-times-2ed88e1743ab.

O'Brien, Dan. 2019. "Break the 'Broken Windows' Spell: The Policing Theory Made Famous in New York City Under Giuliani and Bratton Doesn't Hold Up to Scrutiny." *New York Daily News*, May 26, 2019. https://www.nydailynews.com/opinion/ny-oped-break-the-broken-windows-spell-20190526-ulwcdd7fnjg4fgv6dnskls6vhi-story.html.

O'Connell, Robert. 2022. "In the N.F.L., It's Fourth-and-It-Doesn't-Matter." *New York Times*, February 9, 2022, sec. Sports. https://www.nytimes.com/2022/02/09/sports/football/nfl-analytics-super-bowl.html.

O'Connor, Matt. 2020. "Revolutionary AI-Powered Digital Pathology Algorithm Customizes Cancer Care." *Health Imaging*, July 27, 2020. https://www.healthimaging.com/topics/advanced-visualization/ai-digital-pathology-cancer-care.

O'Keefe, Patrick. 2018. "How P2P Texting Is Revolutionizing Politics." *Political Moneyball* (blog), *Medium*, August 26, 2018. https://medium.com/political-moneyball/how-p2p-texting-is-revolutionizing-politics-bfe697c2abb8.

Olavson, Thomas. 2012. "High-Impact Analytics Teams: Defining Choices and Timeless Lessons—Google." Presentation at the INFORMS Analytics Conference, Huntington Beach, CA, April 2012. Published September 10, 2012. *YouTube*. https://youtu.be/A1piSz1rqbU.

O'Neil, Cathy. 2017. *Weapons of Math Destruction: How Big Data Increases Inequality and Threatens Democracy*. Reprint ed. New York: Crown.

O'Neil, Cathy, and Hanna Gunn. 2020. "Near-Term Artificial Intelligence and the Ethical Matrix." In *Ethics of Artificial Intelligence*, ed. S. Matthew Liao. New York: Oxford University Press.

O'Neill, Patrick Howell, Tate Ryan-Mosley, and Bobbie Johnson. 2020. "A Flood of Coronavirus Apps Are Tracking Us. Now It's Time to Keep Track of Them." *MIT Technology Review*, May 7, 2020. https://www.technologyreview.com/2020/05/07/1000961/launching-mittr-covid-tracing-tracker.

O'Neill, Paul. 2021. Interviews by Zachary Tumin. Appeared originally in William Bratton and Zachary Tumin, *Collaborate or Perish!: Reaching Across Boundaries in a Networked World*. New York: Crown Business.

Osterman, Caroline. 2019. "Blue River Technology: How Robotics and Machine Learning Are Transforming the Future of Farming." *Berkeley Master of Engineering* (blog), April 19, 2019. https://medium.com/the-coleman-fung-institute/blue-river-technology-how-robotics-and-machine-learning-are-transforming-the-future-of-farming-f355398dc567.

Owen, Chris (@ChrisO_wiki). 2022. "For the Attack, a Modified VOG-17 Grenade Was Used. A Fin and a Front Part, Created on a 3D Printer, Are Added. The VOG-17 Is a Soviet-Era 30x120 Mm Fragmentation Grenade with a Claimed Effective Radius of 7m, Covering an Area of about 150 M2. *Twitter*. https://twitter.com/ChrisO_wiki/status/1520561974127603712.

Owen, Laura Hazard. 2021. "Wirecutter, Which Makes Money When You Shop, Is Going Behind *The New York Times*' Paywall." *Nieman Lab* (blog), August 31, 2021. https://www.niemanlab.org/2021/08/wirecutter-which-makes-money-when-you-shop-is-going-behind-the-new-york-times-paywall/.

Oyefusi, Daniel. 2019. "Ravens' John Harbaugh Claims Analytics Back Up Aggressive Decisions vs. Chiefs. Here's What the Numbers Say." *Baltimore Sun*, September 23, 2019. https://www.baltimoresun.com/sports/ravens/bs-sp-ravens-john-harbaugh-analytics-chiefs-20190923-oajueu7eqzcztj5aisldofntu4-story.html.

P2016.org. 2016. "Organization of Bernie 2016—Staff, Advisors and Supporters on Sen. Bernie Sanders' Presidential Campaign." Bernie 2016. Accessed January 28, 2021. http://www.p2016.org/sanders/sandersorg.html.

Padwick, Chris. 2020a. "AI for Agriculture: How PyTorch Enables Blue River's Robots." *Robot Report*, August 6, 2020. https://www.therobotreport.com/ai-for-agriculture-how-pytorch-enables-blue-rivers-robots.

Padwick, Chris. 2020b. "The Future of Farming Is One Giant A/B Test on All the Crops in the World at Once." Interview by Mike Murphy. *Protocol*, August 11, 2020. https://www.protocol.com/the-future-of-farming-is-math.

Pan, Ian, Grayson L. Baird, Simukayi Mutasa, et al. 2020. "Rethinking Greulich and Pyle: A Deep Learning Approach to Pediatric Bone Age Assessment Using Pediatric Trauma Hand Radiographs." *Radiology: Artificial Intelligence* 2, no. 4 (July 2020): 1–9. https://doi.org/10.1148/ryai.2020190198.

Papadopoulos, Anna. 2020. "CEO Spotlight: Joel Rose, CEO of Teach to One." *CEOWORLD Magazine*, February 26, 2020. https://ceoworld.biz/2020/02/26/ceo-spotlight-joel-rose-ceo-of-teach-to-one.

Park, Andrea. 2019. "5 Key Quotes About How AI Will Transform Healthcare." Beckershospitalreview.com, July 1, 2019. https://www.beckershospitalreview.com/innovation/5-key-quotes-about-how-ai-will-transform-healthcare.html?utm_campaign=bhr&utm_source=website&utm_content=related.

Parker, Charlie. 2022. "Russian Battalion Wiped Out Trying to Cross River of Death." *The Times*, May 12, 2022, sec. News. https://www.thetimes.co.uk/article/russian-battalion-devastated-as-it-crosses-river-989vvnj9v.

Patel, Shesh. 2019. "Stress Testing in Production: *The New York Times* Engineering Survival Guide." Presentation at the 2019 Dash Conference in Chelsea, NY, July 17, 2019. Published August 5, 2019. *YouTube*. https://youtu.be/Ga9UxGRgtEE.

Patterson, Jessica. 2017. "How Great Companies Analyse and Act on Data." *Insights/Innovation Media Consulting* (blog), September 3, 2017. https://innovation.media/insights/how-great-companies-analyse-and-act-on-data.

Patton, Desmond Upton. 2020. "Social Work Thinking for UX and AI Design." *Interactions* 27, no. 2 (March–April 2020): 86–89. https://dl.acm.org/doi/10.1145/3380535.

Payan, Tony, and Guadalupe Correa-Cabrera. 2014. "Issue Brief: Energy Reform and Security in Northeastern Mexico." Houston, TX: Rice University's Baker Institute for Public Policy.

PEMEX. 2018. "Annual Report Pursuant to Section 13 or 15(D) of the Securities Exchange Act of 1934." Form 20-F. https://www.sec.gov/Archives/edgar/data/932782/000119312519129698/d632951d20f.htm.

Peters, Adele. 2020. "Zipline Will Bring Its Medical Delivery Drones to the U.S. to Help Fight the Coronavirus." *Fast Company*, March 30, 2020. https://www.fastcompany.com/90483592/zipline-will-use-its-medical-delivery-drones-to-the-u-s-to-help-fight-the-coronavirus.

Peters, Jeremy W. 2011. "*The Times* Announces Digital Subscription Plan." *New York Times*, March 17, 2011. https://www.nytimes.com/2011/03/18/business/media/18times.html.

Peterson, Nolan (@nolanwpeterson). 2022. "For 8 Years, Ukraine's Military Has Transformed, Allowing Front-Line Personnel to Operate Creatively & as Autonomously as Possible. There's a 'Start-up' Mentality Among Many Troops. That's a Big Change from the Strict, Top-down Soviet Chain of Command—Which Russia Still Employs." *Twitter*. https://twitter.com/nolanwpeterson/status/1520686581157449730?lang=en.

Petróleos Mexicanos. 2018. Form 20-F: Annual Report Pursuant to Section 13 or 15(D) Of the Securities Exchange Act of 1934. United States Securities and Exchange Commission. Accessed November 27, 2021. https://www.sec.gov/Archives/edgar/data/932782/000119312519129698/d632951d20f.htm.

Pickett, Terry. 2015. "The Payoff from Precision Agriculture." August 7, 2015. https://johndeerejournal.com/2015/08/the-payoff-from-precision-agriculture/?cid=LNK_JDJ_enUS_ReadMore_TerryPicket.

Pleasance, Chris. 2022. "Ukraine 'Destroys £3million Russian T-90 Tank.'" Mail Online, May 11, 2022. https://www.dailymail.co.uk/news/article-10805727/Ukraine-destroys-4million-Russian-T-90-tank.html.

Podojil, Edward. 2021. "How *The New York Times* Built an End-to-End Cloud Data Platform." *Google Cloud* (blog), January 27, 2021. https://cloud.google.com/blog/products/data-analytics/how-the-new-york-times-build-an-end-to-end-cloud-data-platform.

Podojil, Edward, Josh Arak, and Shane Murray. 2017. "Designing a Faster, Simpler Workflow to Build and Share Analytical Insights." *NYT Open* (blog), *Medium*, May 23, 2017. https://open.nytimes.com/faster-simpler-workflow-analytical-insights-ae6c7055e187.

Politics in the Zeros. 2013. "Scoble Interviews Marc Andreessen on What's Coming," February 22, 2013. https://polizeros.com/2013/02/22/scoble-interviews-marc-andreessen-on-whats-coming/.

Pollock, Lori. 2021. Interviews by Zachary Tumin and Madeleine Want.

Pooley, Eric. 1996. "One Good Apple." *Time*, January 15, 1996. http://content.time.com/time/subscriber/article/0,33009,983960,00.html.

Popke. Michael 2019. "Rail Insider-Data Analytics Is Giving Rail Asset Owners New Ways to Improve Their Predictive Maintenance Practices. Information for Rail Career Professionals from Progressive Railroading Magazine." *Progressive Railroading*, May 2019. https://www.progressiverailroading.com/internet-digital/article/Data-analytics-is-giving-rail-asset-owners-new-ways-to-improve-their-predictive-maintenance-practices—57544.

Potcner, Kevin, and Bill Griffin. 2020. "The Predictive Modeling Workflow." INFORMS Webinar, October 19, 2020. https://www.youtube.com/watch?v=SRRGKAenyUc&list=PLuvtfhwcPzCRcAlS77SSEFh1wtsEMz3s9&index=3&ab_channel=INFORMS.

Pownall, Augusta. 2018. "New York Times Redesigns Website to Catch Up with Mobile Offering." *Dezeen*, August 17, 2018. https://www.dezeen.com/2018/08/17/new-york-times-newspaper-website-redesign-news-design.

Prabhat, Pranay. 2020. "To Serve Better Ads, We Built Our Own Data Program." *NYT Open* (blog), *Medium*, December 17, 2020. https://open.nytimes.com/to-serve-better-ads-we-built-our-own-data-program-c5e039bf247b.

Press, Gil. 2013. "A Very Short History of Data Science." *Forbes*, May 28, 2013. https://www.forbes.com/sites/gilpress/2013/05/28/a-very-short-history-of-data-science/.

Prime Tide Sports. "Why NBA Teams Are Shooting More 3 Pointers." n.d. Accessed November 28, 2021. *YouTube*. https://www.youtube.com/watch?v=uGCRzuWIoN8&ab_channel=PrimeTideSports.

Privacy Sandbox. n.d. "Building a More Private, Open Web." Google. Accessed November 26, 2021. https://privacysandbox.com.

Pulley, Brett. 2011. "New York Times Fixes Paywall to Balance Free and Paid." *Bloomberg*, January 28, 2011. https://www.bloomberg.com/news/articles/2011-01-28/new-york-times-fixes-paywall-glitches-to-balance-free-vs-paid-on-the-web.

Quinn, Colleen. 2021. "3 Reasons You Should Be Using Site Direct vs. Programmatic." Accessed November 24, 2021. https://www.meredithconnecticut.com/blog/3-reasons-you-should-be-using-site-direct-vs.-programmatic.

Ralby, Ian M. 2017. "Downstream Oil Theft: Global Modalities, Trends, and Remedies." Atlantic Council of the United States, and Global Energy Center. 2017. http://www.atlanticcouncil.org/images/publications/Downstream_Oil_Theft_web_0106.pdf.

Ramaswamy, Poornima. 2021. "How Data Can Help Fast Followers Close the Gap Faster." *LinkedIn/Pulse* (blog), February 2, 2021. https://www.linkedin.com/pulse/how-data-can-help-fast-followers-close-gap-faster-poornima-ramaswamy/.

Rao, Leena. 2010. "Former Google CIO Douglas Merrill Wants to Reform Payday Loans with ZestCash." *TechCrunch* (blog), October 12, 2010. https://social.techcrunch.com/2010/10/12/former-google-cio-douglas-merrill-wants-to-reform-payday-loans-with-zestcash.

Rao, Leena. 2012. "ZestFinance Debuts New Data Underwriting Model to Ensure Lower Consumer Loan Default Rates." *TechCrunch* (blog), November 19, 2012. https://social.techcrunch.com/2012/11/19/zestfinance-debuts-new-data-underwriting-model-to-ensure-lower-consumer-loan-default-rates.

Ravishankar, Hariharan, Prasad Sudhakar, Rahul Venkataramani, et al. 2017. "Understanding the Mechanisms of Deep Transfer Learning for Medical Images." ArXiv:1704.06040 [Cs], April 2017. http://arxiv.org/abs/1704.06040.

Ravitch, Diane. 2011. *The Death and Life of the Great American School System: How Testing and Choice Are Undermining Education*. New York: Basic Books.

Rayman, Graham, Jillian Jorgensen, and Larry McShane. 2017. "Hundreds of NYPD Cops Turn Backs to de Blasio in Protest as He Speaks at Funeral for Slain Officer Miosotis Familia." *New York Daily News*, July 11, 2017. https://www.nydailynews.com/new-york/scores-nypd-cops-turn-backs-de-blasio-officer-funeral-article-1.3318292.

Ready, Douglas, Ellen Meier, et al. 2013. *Student Mathematics Performance in Year One Implementation of Teach to One: Math*. November 2013. New York: Center for Technology and School Change at Teachers College, Columbia University.

Reagan, Mitt. 2022. "Do Targeted Strikes Work? The Lessons of Two Decades of Drone Warfare." Modern War Institute. June 2, 2022. https://mwi.usma.edu/do-targeted-strikes-work-the-lessons-of-two-decades-of-drone-warfare/.

Reagan, Mitt. 2022. *Drone Strike–Analyzing the Impacts of Targeted Killing*. 2022 edition. Cham, Switzerland: Palgrave Pivot.

Recruiter.com. 2017. "Outdated Recruiting Methods Don't Work on Millennials." *Fox Business*, October 5, 2017. https://www.foxbusiness.com/features/outdated-recruiting-methods-dont-work-on-millennials.

Redden, Lee. 2017. "Growth Opportunity—How AI Puts Lettuce in Your Salad Bowl." The AI Podcast, Episode 13, April 4, 2017. https://soundcloud.com/theaipodcast/ep-16-growth-opportunity-how-ai-puts-lettuce-in-your-salad-bowl.

Reich, Justin. 2014. "Teach to One Blended Math Study Results Similar to Other Computer-Assisted Math Instruction." *Education Week*, December 5, 2014. https://www.edweek.org/leadership/opinion-teach-to-one-blended-math-study-results-similar-to-other-computer-assisted-math-instruction/2014/12.

Reich, Justin. 2020. *Failure to Disrupt: Why Technology Alone Can't Transform Education*. Cambridge, MA: Harvard University Press.

Reilly, Patrick. 2018. "How IoT Is Making Private Railways Safer." *GovTech*, March 2, 2018. https://www.govtech.com/fs/transportation/How-IoT-Is-Making-Private-Railways-Safer.html.

"Restrictions on Geographic Data in China." 2018. *Wikipedia*. https://en.wikipedia.org/w/index.php?title=Restrictions_on_geographic_data_in_China&oldid=862085456.

Reuter, Timothy, and Peter Liu. 2021. "Medicine from the Sky: Opportunities and Lessons from Drones in Africa." World Economic Forum. https://www.updwg.org/wp-content/uploads/2021/04/WEF_Medicine_from_the_Sky_2021.pdf.

Ridley, Erik L. 2020. "Can AI Interpret Chest X-Rays as Well as Rad Residents?" AuntMinnie.com, October 9, 2020. https://www.auntminnie.com/index.aspx?sec=sup&sub=aic&pag=dis&ItemID=130443.

Ridley, Erik L. 2021a. "AI Applications Progress in Musculoskeletal Applications." AuntMinnie.com, December 1, 2021. https://www.auntminnie.com/index.aspx?sec=rca&sub=rsna_2021&pag=dis&ItemID=134400.

Ridley, Erik L. 2021b. "AI Captures Bone Mineral Density Data on Hip X-Rays." AuntMinnie.com, November 29, 2021. https://www.auntminnie.com/index.aspx?sec=rca&sub=rsna_2021&pag=dis&ItemID=134308.

Ridley, Erik L. 2021c. "Report: Reimbursement Drives Adoption of AI Software for Stroke." AuntMinnie.com, November 15, 2021. https://www.auntminnie.com/index.aspx?sec=sup&sub=aic&pag=dis&ItemID=134111.

Ridley, Erik L. 2021d. "How Will AI Affect Radiologist Productivity?" AuntMinnie.com, November 18, 2021. https://www.auntminnie.com/index.aspx?sec=sup&sub=cto&pag=dis&ItemID=134178.

Ries, Eric. 2011. *The Lean Startup: How Today's Entrepreneurs Use Continuous Innovation to Create Radically Successful Businesses*. Illustrated ed. New York: Currency.

Rifkin, Nathan. 2020a. "How Team Warren Organized Everywhere." *Medium* (blog), July 30, 2020. https://nfrifkin.medium.com/how-team-warren-organized-everywhere-3a08d66fa126.

Rifkin, Nathan. 2020b. "160,000 Organizers." *Distro List* (blog), December 15, 2020. https://nathanrifkin.substack.com/p/160000-organizers.

Rifkin, Nathan. 2020c. "160,000 Organizers: The 2020 Biden-Harris National Distributed Organizing Program." *Medium* (blog), December 15, 2020. https://nfrifkin.medium.com/160-000-organizers-the-2020-biden-harris-national-distributed-organizing-program-bef8b0a6ea88.

Rifkin, Nathan. 2021. Interviews by Zachary Tumin.

Ripston, Ramona. 2007. "Does Bratton Deserve Five More Years?" *Los Angeles Times*, May 6, 2007. https://www.latimes.com/la-op-bratton06may06-story.html.

Rittell, Horst W. J., and Melvin M. Webber. 1973. "Dilemmas in a General Theory of Planning." *Policy Sciences* 4, no. 2 (June 1973): 155–169. https://archive.epa.gov/reg3esd1/data/web/pdf/rittel%2bwebber%2bdilemmas%2bgeneral_theory_of_planning.pdf.

Roberts, Paul. 2013. "At MIT Conference, Warnings of Big Data Fundamentalism." The Security Ledger with Paul F. Roberts, October 10, 2013. https://securityledger.com/2013/10/at-mit-conference-warnings-of-big-data-fundamentalism/.

Rockoff, Jonah E. 2015. *Evaluation Report on the School of One i3 Expansion*. September 2015. New York: Columbia Business School. https://www.classsizematters.org/wp-content/uploads/2019/02/Rockoff-evaluation-of-the-school-of-one-Sept.-2015.pdf.

Rockwell, Nick. 2017a. "Develop Your Culture Like Software." *NYT Open* (blog), *Medium*, November 26, 2017. https://open.nytimes.com/develop-your-culture-like-software-a1a3c1acfd6e.

Rockwell, Nick. 2017b. "Inside Election Night at the NYT." Talk at the 2017 Altitude NYC Summit, New York, March 21, 2017. Published April 7, 2017. *Vimeo*. https://vimeo.com/212296243.

Rockwell, Nick. 2017c. "The Resistance to Serverless." *Medium* (blog), October 7, 2017. https://nicksrockwell.medium.com/the-futile-resistance-to-serverless-9f0303ba2b24.

Rockwell, Nick. 2017d. "What If Serverless Was Real?—Nick Rockwell (*The New York Times*)." Presentation at the O'Reilly Velocity Conference 2017, New York, October 2017. Published October 20, 2017. *YouTube*. https://youtu.be/6zNpkZF9z9U.

Rockwell, Nick. 2018. "Cindy Taibi Named Chief Information Officer of *The New York Times*." *NYT Open* (blog), *Medium*, April 19, 2018. https://open.nytimes.com/cindy-taibi-named-chief-information-officer-of-the-new-york-times-fe3f74cfa2bf.

Rockwell, Nick. 2019a. "Fireside Chat: Nick Rockwell, CTO of *The New York Times* (FirstMark's Data Driven NYC)." Interview by Matt Turck at the 2019 Data Driven NYC in New York. Published January 16, 2019. *YouTube*. https://youtu.be/PxYlyv_HM_8.

Rockwell, Nick. 2019b. "News in the Age of Algorithmic Recommendation: *The New York Times*." Presentation at the 2019 Data Council Conference, New York. Published November 20, 2019. *YouTube*. https://youtu.be/rgIYxpjXpPc.

Rockwell, Nick. 2019c. "New York Times' CTO Nick Rockwell on How Tech Drives Subscription Goals." Interview by Pia Frey. Published April 24, 2019. OMR Media podcast. https://omrmedia.podigee.io/24-mit-nick-rockwell.

Rockwell, Nick. 2020. "Looking Back on Four Years at *The Times*." *Start It Up* (blog), *Medium*, April 5, 2020. https://medium.com/swlh/looking-back-on-four-years-at-the-times-e158ec3a5936.

Rockwell, Nick. 2021. Interviews by Zachary Tumin.

Rockwell, Nick, and James Cunningham. 2018. "The Evolution of *The New York Times* Tech Stack." Interview by Yonas Beshawred. Published April 26, 2018. SoundCloud podcast. https://stackshare.io/posts/evolution-of-new-york-times-tech-stack.

Rockwell, Nick, and Chris Wiggins. 2019. "Open Questions: Carlos A. Gomez-Uribe." Medium, January 15, 2019. https://towardsdatascience.com/open-questions-carlos-a-gomez-uribe-980c1af1195c.

Rodriguez, Mauricio. 2019. "Analytics: Too Late for Cowboys to Buy In?" *Inside The Star*, December 1, 2019. https://insidethestar.com/analytics-too-late-for-cowboys-to-buy-in.

Rogati, Monica (@mrogati). 2014. "My favorite data science algorithm is division. Seriously: over-represented X in group Y is a beat blog-post generator and cheap classifier." *Twitter*. June 25, 2014. https://twitter.com/mrogati/status/481927908802322433.

Rogati, Monica. 2017. "The AI Hierarchy of Needs." *Hackernoon* (blog). June 12, 2017. https://hackernoon.com/the-ai-hierarchy-of-needs-18f111fcc007.

Rogers, Everett M. 2003. *Diffusion of Innovations*. New York: Free Press.

Rose, Joel. 2020a. "The Grade-Level Expectations Trap." *Education Next* (blog), May 12, 2020. https://www.educationnext.org/grade-level-expectations-trap-how-lockstep-math-lessons-leave-students-behind.

Rose, Joel. 2020b. "Mike Bloomberg Can't Shake the Legacy of Stop-and-Frisk Policing in New York." *NPR*, February 25, 2020. https://www.npr.org/2020/02/25/809368292/the-legacy-of-stop-and-frisk-policing-in-michael-bloombergs-new-york.

Rose, Joel. 2021. "Addressing Significant Learning Loss in Mathematics During COVID-19 and Beyond." *Education Next* (blog), January 29, 2021. https://www.educationnext.org/addressing-significant-learning-loss-in-mathematics-during-covid-19-and-beyond.

Rosenberg, Bernard, and Ernest Goldstein. 1982. *Creators and Disturbers: Reminiscences by Jewish Intellectuals of New York*. New York: Columbia University Press.

Roth, Brian, Anantaram Balakrishnan, Pooja Dewan, et al. 2018. "Crew Decision Assist: System for Optimizing Crew Assignments at BNSF Railway." *Interfaces* 48, no. 5 (December 2018): 436–448. https://doi.org/10.1287/inte.2018.0963.

Rotman, David. 2020. "Why Tech Didn't Save Us from COVID-19." *MIT Technology Review*, June 17, 2020. https://www.technologyreview.com/2020/06/17/1003312/why-tech-didnt-save-us-from-covid-19.

Rouhiainen, Lasse. 2019. "How AI and Data Could Personalize Higher Education." *Harvard Business Review*, October 14, 2019. https://hbr.org/2019/10/how-ai-and-data-could-personalize-higher-education.

Rucker, Patrick, and Armando Tovar. 2010. "Oil Blast Causes Inferno in Mexican Town, 28 Dead." *Reuters*, December 20, 2010, sec. World News. https://www.reuters.com/article/us-mexico-explosion-idUSTRE6BI1DT20101220.

Rudder, Christian. 2015. *Dataclysm: Who We Are (When We Think No One's Looking)*. London: Fourth Estate.

Rumelt, Richard. 2011. *Good Strategy/Bad Strategy: The Difference and Why It Matters*. Illustrated ed. New York: Currency.

Russell, Jon. 2019. "India's ZestMoney Raises $20M to Grow Its Digital Lending Service." *TechCrunch* (blog), April 22, 2019. https://social.techcrunch.com/2019/04/22/zestmoney-raises-20m.

Ryan-Mosley, Tate. 2020. "The Technology That Powers the 2020 Campaigns, Explained." *MIT Technology Review*, September 28, 2020. https://www.technologyreview.com/2020/09/28/1008994/the-technology-that-powers-political-campaigns-in-2020-explained.

Sailthru. 2018. "Personalization Failure, New York Times Paywall." *Sailthru* (blog), January 2, 2018. https://www.sailthru.com/marketing-blog/price-personalization-failures-paywall-changes-new-york-times-wired.

Samuelson, William, and Richard Zeckhauser. 1988. "Status Quo Bias in Decision Making." *Journal of Risk and Uncertainty* 1, no. 1 (March 1998): 7–59. https://www.jstor.org/stable/41760530.

Sandberg, C., J. Holmes, K. McCoy, and H. Koppitsch. 1989. "The Application of a Continuous Leak Detection System to Pipelines and Associated Equipment." *IEEE Transactions on Industry Applications* 25, no. 5 (September 1989): 906–909. https://doi.org/10.1109/28.41257.

Sankin, Aaron. 2016. "This Text Messaging App Is Bernie Sanders's Secret Weapon." *Daily Dot*, April 18, 2016. https://www.dailydot.com/debug/hustle-app-bernie-sanders-texting.

Sashihara, Stephen. 2011. *The Optimization Edge: Reinventing Decision Making to Maximize All Your Company's Assets*. New York: McGraw-Hill Education.

Sashihara, Stephen. 2020. "The Princeton 20 for AI Projects: Framework to Manage Project Risks & Successfully Deploy Solutions." INFORMS Webinar. Published September 10, 2020. *YouTube*. https://www.youtube.com/watch?v=fgmoBM-Niks&list=PLuvtfhwcPzCRcAlS77SSEFh1wtsEMz3s9&index=3&ab_channel=INFORMS.

Sault, Spring. 2019. "How Many Minutes of Game Play Does the Average NFL Game Have?" *Texas Hill Country* (blog), October 1, 2019. https://texashillcountry.com/minutes-play-average-nfl-game/.

Schatz, Aaron. 2022. "Staley Leads NFL in Re-Calibrated Aggressiveness Index: Football Outsiders." FootballOutsiders.com, February 28, 2022. https://www.footballoutsiders.com/stat-analysis/2022/staley-leads-nfl-re-calibrated-aggressiveness-index.

Schaufeli, Wilmar B. 2012. "The Measurement of Work Engagement." In *Research Methods in Occupational Health Psychology: Measurement, Design and Data Analysis*, ed. Robert R. Sinclair, Mo Wang, and Lois E. Tetrick, 138–153. New York: Routledge. https://doi.org/10.4324/9780203095249.

Schein, Aaron. 2021. "Assessing the Effects of Friend-to-Friend Texting on Turnout in the 2018 US Midterm Elections." April 11. https://www.youtube.com/watch?v=bi3VDktL7Y4.

Schein, Edgar H. 1996. "Three Cultures of Management: The Key to Organizational Learning." *MIT Sloan Management Review*, October 15, 1996. http://sloanreview.mit.edu/article/three-cultures-of-management-the-key-to-organizational-learning.

Schein, Edgar H., and Peter A. Schein. 2016. "How Leaders Embed and Transmit Culture." In *Organizational Culture and Leadership*. 5th ed., 228–253. Hoboken, NJ: Wiley.

Schellmann, Hilke. 2021. "Auditors Are Testing Hiring Algorithms for Bias, but There's No Easy Fix." *MIT Technology Review*, February 11, 2021. https://www.technologyreview.com/2021/02/11/1017955/auditors-testing-ai-hiring-algorithms-bias-big-questions-remain.

Schenkler, Martin. 2021. Interview by Zachary Tumin.

Schiller, Joe. 2019. "Late for Work 11/25: How John Harbaugh and the Ravens Gained an Analytical Edge." *Baltimore Ravens*, November 25, 2019. https://www.baltimoreravens.com/news/late-for-work-11-25-how-john-harbaugh-and-the-ravens-gained-an-analytical-edge.

Schipper, Burkhard C., and Hee Yeul Yoo. 2019. "Political Awareness, Microtargeting of Voters, and Negative Electoral Campaigning." *Quarterly Journal of Political Science* 14, no. 1 (2019): 41–88. https://doi.org/10.1561/100.00016066.

Schmidt, David. 2021. "SailGP Teams, Back at Full Strength, Power to the $1 Million Prize." *New York Times*, October 8, 2021. https://www.nytimes.com/2021/10/08/sports/sailing/sailgp-catamaran.html.

Schmidt, Eric. 2015. "Blitzscaling 08: Eric Schmidt on Structuring Teams and Scaling Google." Interview by Reid Hoffman on October 15, 2015. Published October 23, 2015. *YouTube*. https://youtu.be/hcRxFRgNpns.

Schneider Electric. 2018. "Solutions for Gas Management System." Schneider Electric. https://download.schneider-electric.com/files?p_enDocType=Customer+success+story&p_File_Name=SCADA+Industrial.pdf&p_Doc_Ref=SCADA+International.

Schön, Donald A. 1987. *Educating the Reflective Practitioner: Toward a New Design for Teaching and Learning in the Professions*. New York: Jossey-Bass.

Schrage, Michael. 2020. *Recommendation Engines*. Cambridge, Massachusetts: The MIT Press.

Schreiber, Simon. 2020. "How to Measure AI Product Performance the Right Way." *Start It Up* (blog), *Medium*, June 30, 2020. https://medium.com/swlh/how-to-measure-ai-product-performance-the-right-way-2d6791c5f5c3.

Schrimpf, Paul. 2021. "Deere Launches See & Spray Select Technology." *PrecisionAg* (blog), March 2, 2021. https://www.precisionag.com/market-watch/deere-see-spray-select-release/.

Schwab, Klaus. 2015. "The Fourth Industrial Revolution: What It Means and How to Respond." *Foreign Affairs*, December 12, 2015. https://www.weforum.org/agenda/2016/01/the-fourth-industrial-revolution-what-it-means-and-how-to-respond.

Scire, Sarah. 2020a. "How *The New York Times* Prepared for the Ultimate Stress Test—the 2020 Election." *Nieman Lab* (blog), December 3, 2020. https://www.niemanlab.org/2020/12/how-the-new-york-times-prepared-for-the-ultimate-stress-test-the-2020-election.

Scire, Sarah. 2020b. "Outgoing *New York Times* CEO Mark Thompson Thinks There Won't Be a Print Edition in 20 Years." *Nieman Lab* (blog), August 11, 2020. https://www.niemanlab.org/2020/08/outgoing-new-york-times-ceo-mark-thompson-thinks-there-wont-be-a-print-edition-in-20-years.

Scire, Sarah. 2020c. "Readers Reign Supreme, and Other Takeaways from *The New York Times* End-of-Year Earnings Report." *Nieman Lab* (blog), February 6, 2020. https://www.niemanlab.org/2020/02/readers-reign-supreme-and-other-takeaways-from-the-new-york-times-end-of-year-earnings-report.

Sculley, D., Gary Holt, Daniel Golovin, et al. 2015. "Hidden Technical Debt in Machine Learning Systems." In *Proceedings of the 28th International Conference on Neural Information Processing Systems—Volume* 2, 2503–11. NIPS'15. Cambridge, MA: MIT Press. https://papers.nips.cc/paper/2015/file/86df7dcfd896fcaf2674f757a2463eba-Paper.pdf.

Sebastian-Coleman, Laura. 2010. "Data Quality Assessment Framework." Presentation at the Fourth MIT Information Quality Industry Symposium, Cambridge, MA, July 2010. http://mitiq.mit.edu/IQIS/Documents/CDOIQS_201077/Papers/03_08_4B-1.pdf.

Sebastian-Coleman, Laura. 2013. *Measuring Data Quality for Ongoing Improvement: A Data Quality Assessment Framework*. Waltham, MA: Morgan Kaufmann.

Seegene. n.d. "Rising to Global Challenges." Access Seegene. Accessed November 18, 2021. https://sponsorcontent.cnn.com/int/seegene/rising-to-global-challenges.

Semple, Kirk. 2017. "In Mexico, an Epidemic of Fuel Thefts Becomes a Crisis." *New York Times*, April 26, 2017, sec. World. https://www.nytimes.com/2017/04/26/world/americas/mexico-fuel-theft-crisis.html.

Seifert, Kevin. 2021. "Fourth-down Offense Is at an All-Time High: Why NFL Teams Are Following the Analytics and Going for It." ESPN.Com. September 24, 2021. https://www.espn.com/nfl/insider/insider/story/_/id/32264153/what-changed-fourth-why-nfl-headed-golden-era-analytically-driven-four-offense.

Seto, J.L. 2022. "Lack of EV Charging Stations Creates Range Anxiety in the West." *MotorBiscuit* (blog). June 10, 2022. https://www.motorbiscuit.com/lack-ev-charging-stations-creates-range-anxiety-west/.

Setty, Prasad. 2014. "HR Meets Science at Google." Speech at the 2014 Re:Work with Google event. Published November 10, 2014. *YouTube*. https://youtu.be/KY8v-O5Buyc.

Shaheen, Alexandra, Megan Araula, and Shawn Bower. 2021a. "Failover Plans, Outage Playbooks and Resilience Gaps." *NYT Open* (blog), *Medium*, July 30, 2021. https://open.nytimes.com/failover-plans-outage-playbooks-and-resilience-gaps-35047aed6213.

Shaheen, Alexandra, Megan Araula, and Shawn Bower. 2021b. "How *The New York Times* Assesses, Tests, and Prepares for the (UN)Expected News Event." *Nieman Lab* (blog), July 2021. https://www.niemanlab.org/2021/07/how-the-new-york-times-assesses-tests-and-prepares-for-the-unexpected-news-event.

Shaheen, Alexandra, Megan Araula, and Suman Roy. 2020. "How *The New York Times* Technology Teams Prepared for the 2020 Election." Interview by Sara Bures. *NYT Open* (blog), *Medium*, November 20, 2020. https://open.nytimes.com/how-the-new-york-times-technology-teams-prepared-for-the-2020-election-3928ce7f923c.

Sharma, Shradha and Jarshad NK. 2022. "Banking on the Bottom of the Pyramid." *YourStory.Com* (blog). July 29, 2022. https://yourstory.com/2022/07/financial-inclusion-unbanked-fia-global-seema-prem/amp.

Shashkevich, Alex. 2017. "10 Notable Books Published by Stanford University Press." *Stanford News*, November 9, 2017. https://news.stanford.edu/2017/11/09/10-notable-books-published-stanford-university-press/.

Shaw, Kathryn, and Debra Schifrin. 2015. "LinkedIn and Modern Recruiting (A)." HR41A-PDF-ENG, Palo Alto, CA: Stanford Graduate School of Business. https://hbsp.harvard.edu/product/HR41A-PDF-ENG.

Shea, Stephen. n.d. "The 3-Point Revolution." ShotTracker. Accessed November 28, 2021. https://shottracker.com/articles/the-3-point-revolution.

Shellenbarger, Sue. 2019. "The Dangers of Hiring for Cultural Fit." *Wall Street Journal*, September 23, 2019. https://www.wsj.com/articles/the-dangers-of-hiring-for-cultural-fit-11569231000.

Shirky, Clay. 2009. *Here Comes Everybody: The Power of Organizing Without Organizations*. Reprint ed. New York: Penguin.

Shomikdutta. 2021. "How Tech Helped Democrats Win: By the Numbers." Higher Ground Labs, March 31, 2021. https://highergroundlabs.com/how-tech-helped-democrats-win-by-the-numbers/.

Shrestha, Yash Raj, Shiko M. Ben-Menahem, and Georg von Krogh. 2019. "Organizational Decision-Making Structures in the Age of Artificial Intelligence." *California Management Review* 61, no. 4 (2019): 66–83.

Shulga, Dima. 2018. "5 Reasons 'Logistic Regression' Should Be the First Thing You Learn When Becoming a Data Scientist." *Medium*, Towards Data Science, April 21, 2018. https://towardsdatascience.com/5-reasons-logistic-regression-should-be-the-first-thing-you-learn-when-become-a-data-scientist-fcaae46605c4.

Shulman, Robyn D. 2017. "The Top 5 Reasons EdTech Startups Fail and How to Avoid Them." *Forbes*, June 13, 2017. https://www.forbes.com/sites/robynshulman/2017/06/13/the-top-5-reasons-edtech-startups-fail-and-how-to-avoid-them.

Sicha, Choire. 2022. "Journalism's Twitter Problem Is the Journalists." Intelligencer, April 7, 2022. https://nymag.com/intelligencer/2022/04/new-york-times-journalists-are-terrible-on-twitter.html.

Siddhartha, Sasha, and Nick Rockwell. 2021. "The Secret Advantage of Goal Setting with Sasha Siddhartha, Co-founder and CTO of Thrive Market and Nick Rockwell, Senior VP of Engineering and Infrastructure at Fastly." In *CIO Classified.* Published February 24, 2021. Spotify podcast. https://ciopod.com/podcasts/the-secret-advantage-of-goal-setting.

Silverman, Steve. 2018. "How Much Time Is Played During a Football Game?" *SportsRec*, December 5, 2018. https://www.sportsrec.com/6879403/how-much-time-is-played-during-a-football-game.

Simonite, Tom. 2020. "A Clever Strategy to Distribute Covid Aid—with Satellite Data." *Wired*, December 17, 2020. https://www.wired.com/story/clever-strategy-distribute-covid-aid-satellite-data.

Singer, P. W. 2009. *Wired for War: The Robotics Revolution and Conflict in the 21st Century.* Reprint ed. New York: Penguin.

Singleton, Ben. 2020–2021, various dates. Interviews by Zachary Tumin and Madeleine Want.

Sisak, Michael R. 2019. "Modern Policing: Algorithm Helps NYPD Spot Crime Patterns." *NBC Connecticut*, March 10, 2019. https://www.nbcconnecticut.com/news/national-international/modern-policing-algorithm-helps-nypd-spot-crime-patterns/172511.

Sisak, Michael R. 2021. "Monitor: NYPD Officers Underreporting Use of Stop-and-Frisk." *AP News*, September 1, 2021. https://apnews.com/article/business-race-and-ethnicity-police-reform-nyc-state-wire-273a0d430d3d4ccfe172ba367ac3d44f.

Sittig, Dean, Joan Ash, Jiajie Zhang, Jerome Osheroff, and M. Shabot. 2006. "Lessons from 'Unexpected Increased Mortality After Implementation of a Commercially Sold Computerized Physician Order Entry System.'" *Pediatrics* 118 (September 2006): 797–801. https://doi.org/10.1542/peds.2005-3132.

Skillroads. n.d. "AI and Facial Recognition Are Game Changers for Recruitment." *Skillroads.* Accessed November 22, 2021. https://skillroads.com/blog/ai-and-facial-recognition-are-game-changers-for-recruitment.

Skomoroch, Peter. 2019. "Product Management for AI." Presentation at the Rev 2 Data Science Leaders' Summit, New York, May 2019. Published June 21, 2019. https://www.slideshare.net/pskomoroch/product-management-for-ai.

Skomoroch, Peter, and Mike Loukides. 2020. "What You Need to Know About Product Management for AI." *Radar* (blog), *O'Reilly*, March 31, 2020. https://www.oreilly.com/radar/what-you-need-to-know-about-product-management-for-ai.

Smith, Brittany. 2019. "Behind the Scenes with U.S. SailGP Team: Tech, Training, and Teamwork." *Men's Journal* (blog), June 7, 2019. https://www.mensjournal.com/health-fitness/mens-journal-x-u-s-sailgp-team-strength-conditioning-workout/.

Smith, Chris. 1996. "The NYPD Guru." *New York Magazine*, April 2, 1996.

Smith, Chris. 2018. "The Controversial Crime-Fighting Program That Changed New York Forever." *New York Magazine Intelligencer*, March 2, 2018. https://nymag.com/intelligencer/2018/03/the-crime-fighting-program-that-changed-new-york-forever.html.

Smith, Gerry. 2017. "NY Times Scales Back Free Articles to Get More Subscribers." *Bloomberg*, December 1, 2017. https://www.bloomberg.com/news/articles/2017-12-01/n-y-times-scales-back-free-articles-to-get-readers-to-subscribe.

Smith, Hedrick. 2005. "District-Wide Reform: Anthony Alvarado Interview." Making Schools Work with Hedrick Smith, PBS. Accessed November 20, 2021. http://www.pbs.org/makingschoolswork/dwr/ca/alvarado.html.

Smith, Shelley. 2021. "Week 5 Takeaways and Big Questions: Chaos in Packers-Bengals, Lions Heartbreak, Cardinals 5–0." *ESPN*, October 10, 2021. https://www.espn.com/nfl/story/_/id/32355896/nfl-week-5-takeaways-learned-big-questions-every-game-future-team-outlooks.

Snyder, Gabriel. 2017. "How *The New York Times* Is Clawing Its Way into the Future." *Wired*, February 12, 2017. https://www.wired.com/2017/02/new-york-times-digital-journalism.

Sohn, Soopum. 2020. "The Real Reason South Korea Was Able to Stop COVID-19." *GEN* (blog), *Medium*, May 19, 2020. https://gen.medium.com/the-real-reason-south-korea-was-able-to-stop-covid-19-2742afaea400.

Sohrabi, Shirin, Anton Riabov, et al. 2018. "An AI Planning Solution to Scenario Generation for Enterprise Risk Management." In *AAAI 2018*. https://research.ibm.com/publications/an-ai-planning-solution-to-scenario-generation-for-enterprise-risk-management.

Somaiah, Jasmine. 2018. "Everything You Need to Know About Deep Canvassing." *CallHub* (blog), June 25, 2018. https://callhub.io/deep-canvassing.

Sousa, William H., and George L. Kelling. 2010. "Police and Reclamation of Public Places a Study of MacArthur Park in Los Angeles." *International Journal of Police Science and Management* 12, no. 1 (2010): 41–54. https://doi.org/10.1350/ijps.2010.12.1.156.

Sparrow, Malcolm K. 2015. *Measuring Performance in a Modern Police Organization*. New Perspectives in Policing Bulletin. Washington, DC: US Department of Justice, National Institute of Justice. NCJ 248476. https://www.ojp.gov/pdffiles1/nij/248476.pdf.

Sparrow, Malcolm K. 2016. *Handcuffed: What Holds Policing Back, and the Keys to Reform*. Washington, DC: Brookings Institution Press.

Spayd, Liz. 2017. "A 'Community' of One: *The Times* Gets Tailored." *New York Times*, March 18, 2017. https://www.nytimes.com/2017/03/18/public-editor/a-community-of-one-the-times-gets-tailored.html.

Spear, Steven J. 2010. *The High-Velocity Edge: How Market Leaders Leverage Operational Excellence to Beat the Competition*. New York: McGraw-Hill Education.

Springer, Bill. 2022. "Team Australia Wins SailGP Championship—and $1 Million Prize—in San Francisco." *Forbes*, March 28, 2022. https://www.forbes.com/sites/billspringer/2022/03/28/team-australia-wins-sailgp-championship-and-1-million-prize-in-san-francisco/.

Stern, Sebastian, Andreas Behrendt, Elke Eisenschmidt, et al. 2017. *The Rail Sector's Changing Maintenance Game: How Rail Operators and Rail OEMs Can Benefit from Digital Maintenance Opportunities. Digital McKinsey*, December 2017. https://www.mckinsey.com/~/media/mckinsey/industries/public%20and%20social%20sector/our%20insights/the%20rail%20sectors%20changing%20maintenance%20game/the-rail-sectors-changing-maintenance-game.pdf.

Sternlicht, Alexandra. 2020. "South Korea's Widespread Testing and Contact Tracing Lead to First Day with No New Cases." *Forbes*, April 30, 2020. https://www.forbes.com/sites/alexandrasternlicht/2020/04/30/south-koreas-widespread-testing-and-contact-tracing-lead-to-first-day-with-no-new-cases.

Stevenson, Howard H. 2006. "A Perspective on Entrepreneurship." Harvard Business School Background Note 384–131, October 1983. Revised April 2006.

Stokenberga, Aiga, and Maria Catalina Ochoa. 2021. "Unlocking the Lower Skies: The Costs and Benefits of Deploying Drones Across Use Cases in East Africa." The World Bank. https://doi.org/10.1596/978-1-4648-1696-3.

Strobel, Gordon and Warren P. Lubold. 2019. "Secret U.S. Missile Aims to Kill Only Terrorists, Not Nearby Civilians." *Wall Street Journal*, May 9, 2019. https://www.wsj.com/articles/secret-u-s-missile-aims-to-kill-only-terrorists-not-nearby-civilians-11557403411.

Stuart, Mark, David Angrave, Andy Charlwood, et al. 2016. "HR and Analytics: Why HR Is Set to Fail the Big Data Challenge." *Human Resource Management Journal* 26, no. 1 (2016): 1–11.

Sull, Donald, and Kathleen M. Eisenhardt. 2012. "Simple Rules for a Complex World." *Harvard Business Review* 90, no. 9 (September): 68–75. https://hbr.org/2012/09/simple-rules-for-a-complex-world.

Sull, Donald, and Kathleen M. Eisenhardt. 2016. *Simple Rules: How to Thrive in a Complex World*. Boston: Mariner.

Sullivan, Margaret. 2014. "Is There Any Escape from 'Recommended for You'?" *Public Editor's Journal* (blog), *New York Times*, February 21, 2014. https://publiceditor.blogs.nytimes.com/2014/02/21/is-there-any-escape-from-recommended-for-you.

Sverdlik, Yevgeniy. 2017. "*The New York Times* to Replace Data Centers with Google Cloud, AWS." *Data Center Knowledge*, April 18, 2017. https://www.datacenterknowledge.com/archives/2017/04/18/the-new-york-times-to-replace-data-centers-with-google-cloud-aws.

Svingen, Børge. 2017. "Publishing with Apache Kafka at *The New York Times*." *NYT Open* (blog), *Medium*, September 6, 2017. https://open.nytimes.com/publishing-with-apache-kafka-at-the-new-york-times-7f0e3b7d2077.

Swisher, Kara. 2021. "Another Big Step Toward Digitizing Our Lives." *New York Times*, March 19, 2021. https://www.nytimes.com/2021/03/19/opinion/NFTs-beeple-crypto.html.

Taibi, Cindy, and Brian Hamman. 2019. "New York Times Fireside Chat: The Changing Role of IT." Interview by Larry Dignan. Published July 30, 2019. *YouTube*. https://youtu.be/L3sfA2q9wiA.

Tameez, Hannaa'. 2021. "*The New York Times*' New Slack App Aims to Deliver (Non-Depressing) Times Stories to You While You Work." *Nieman Lab* (blog), February 18, 2021. https://www.niemanlab.org/2021/02/the-new-york-times-new-slack-app-aims-to-deliver-non-depressing-times-stories-to-you-while-you-work.

Tarnoff, Ben, and Moira Weigel. 2018. "Why Silicon Valley Can't Fix Itself." *Guardian*, May 3, 2018. https://www.theguardian.com/news/2018/may/03/why-silicon-valley-cant-fix-itself-tech-humanism.

Tarsney, Catherine. 2021. "Knock, Knock." *DNC Tech Team* (blog), *Medium*, January 28, 2021. https://medium.com/democratictech/knock-knock-5cfbe84c3b25.

Teachout, Zephyr, and Thomas Streeter. 2007. *Mousepads, Shoe Leather, and Hope: Lessons from the Howard Dean Campaign for the Future of Internet Politics*. Boulder, CO: Routledge.

Teach to One. 2020a. "How 'Teach to One' Personalizes Math Learning." *Medium* (blog), May 16, 2020. https://teachtoone.medium.com/how-teach-to-one-personalizes-math-learning-a8d7a0a8c977.

Teach to One. 2020b. "New Orleans Educators Share How Teach to One Helps Save Students from the Iceberg Problem." *Medium* (blog), June 22, 2020. https://teachtoone.medium.com/new-orleans-educators-share-how-teach-to-one-helps-save-students-from-the-iceberg-problem-450cc0c6cbb6.

Teach to One. 2020c. "NWEA's MAP Growth and New Classrooms' Teach to One Roadmaps Launch New Partnership." *PR Newswire*, October 19, 2020. https://www.prnewswire.com/news-releases/nweas-map-growth-and-new-classrooms-teach-to-one-roadmaps-launch-new-partnership-301155010.html.

Teach to One. 2020d. "Star Charts, Students, and Staying Connected at Sacajawea Middle School with Teach to One: Math." *Medium* (blog), June 22, 2021. https://teachtoone.medium.com/star-charts-students-and-staying-connected-at-sacajawea-middle-school-with-teach-to-one-math-9cba2e595974.

Teach to One. 2020e. "Teach to One Addresses Educational Gaps with a Personalized Learning Experience." *Medium* (blog), May 18, 2020. https://teachtoone.medium.com/teach-to-one-addresses-educational-gaps-with-a-personalized-learning-experience-5b349ab6c624.

Teach to One. 2020f. "Teach to One's Advanced Math Algorithms Map Personalized Learning Routes for All Students." *Medium* (blog), October 24, 2020. https://teachtoone.medium.com/teach-to-ones-advanced-math-algorithms-map-personalized-learning-routes-for-all-students-827641045ca.

Tech for Campaigns. "2020 Political Texting Report" n.d. Accessed November 29, 2021. https://www.techforcampaigns.org/impact/2020-texting-analysis.

Tegelberg, Erland. 2018. "Effective Asset Management and Exciting New Big Data Sources." Presentation at the Big Data in Railway Maintenance Planning Conference, University of Delaware, December 2018.

Telenko, Trent (@TrentTelenko). 2022. "If You Want to Understand Why Russian Missiles Are Not Always Able to Kill Ukrainian Drones, This Map Is a Very Good Start in Understanding Why. If Your Radio Energy Direction Finding Gear Locates an Enemy Radar Accurately. Good Digital Terrain Maps Can Predict Where 1/." *Twitter*. https://twitter.com/TrentTelenko/status/1507837884266856449.

Tetlock, Philip E., and Dan Gardner. 2016. *Superforecasting: The Art and Science of Prediction*. New York: Crown.

Thaler, Richard H. 1980. "Toward a Positive Theory of Consumer Choice." *Journal of Economic Behavior and Organization* 1 (1980): 39–60.

Thaler, Richard H., and Cass Sunstein. 2009. *Nudge: Improving Decisions About Health, Wealth, and Happiness*. New York: Penguin.

Thomas, Nellwyn. 2020. "One Year as DNC CTO." *DNC Tech Team* (blog), *Medium*, June 11, 2020. https://medium.com/democratictech/one-year-as-dnc-cto-76fa813b7836.

Thomas, Nellwyn. 2021a. "Breaking the Boom & Bust Cycle: What Made the Difference in DNC." *DNC Tech Team* (blog), *Medium*, January 25, 2021. https://medium.com/democratictech/breaking-the-boom-and-bust-cycle-955643e48949.

Thomas, Nellwyn. 2022. "We're Bringing Advanced Data Infrastructure to Thousands of Midterm Campaigns." *DNC Tech Team* (blog). March 23, 2022. https://medium.com/democratictech/were-bringing-advanced-data-infrastructure-to-thousands-of-midterm-campaigns-2bfb32f400bc.

Thomas, Nellwyn. 2021b. Interviews by Zachary Tumin and Madeleine Want.

Thomas, Owen. 2010. "Ex-Googler Douglas Merrill Takes on Payday Lenders with ZestCash." *VentureBeat* (blog), October 12, 2010. https://venturebeat.com/2010/10/12/douglas-merrill-zestcash.

Thomas, Rachel. 2022. "7 Great Lightning Talks Related to Data Science Ethics." Fast.AI, March 14, 2022. https://www.fast.ai/2022/03/14/ADSN-ethics/.

Thomke, Stefan, and Jim Manzi. 2014. "The Discipline of Business Experimentation." *Harvard Business Review*, December 10, 2014. https://hbr.org/2014/12/the-discipline-of-business-experimentation.

Thompson, Timothy. 2018. "Utilizing Artificial Intelligence on ERP Data to Increase Rolling Stock Maintenance Efficiency." Presentation at the Big Data in Railway Maintenance Planning Conference, University of Delaware, December 2018.

Tiersky, Howard. 2017. "*The New York Times* Is Winning at Digital." *CIO Magazine*, June 8, 2017. https://www.cio.com/article/3199604/the-new-york-times-is-winning-at-digital.html.

Timberg, Craig, Drew Harwell, and Spencer S. Hsu. 2021. "Police Let Most Capitol Rioters Walk Away. But Cellphone Data and Videos Could Now Lead to More Arrests." *Washington Post*, January 8, 2021. https://www.washingtonpost.com/technology/2021/01/08/trump-mob-tech-arrests.

TNTP and Zearn. 2021. *Accelerate, Don't Remediate: New Evidence from Elementary Math Classrooms*. May 23, 2021. New York: TNTP. https://tntp.org/publications/view/teacher-training-and-classroom-practice/accelerate-dont-remediate.

Togerson, Derek. 2022. "Exploring NFL Analytics: The Brandon Staley Conundrum." *NBC 7 San Diego* (blog), January 24, 2022. https://www.nbcsandiego.com/news/local/exploring-nfl-analytics-the-brandon-staley-conundrum/2847169/.

Toor, Amar. 2016. "This Startup Is Using Drones to Deliver Medicine in Rwanda." *Verge*, April 5, 2016. https://www.theverge.com/2016/4/5/11367274/zipline-drone-delivery-rwanda-medicine-blood.

Topol, Eric. 2019a. *Deep Medicine: How Artificial Intelligence Can Make Healthcare Human Again*. Illustrated ed. New York: Basic Books.

Topol, Eric J. 2019b. "High-Performance Medicine: The Convergence of Human and Artificial Intelligence." *Nature Medicine* 25, no. 1 (2019): 44–56. https://doi.org/10.1038/s41591-018-0300-7.

Topol, Eric. 2020. "The Future of Clinical Machine Intelligence: A Multi-Disciplinary Discussion." Presentation at the Stanford AIMI Symposium, Stanford University, August 5, 2020. https://aimi.stanford.edu/news-events/aimi-symposium/2020-symposium/agenda.

Tracy, Marc. 2020. "*The New York Times* Tops 6 Million Subscribers as Ad Revenue Plummets." *New York Times*, May 6, 2020. https://www.nytimes.com/2020/05/06/business/media/new-york-times-earnings-subscriptions-coronavirus.html.

Trautman, Erik. 2018. "The Virtuous Cycle of AI Products." *Erik Trautman*, June 12, 2018. https://www.eriktrautman.com/posts/the-virtuous-cycle-of-ai-products.

Tse, Kane. 2017. "Direct Versus Programmatic Sales: Which Is Better For Your Website?" *WIRED_MESH* (blog), January 30, 2017. https://medium.com/wired-mesh/direct-vs-programmatic-sales-which-is-better-for-your-website-9fda0ea3a65a.

Tufekci, Zeynep. 2014. "Engineering the Public: Big Data, Surveillance and Computational Politics." *First Monday* 19, no. 7 (July 2014). https://doi.org/10.5210/fm.v19i7.4901.

Tumin, Zachary, and Tad Oelstrom. 2008. "Unmanned and Robotic Warfare: Issues, Options, and Futures." Unpublished manuscript. Summary from the Harvard Executive Session of June 2008. Harvard University, John F. Kennedy School of Government.

Tushman, Michael L., Wendy K. Smith, and Andy Binns. 2011. "The Ambidextrous CEO." *Harvard Business Review*, June 1, 2011. https://hbr.org/2011/06/the-ambidextrous-ceo.

Tzachor, Asaf, Jess Whittlestone, Lalitha Sundaram, and Seán Ó hÉigeartaigh. 2020. "Artificial Intelligence in a Crisis Needs Ethics with Urgency." *Nature Machine Intelligence* 2 (June 2020): 1–2. https://doi.org/10.1038/s42256-020-0195-0.

Useem, Michael, Prasad Setty, Sean Waldheim, et al. 2015. "Panel Discussion: The Biggest Mistakes I've Made in People Analytics." Presentation at the Wharton People Analytics Conference in Philadelphia. Published December 13, 2015. *YouTube*. https://www.youtube.com/watch?v=BoLtjlhveCw.

van Elsland, Sabine L., and Deborah Evanson. 2020. "Identifying Clusters Central to South Korea's COVID-19 Response." *Imperial College London News*, May 29, 2020. https://www.imperial.ac.uk/news/197837/identifying-clusters-central-south-koreas-covid-19.

Velsey, Kim. 2021. "Why Did So Many Homeowners Sell to Zillow This Summer? Because It Overpaid." *Curbed*, November 4, 2021. https://www.curbed.com/2021/11/zillows-i-buying-algorithm-house-flipping.html.

Vlastelica, John. 2021. "Diversity Recruiting Efforts Won't Significantly Improve Until We Address These 2 Barriers." *LinkedIn Talent Blog* (blog), January 12, 2021. https://www.linkedin.com/business/talent/blog/talent-acquisition/remove-these-barriers-to-improve-diversity-recruiting.

Waddell, Kaveh. 2018. "Today's Best Artificial Intelligence Advancements Are One-Trick Ponies." *Axios*, October 17, 2018. https://www.axios.com/2018/10/17/narrow-ai-small-data.

Wadlow, Tom, and James Pepper. 2018. "Technology and the 2017–2021 Business Plan." Mexico City: PEMEX. https://issuu.com/businessreviewasia/docs/bro-pemex?e=26716553/60843559.

Wageman, Ruth. 1997. "Critical Success Factors for Creating Superb Self-Managing Teams." *Organizational Dynamics* 26, no. 1 (1997). https://doi.org/10.1016/S0090-2616(97)90027-9.

Walder, Seth. 2021. "Which NFL Teams Are Most, Least Analytics-Advanced? We Surveyed 22 Team Staffers." ESPN.com, October 6, 2021. https://www.espn.com/nfl/story/_/id/32338821/2021-nfl-analytics-survey-most-least-analytically-inclined-teams-future-gm-candidates-more.

*Wall Street Journal*. 2012. "The Daily Start-Up: ZestCash Nabs $73M To Disrupt Payday Loans." January 19, 2012. https://www.wsj.com/articles/BL-VCDB-11816.

Wang, Yuan, Ping Wang, Xin Wang, and Xiang Liu. 2018. "Position Synchronization for Track Geometry Inspection Data via Big-Data Fusion and Incremental Learning." *Transportation Research Part C: Emerging Technologies* 93 (August 2018): 544–565. https://doi.org/10.1016/j.trc.2018.06.018.

Want, Madeleine. 2020–2021. Interviews by Zachary Tumin.

Warzel, Charlie, and Stuart A. Thompson. 2021. "They Stormed the Capitol. Their Apps Tracked Them." *New York Times*, February 5, 2021. https://www.nytimes.com/2021/02/05/opinion/capitol-attack-cellphone-data.html.

Waters, Richard. 2018. "Why We Are in Danger of Overestimating AI." *Financial Times*, February 4, 2018. https://www.ft.com/content/4367e34e-db72-11e7-9504-59efdb70e12f.

Watson, Ivan, Sophie Jeong, Julia Hollingsworth, and Tom Booth. 2020. "How This South Korean Company Created Coronavirus Test Kits in Three Weeks." *CNN*, March 13, 2020. https://www.cnn.com/2020/03/12/asia/coronavirus-south-korea-testing-intl-hnk/index.html.

Webber, Jude. 2017. "PEMEX Chases Fuel Thieves with Big Data." May 24, 2017. https://www.ft.com/content/8d830426-3f01-11e7-9d56-25f963e998b2.

Weekman, Kelsey. 2020. "Grubhub Was Getting 6,000 Orders A Minute During Its Promo Today That Left Restaurant Workers Stressed and Customers Hangry." BuzzFeed News. May 17, 2020. https://www.buzzfeednews.com/article/kelseyweekman/grubhub-free-lunch-nyc-promo-chaos.

Weichselbaum, Simone. 2016. "The End of the Bratton Era." *Marshall Project*, August 2, 2016. https://www.themarshallproject.org/2016/08/02/the-end-of-the-bratton-era.

Weick, Karl E., Kathleen M. Sutcliffe, and David Obstfeld. 1999. "Organizing for High Reliability: Processes of Collective Mindfulness." In *Research in Organizational Behavior*, vol. 21, ed. R. I. Sutton and B. M. Staw, 81–123. Cambridge, MA: Elsevier Science.

Weintraub, Joshua. 2021. Interviews by Zachary Tumin.

Weisburd, David, Stephen D. Mastrofski, Rosann Greenspan, and James J. Willis. 2004. *The Growth of CompStat in American Policing*. Washington, DC: Police Foundation Reports. https://www.policefoundation.org/wp-content/uploads/2015/06/Weisburd-et-al.-2004-The-Growth-of-Compstat-in-American-Policing-Police-Foundation-Report_0.pdf.

Weiss, Mitchell. 2017. "Public Entrepreneurship." Harvard Business School Course Overview Note 818-006, July 2017. Revised July 2021.

Weiss, Mitchell. 2021. *We the Possibility: Harnessing Public Entrepreneurship to Solve Our Most Urgent Problems*. Boston: Harvard Business Review Press.

Weiss, Mitchell, and Benjamin Henkes. 2019. "An (Abbreviated) Perspective on Entrepreneurship." Harvard Business School Background Note 820-083, December 2019.

Weiss, Mitchell B., and Sarah Mehta. 2020. "TraceTogether." Harvard Business School Case 820-111, June 29, 2020. Revised July 2020. https://www.hbs.edu/faculty/Pages/item.aspx?num=58357.

Whittaker, James, Jason Arbon, and Jeff Carollo. 2012. *How Google Tests Software*. Upper Saddle River, NJ: Addison-Wesley Professional.

Whittle, Richard. 2015. *Predator: The Secret Origins of the Drone Revolution*. New York: Picador.

Whittle, Richard. 2018. "Predator Started Drone Revolution, and Made Military Innovation Cool." *Breaking Defense*, March 9, 2018. https://breakingdefense.sites.breakingmedia.com/2018/03/predator-started-drone-revolution-and-made-military-innovation-cool.

Wiggins, Chris. 2015. "Keynote: Data Science at *The New York Times*." Presentation at the 2015 SciPy Conference, Austin, TX, July 2015. Published July 9, 2015. *YouTube*. https://youtu.be/MNosMXFGtBE.

Wiggins, Chris. 2019. "Data Science at *The New York Times*." Presentation at the Rev 2 Data Science Leaders' Summit, New York, May 2019. Published July 9, 2019. Video. https://blog.dominodatalab.com/data-science-at-the-new-york-times.

Wiggins, Chris, and Anne Bauer. 2020. "Data Science at *The New York Times*: A Mission-Driven Approach to Personalizing the Customer Journey." Presentation at the 2019 PyData Conference, New York, November 2019. Published January 2, 2020. *YouTube*. https://youtu.be/no7q-rZTLTw.

Wikipedia. "Douglas Merrill." Last modified October 24, 2020. https://en.wikipedia.org/w/index.php?title=Douglas_Merrill&oldid=985137850.

Willens, Max. 2017. "*The New York Times* Now Has 13 Million Subscriptions to 50 Email Newsletters." *Digiday* (blog), May 30, 2017. https://digiday.com/media/new-york-times-now-13-million-subscribers-50-email-newsletters.

Willens, Max. 2018a. "From Hard Paywalls to Meters to Dynamic Paywalls: Why New York Media Is Taking a Flexible Approach to Subscriptions." *Digiday* (blog), November 13, 2018. https://digiday.com/media/new-york-media-paywall-subscriptions-flexible.

Willens, Max. 2018b. "How *The New York Times* Is Using Interactive Tools to Build Loyalty (and Ultimately Subscriptions)." *Digiday* (blog), January 24, 2018. https://digiday.com/media/new-york-times-enlisting-interactive-news-desk-subscription-drive.

Williams, Joe. 2021. "Andrew Ng Thinks Your Company Is Doing AI Wrong." *Protocol*, May 18, 2021. https://www.protocol.com/enterprise/andrew-ng-ai-strategy.

Williams, Josh, and Tiff Fehr. 2021. "Tracking COVID-19 from Hundreds of Sources, One Extracted Record at a Time." *NYT Open* (blog), *Medium*, June 17, 2021. https://open.nytimes.com/tracking-covid-19-from-hundreds-of-sources-one-extracted-record-at-a-time-dd8cbd31f9b4.

Williams, Trefor P., and John F. Betak. 2018. "Using Text and Data Analytics to Study Railroad Operations." Presentation at the Big Data in Railway Maintenance Planning Conference, University of Delaware, December 2018.

Wise, Dean. 2021. Interviews by Zachary Tumin.

Witt, Stephen. 2022. "The Turkish Drone That Changed the Nature of Warfare." *The New Yorker*, May 9, 2022. https://www.newyorker.com/magazine/2022/05/16/the-turkish-drone-that-changed-the-nature-of-warfare.

Woo, Stu. 2017. "What's Better in the Classroom—Teacher or Machine?" *Wall Street Journal*, January 30, 2017. https://www.wsj.com/articles/whats-better-in-the-classroomteacher-or-machine-1485772201.

Wood, Daniel B. 2007. "William Bratton: Lauded Chief of Troubled LAPD." *Christian Science Monitor*, July 6, 2007. https://www.csmonitor.com/2007/0706/p01s05-usju.html.

Woods, Dan. 2021. "Artificial Intelligence on the Biden Campaign." *Medium* (blog), February 4, 2021. https://danveloper.medium.com/artificial-intelligence-on-the-biden-campaign-e704a656d956.

Woods, Tim. 2016. "8 Principles of the Innovator's Solution." *Hype* (blog), April 10, 2016. https://blog.hypeinnovation.com/8-principles-of-the-innovators-solution.

Worldometers.info. 2022. "Coronavirus Death Toll and Trends." *Worldometers*, May 11, 2022. https://www.worldometers.info/coronavirus/coronavirus-death-toll/.

Wornhoff, Nina. 2021. Interviews by Zachary Tumin and Madeleine Want.

Wortman, Joan Pearson Heck. 2019. "Record Linkage Methods with Applications to Causal Inference and Election Voting Data." PhD diss., Duke University, 2019. https://dukespace.lib.duke.edu/dspace/handle/10161/18657.

Wortman, Joan Pearson Heck, and Jerome P. Reiter. 2018. "Simultaneous Record Linkage and Causal Inference with Propensity Score Subclassification." *Statistics in Medicine* 37, no. 24 (2018)): 3533–3546. https://doi.org/10.1002/sim.7911.

Wurster, Thomas S., Derek Kennedy, and Jason Zajac. 2015. "Insight to Outcome: A Note on Strategy Development and Implementation." Stanford Graduate School of Business Case SM238, 2015.

Xu, P., Q. Sun, R. Liu, and F. Wang. 2011. "A Short-Range Prediction Model for Track Quality Index." *Proceedings of the Institution of Mechanical Engineers, Part F: Journal of Rail and Rapid Transit* 225, no. 3 (2–11): 277–285. https://doi.org/10.1177/2041301710392477.

Yale Institution for Social and Policy Studies. n.d. "Lessons from GOTV Experiments." Field Experiments Initiative. Accessed November 23, 2021. https://isps.yale.edu/node/16698.

Yampolskiy, Roman. n.d. "Artificial Intelligence Incident Database (for Incident Number 43 (Incident Date 1998-03-05)." Partnership on AI. Accessed October 24, 2021. incidentdatabase.ai/cite/43.

Yee, Kate Madden. 2019. "AI May Help Women with Ductal Hyperplasia Avoid Surgery." AuntMinnie.com, March 15, 2019. https://www.auntminnie.com/index.aspx?sec=sup&sub=aic&pag=dis&ItemID=124904.

Yee, Kate Madden. 2021. "Deep Learning: A New Frontier for Musculoskeletal MRI." AuntMinnie.com, September 6, 2021. https://www.auntminnie.com/index.aspx?sec=sup&sub=aic&pag=dis&ItemID=133396.

Yemen, Gerry, and Martin N. Davidson. 2009. "Diversity at JPMorgan Chase: Right Is Good Enough for Me (A) and (B)." Darden Business Publishing, Teaching Note UVA-OB-0975TN.

Yew, Lauren. 2021. "4 Steps to Win Advocates and Implement a Technical Change." *NYT Open* (blog), *Medium*, May 13, 2021. https://open.nytimes.com/4-steps-to-win-advocates-and-implement-a-technical-change-b2a9b922559b.

Yin, Jiateng, and Wentian Zhao. 2016. "Fault Diagnosis Network Design for Vehicle On-Board Equipments of High-Speed Railway: A Deep Learning Approach." *Engineering Applications of Artificial Intelligence* 56 (November 2016): 250–259. https://doi.org/10.1016/j.engappai.2016.10.002.

Young, Michael, and Anne Bauer. 2021. Interview by Zachary Tumin and Madeleine Want.

Yu, Allen. 2019. "How Netflix Uses AI and Machine Learning." *Becoming Human* (blog), *Medium*, February 27, 2019. https://becominghuman.ai/how-netflix-uses-ai-and-machine-learning-a087614630fe.

Zarembski, Allan M. 2014. "Some Examples of Big Data in Railroad Engineering." 2014 IEEE International Conference on Big Data (Big Data) (2014): 96–102. doi:10.1109/BigData.2014.7004437.

Zarembski, Allan M. 2019. "Big Data Drives Big Results." *Railway Age*, February 11, 2019. https://www.railwayage.com/news/big-data-drives-big-results.

Zarembski, Allan M., Daniel Einbinder, and Nii Attoh-Okine. 2016. "Using Multiple Adaptive Regression to Address the Impact of Track Geometry on Development of Rail Defects." *Construction and Building Materials* 127 (November 2016): 546–555. https://doi.org/10.1016/j.conbuildmat.2016.10.012.

Zax, David. 2012. "Fast Talk: How a Former Google Exec Plans to Transform Loans." *Fast Company*, February 2, 2012. https://www.fastcompany.com/1813256/fast-talk-how-former-google-exec-plans-transform-loans.

Zhou, Chenyi, Liang Gao, Hong Xiao, and Bowen Hou. 2020. "Railway Wheel Flat Recognition and Precise Positioning Method Based on Multisensor Arrays." *Applied Sciences* 10, no. 4 (2020): 1297. https://doi.org/10.3390/app10041297.

Zinkevich, Martin. 2021. "Rules of Machine Learning." *Google Developers*, Last modified September 27, 2021. https://developers.google.com/machine-learning/guides/rules-of-ml.

Zipline. n.d. "How It Works." Accessed December 16, 2020. https://flyzipline.com/how-it-works.

Zipline. 2021a. *Protecting Ghana's Election: Instant Agility with Zipline's Autonomous Delivery Network*. February 2021. https://assets.ctfassets.net/pbn2i2zbvp41/3yrQaMNdJ1u1J2aSEucjzt/4412ea5d12896d15b7eb41a2212d0295/Zipline_Ghana_PPE_Global_Healthcare_Feb-2021.pdf.

Zipline. 2021b "Walmart Teams with Zipline to Launch Autonomous Aircraft Delivery Service in Northwest Arkansas." *GlobeNewswire*, November 18, 2021. https://www.globenewswire.com/news-release/2021/11/18/2337193/0/en/Walmart-Teams-with-Zipline-to-Launch-Autonomous-Aircraft-Delivery-Service-in-Northwest-Arkansas.html.

Zipline. 2021c. "Zipline Celebrates Five Years in Flight." Flyzipline.com. Last modified October 28, 2021. https://flyzipline.com/press/zipline-celebrates-five-years-in-flight.

# INDEX